Must Have for the Morning Session……..

Civil PE Morning Breadth Exam Handbook

All Modules Covered…..

Theory is explained in a very simple manner and loaded with problems and solutions. Solutions are given in a step by step manner.

This book will cover all subject matter that you will encounter during the morning exam.

- WATER RESOURCES/ENVIRONMENTAL
- GEOTECHNICAL
- TRANSPORTATION
- CONSTRUCTION AND
- STRUCTURAL

Author's Other Books:

Civil PE Morning Breadth Exam Handbook –

Theory is explained in a very easy manner and loaded with problems and solutions. Solutions are given in a step by step manner.

Three Sample Exams for the Civil FE Exam, Ruwan Rajapakse. Good book for PE students as well.

This book has many problems that would be useful to PE students as well. Also FE supplied reference handbook by NCEES is a good guide to take to the PE exam.

Civil PE Construction Module, Fifth Edition

(Construction Depth, but good for other students as a reference)

Civil PE Construction Module Practice Problems, Second Edition

These two books cover 90% of construction module depth exam

Civil PE Geotechnical Module (Depth)

Do you know that the minor subject of the Construction Module is Geotechnical? Structural and other modules also have Geotechnical questions.

PRACTICE, PRACTICE, PRACTICE.......KEY TO EXAM SUCCESS

Four Sample Exams for the Civil PE Exam All Modules, Second Edition

by Ruwan Rajapakse, PE, CCM, CCE, AVS

Printed in the United States of America

Library of Congress cataloging in Publication Data

ISBN 1939493056
ISBN: 978-1-939493-05-7

ISBN 10: 1-939493-05-6

ISBN 13: 978-1-939493-05-7
April 17, 2014

www.geotechweb.com

UNITS:

fps Units	SI Units
Length	
1 ft = 0.3048 m	1 m = 3.28084 ft
1 inch = 2.54 cm	
Pressure	
1 ksf = 1,000 psf	1 Pascal = 1 N/m^2
1 ksf = 0.04788 MPa	1 MPa = 20.88543 ksf
1 ksf = 47,880.26 Pascal	1 MPa = 145.0377 psi
1 ksf = 47.88 kPa	1 kPa = 0.020885 ksf
1 psi = 6,894.757 Pascal	1 kPa = 0.1450377 psi
1 psi = 6.894757 kPa	1 bar = 100 kPa
1 psi = 144 psf	
Area	
1 ft^2 = 0.092903 m^2	1 m^2 = 10.76387 ft^2
1 ft^2 = 144 in^2	1 Acre = 43,560 sq. ft
Volume	
1 ft^3 = 0.028317 m^3	1 m^3 = 35.314667 ft^3
1 gallon = 8.34 lbs	1 cu. ft = 7.48 gallons
Density	
1 lbs/ft^3 = 157.1081 N/m^3	1 kN/m^3 = 6.3658 lbs/ft^3
1 lbs/ft^3 = 0.1571081 kN/m^3	
Weight:	
1 kip = 1,000 lbs	1 kg = 9.80665 N
1 lb = 0.453592 kg	1 kg = 2.2046223 lbs
1 lb = 4.448222 N	1 N = 0.224809 lbs
1 ton (short) = 2,000 lbs	1 N = 0.101972 kg
1 ton = 2 kips	1 kN = 0.224809 kips

DENSITY OF WATER

1 g per cubic centimeter = 1,000 g per liter = 1,000 kg/m^3

= 62.42 pounds per cu. feet

Exam Tips:

Concepts and Computations:

Civil PE construction module contains questions that require the knowledge of concepts and at the same time which require the ability to conduct simple computations. This book has provided worked examples that are similar in nature to the examination questions.

Calculators: This is the most important item that you take to the exam. It is extremely important that you have a good calculator and very familiar with its functions.

What calculators are allowed? *(Source: NCEES website)*

Casio:	*All fx-115 models.*
HP:	*HP33s and HP 35s models.*
TI:	*All TI-30X and TI-36X models.*

Frequently used functions in the exams:

- Degree to radian conversion and vice versa
- log, ln, e^x, inv log
- Sin, cos, tan
- $\sin^{-1}$, $\cos^{-1}$ and $\tan^{-1}$

Solve Function in Calculators: There may be couple of questions that you may need the "solve" function. Hence it is important to learn this function in your calculator.

For an instance if you need to solve an equation such as $x^3 + 30x + \log x = 0$, using the solve function would save you some vital minutes.

TABLE OF CONTENTS

Problem 1.21) Shear force diagrams and finding the maximum shear force of a beam

Problem 1.22) Bending moment diagrams and finding the maximum bending moment

Problem 1.23): Finding the Euler buckling load of a column

Problem 1.24): Sand liquefaction problem

Problem 1.25) Natural period of buildings as related to seismic engineering.

Problem 1.26) Spectral acceleration of buildings as related to earthquakes.

CONSTRUCTION:

Problem 1.27): Successor – predecessor relationships in activity on node networks

Problem 1.28) Swell factor and shrinkage factor basics

Problem 1.29): Removing tree stumps using tractors

Problem 1.30): SMAW welding

Problem 1.31) Crew hour rate and labor hour rate

Problem 1.32) Finding the cost of labor for formwork

Geotechnical:

Problem 1.33): Finding the bearing capacity of a footing using Terzaghi bearing capacity equation

Problem 1.34): Soil phase relationship equations

Problem 1.35) Modified Proctor and dry density computation

Problem 1.36) Lateral earth pressure coefficients (active at rest and passive)

Problem 1.37): Retaining wall stability computation

Problem 1.38) Factor of safety against overturning in retaining walls

WASTEWATER: (WATER RESOURCES)

Problem 1.39): Wastewater processing

Problem 1.40) Primary clarifier (Primary sedimentation tank) design

SAMPLE EXAM 2

WATER RESOURCES:

Problem 2.1) Finding the hydraulic radius of a trapezoidal channel.

Problem 2.2) Use of Manning formula for a trapezoidal channel.

Problem 2.3) Computing energy dissipation during a hydraulic jump

Problem 2.4: Finding the pressure at a point in pressure conduits ignoring frictional head loss.

Problem 2.5: Use of Hazen Williams formula to compute head loss due to friction

Problem 2.6: Velocity computation in a Pitot tubes

Problem 2.7): Time of concentration in a drainage basin

Problem 2.8) Rational formula when two drainage areas are discharging

Problem 2.9) Hydrograph study (Net precipitation and runoff)

Problem 2.10) Unit hydrographs - Finding the maximum flow in a drain pipe for a unit hydrograph.

TRANSPORTATION:

Problem 2.11) Accidental analysis using dynamic formulas.

Problem 2.12) Stopping sight distance of a road is 460 ft. Find the horizontal sight offset required for the horizontal curve.

Problem 2.13): A traverse is done as shown. What is the total error of internal angles?

Problem 2.14) Vehicle travel

Problem 2.15): Finding the PT station of a horizontal curve

Problem 2.16): Finding elevations in vertical curves.

Problem 2.17) Superelevation requirements in horizontal curves

Problem 2.18) Normal crown to superelevation transition

STRUCTURAL:

Problem 2.19) Finding the deflection at the end of a hollow rectangular section.

Problem 2.20): Seismic design categories

Problem 2.21): Generalized spectral response curves in seismic engineering

Problem 2.22) Moment frame buildings

Problem 2:23): Finding the nominal moment of a concrete beam.

Problem 2:24) Finding the maximum steel allowed in concrete beams as per ACI 318.

Problem 2:25) Shear force diagram of a beam loaded with a uniform load and concentrated load

Problem 2:26) Finding the maximum bending moment of a beam

WATER RESOURCES - WASTEWATER:

Problem 2.27): Moles and molecular weight of chemicals

Problem 2.28): BOD computation in wastewater samples

Problem 2.29) Activated sludge process

GEOTECHNICAL:

Problem 2.30): Soil phase relationships and total density of soils

Problem 2.31) Finding the lateral earth pressure. (at rest situation)

Problem 2.32): Standard penetration test (SPT).

Problem 2.33): Soil phase relationships (void ratio and porosity).

Problem 2.34): Pile driven in clay soils.

Problem 2.35): Consolidation test of a normally consolidated soil

CONSTRUCTION:

Problem 2.36) Cut and fill problem integrated with swell factors and shrinkage factors

Problem 2:37) Steel construction estimating

Solution 2.38): Construction loads

Problem 2:39) Finding the number of bricks required for a wall.

Problem 2:40) Finding the volume of mortar required for a wall

SAMPLE EXAM 3

WATER RESOURCES:

Problem 3.1) Open channel flow computation in complex channels

Problem 3.2) Hydraulic efficiency of open channels

Problem 3.3) Finding the Froude Number of a Trapezoidal Channel

Problem 3.4) Finding the critical depth of a trapezoidal Channel

Problem 3.5) Supercritical/subcritical flow during hydraulic jumps

Problem 3.6) Uniform flow and steady flow in open channels

Problem 3.7: Friction head loss computation using Darcy equation

Problem 3.8: Flow in Pitot tubes

Problem 3.9: Finding flow in Venturi meters

Problem 3.10) Rainfall intensity, time of concentration and flow in two drainage basins

Problem 3.11) Runoff coefficient, hydrologically most distant point and discharge

WASTEWATER:

Problem 3.12) Design of circular primary sedimentation tanks

TRANSPORTATION:

Problem 3.13) Heavy vehicle factor computation in traffic flow

Problem 3.14) Trips generation in a city

Problem 3.15) AADT, K factor and directional factor

Problem 3.16) PHF (Peak hour factor), free flow speed and LOS

Problem 3.17) Degree of curve and station computation in horizontal curves

Problem 3.18): Design of sag vertical curves

Problem 3.19) Design of horizontal curves with a superelevation.

Problem 3.20): Tangent runout and tangent runoff in horizontal curves

CONSTRUCTION:

Problem 3.21) Elevation computation using level surveys

Problem 3.22): Azimuth and bearing computations

Problem 3.23): Activity on node diagrams and successor relationships

Problem 3.24: Formwork estimating for a footing

Problem 3.25): Contractor safety – Incidence rate

Problem 3.26) Traverse surveys (Latitude and departure computation)

Problem 3.27): Activity on node network. ES, LS, EF, LF computation. Forward pass and backward pass.

Problem 3.28): Total float computation in activity on node networks.

Problem 3.29): Free float computation in activity on node networks.

STRUCTURAL:

Problem 3.30) Shear force diagrams

Problem 3.31): Bending moment diagrams

Problem 3.32): Location of centroid of a T-beam.

Problem 3.33): Finding the nominal moment of a concrete T-beam

Problem 3.34): Finding the total force in reinforcements in a concrete beam

GEOTECHNICAL:

Problem 3.35): Soil phase relationships - Specific gravity, water content and degree of saturation

Problem 3.36): Shelby tubes in geotechnical engineering

Problem 3.37): RQD computation in rock cores

Problem 3.38): Sieve analysis curve and D_{30}

Problem 3.39): Normally consolidated soils and overconsolidated soils.

Problem 3.40): Well graded gravel and poorly graded gravels

SAMPLE EXAM 4

WATER RESOURCES:

Problem 4.1): Hydraulic jump and the Froude number

Problem 4.2) Broad crested weir computations

Problem 4.3) Open channel flow computation in circular channels

Problem 4.4) Open channel flow computation in Triangular channels

Problem 4.5): Finding the total head added by a pump

Problem 4.6): Computing flow in parallal pipe systems

Problem 4.7): Finding the horse power of a pump.

Problem 4.8): Pressure conduits and minor head losses

Problem 4.9) Application of rational formula for two drainage basins

Problem 4.10) Flow computation using Trapezoidal hydrographs

STRUCTURAL:

Problem 4.11) Occupancy category with regard to seismic design

Solution 4.12): Site classes in seismic design

Problem 4.13) Seismic response coefficient (C_s) of a building and seismic force

Problem 4.14): Finding the area of steel required for a balanced section

Problem 4.15): Finding the nominal moment of a concrete T-beam

Problem 4.16): Moment connections and shear connections in structural steel

Problem 4.17): Finding the moment of inertia (I) thru the centroid of a T- beam

Problem 4.18): Column buckling analysis.

TRANSPORTATION:

Problem 4.19): Horizontal curves, arc method and station computations

Problem 4.20) Finding the deflection angle of a horizontal curve

Problem 4.21) Computing base free flow speed of a highway

Problem 4.22) Computing design traffic volume of a highway

Problem 4.23): Sag vertical curve going under an overpass

CONSTRUCTION:

Problem 4.24) surveying computations

Problem 4.25): Finding the azimuth of a line

Problem 4.26) Finding the duration of a project using an activity on arrow diagram

Problem 4.27) Finding the early start and late start of activities in activity on arrow diagram

Problem 4.28) Finding the total float of activities in an activity on arrow diagram

Problem 4.29) Cement types and their purpose

Problem 4.30): Cold weather concreting

Problem 4.31): Mass haul diagrams

Problem 4.32): Cost estimating (Labor hour method)

GEOTECHNICAL:

Problem 4.33): Finding vertical effective stress in a site

Problem 4.34): Secondary compression of clay soils

Problem 4.35): Consolidation test in overconsolidated soils

Problem 4.36): Recompression curves and Virgin curves

Problem 4.37): Shallow foundation design

Problem 4.38): Finding the length of the footing required to carry a given load (Terzaghi bearing capacity equation is used).

Problem 4.39): Soil porosity, specific gravity, moisture content and degree of saturation

Problem 4.40): Stress distribution under footings.

SAMPLE EXAM 1

Problem 1.1): Find the wetted perimeter of the channel shown. Side slopes are 3 to 1 as shown.

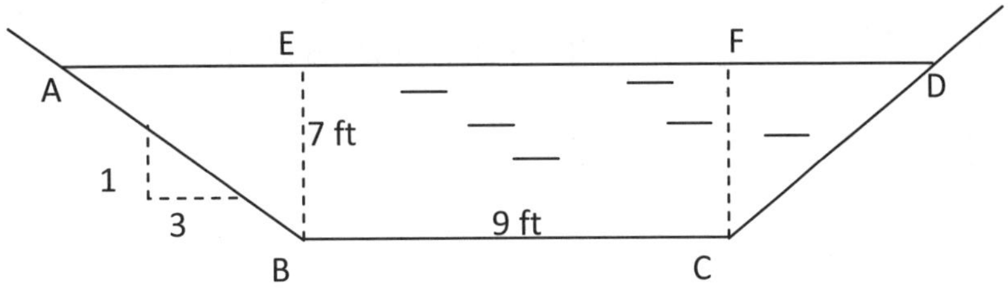

A) 23.5   B) 44.8   C) 53.2   D) 43.2

Problem 1.2) Water is flowing thru a rectangular channel at a flow rate of 800 cu. ft/sec. The depth of the channel is 6 ft and the width is 10 ft. What is the slope of the channel if Manning's coefficient is 0.02?

A) 0.8%   B) 0.3%   C) 0.15%   D) 0.09%

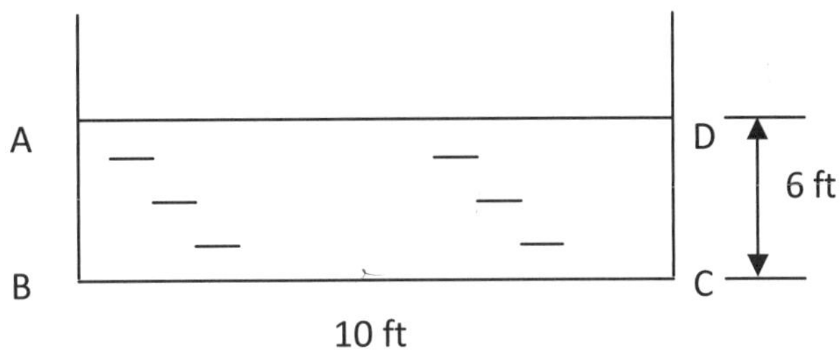

Problem 1.3) Find the Froude number of the channel shown if the flow is 13 cu.ft/sec. Channel width is 11 ft and channel depth is 4.3 ft.

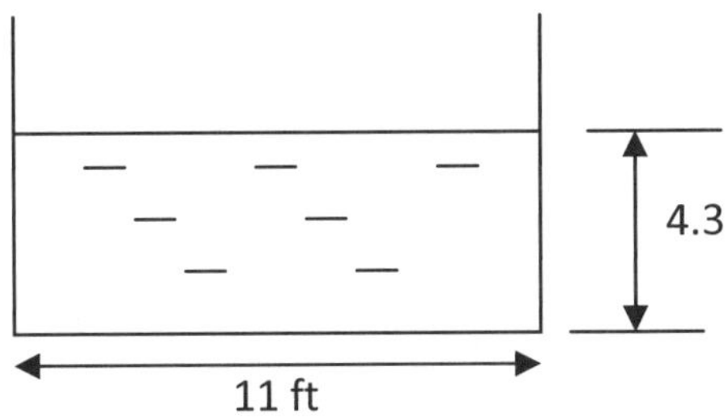

A) 0.034   B) 0.024   C) 0.765   D) 0.876

Problem 1.4) Find the critical depth of the channel shown if the flow is 17.5 cu.ft/sec. Channel width is 11 ft.

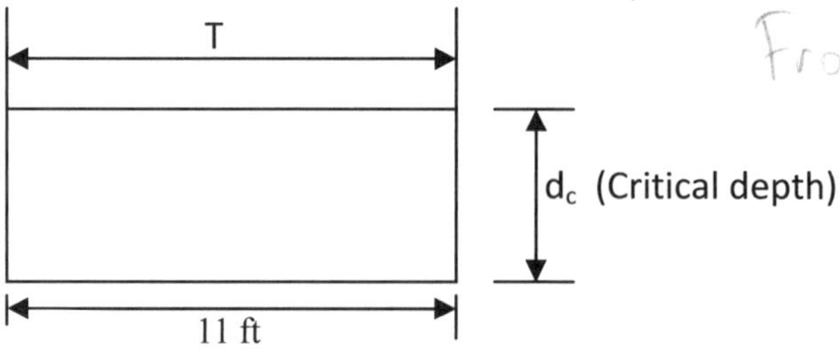

A) 0.428 ft B) 0.254 ft C) 0.654 ft D) 0.876 ft

Problem 1.5) A rectangular channel is undergoing a hydraulic jump. Upstream depth of the channel is 0.4 ft and velocity of the channel is 7.0 ft/sec. What is the depth after the hydraulic jump?

A) 1.23 ft B) 2.84 ft C) 0.92 ft D) 0.21 ft

Problem 1.6): Find the velocity at point B in the pipe shown. Ignore the head loss due to friction.

A) 32.8 ft/sec B) 28.8 ft/sec C) 13.9 ft/sec D) 65.8 ft/sec

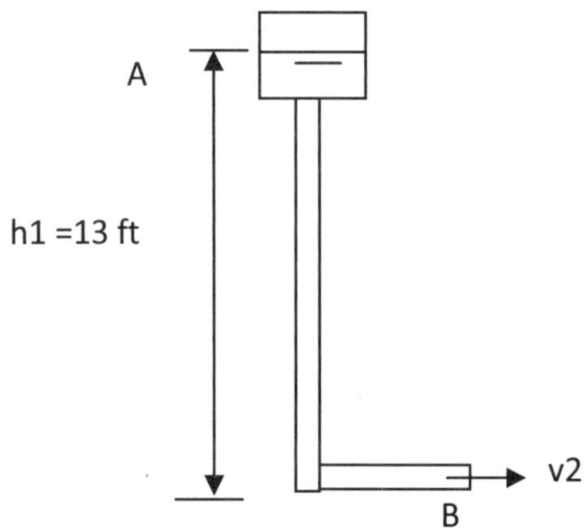

Problem 1.7): Pressure inside a pipe is 1 psi. What is the pressure head in ft of water?
A) 3.8 ft B) 2.3 ft C) 5.8 ft D) 9.7 ft

Problem 1.8): Net precipitation (after deducting for infiltration) of a watershed area was found to be 2.45 inches. The main culvert located at the most downstream end of the watershed area had 40,210 cu. ft of water passing thru it during the storm. Find the total acreage of the watershed area.

A) 2.34 Acres B) 12.76 Acres C) 4.52 Acres D) 8.76 Acres

Problem 1.9): 20 year and 50 year storms are shown in the figure.

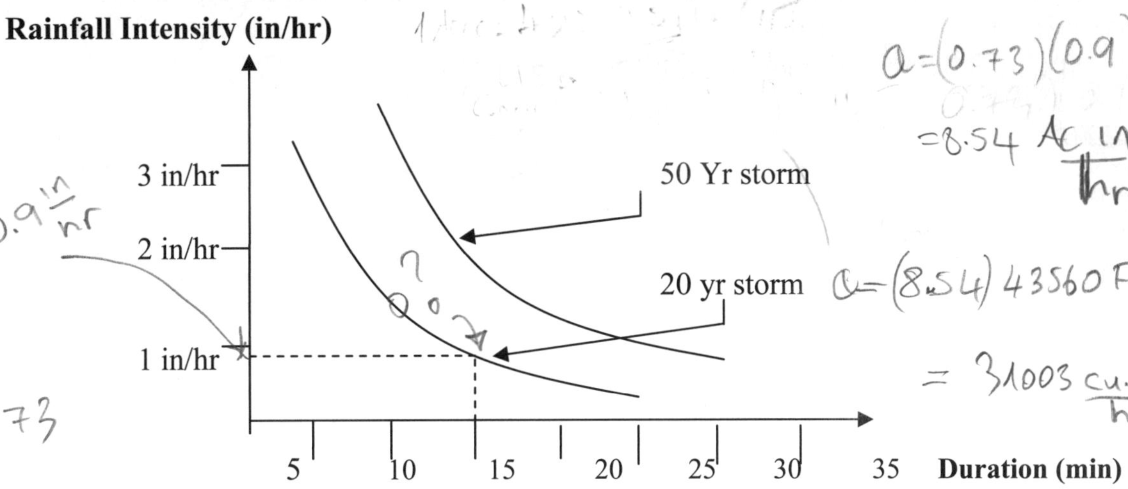

Drainage basin has an area of 13 acres. Time of concentration for the drainage basin was found to be 15 min. What is the design flow if the decision was made to design for a 20 year rainfall? Runoff coefficient (C) of the drainage area is 0.73.

A) 4.56 cu.ft/sec B) 8.61 cu.ft/sec C) 6.90 cu.ft/sec D) 12.78 cu.ft/sec

Problem 1.10): Hydrograph of a drain pipe is shown below. Find the maximum flow in the drain pipe (cu.ft/sec) if the net precipitation is 3.5 inches. Area of the drainage basin is 4 acres.

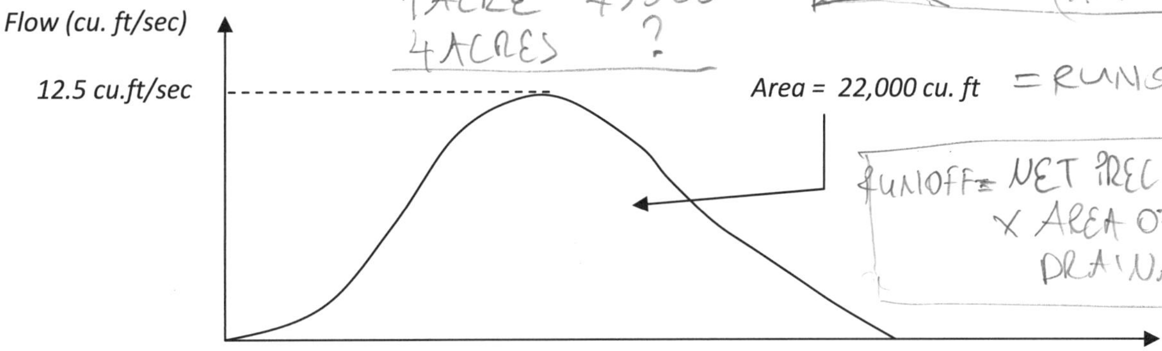

A) 12.6 cu.ft/sec B) 3.6 cu.ft/sec C) 28.9 cu.ft/sec D) 23.7 cu.ft/sec

Problem 1.11) A truck was left on top of a hill without brakes. The truck rolls down the hill for 500 ft. The friction coefficient between tires and the road is 0.02. The grade of the hill is 2.5%. What is the speed of the vehicle after rolling down 500 ft.

A) 21.3 mph   B) 12.7 mph   C) 34.5 mph   D) 19.2 mph

Problem 1.12) A car is travelling downhill in a 3% grade and collides with a truck. Impact speed was estimated to be 35 mph. Brake marks were visible for 210 feet. Coefficient of friction is 0.35. What is the speed of the car before applying brakes?

A) 67.9 mph   B) 56.9 mph   C) 45.6 mph   D) 32.1 mph

Problem 1.13) A total station is used to find the height of a building as shown in the figure. Find the height of the building (AB) from the information given. Angle ACD = 82^0 13' 11", Angle BCD = 65^0 45' 12"
Length DC = 165.9 ft

A) 654.4 ft   B) 274.9 ft   C) 845.9 ft   D) 454,9 ft

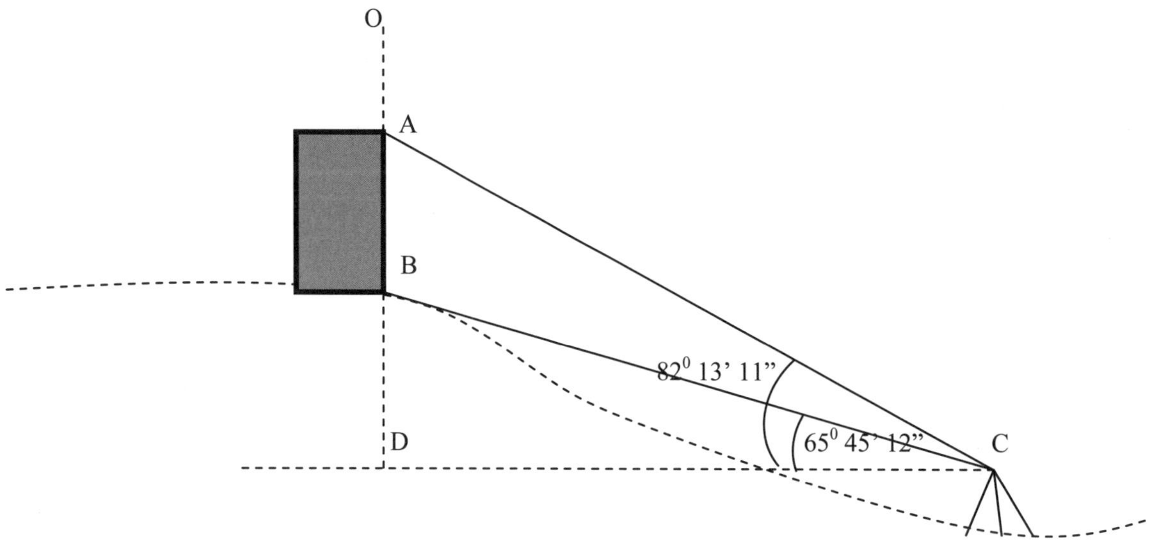

Problem 1.14) Intersection angle of a horizontal curve is 144^0 12' 14". Radius of the curve is 531.3 ft. Find the area of the hatched section.

A) 1.544 acres   B) 3.549 acres   C) 8.155 acres   D) 9.534 acres

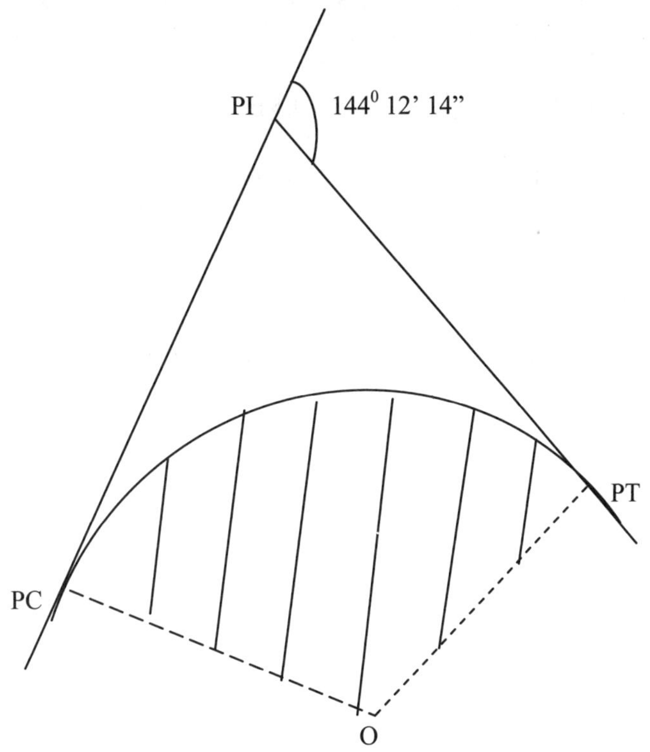

Problem 1.15) Density of traffic is found to be 230 vehicles per mile. Speed of vehicles is measured to be 60 mph. What is the flow or volume of traffic?

A) 4,500 veh/hr B) 28,202 veh/hr C) 13,800 veh/hr D) 29,670 veh/hr

Problem 1.16): Stopping sight distance of a road is 350 ft. The engineers are required to design a crest curve with an approaching gradient of 1.5% and going away gradient of 2.0%. Find the length of the crest curve needed.

A) 83.4 ft B) 198.7 ft C) 235.9 ft D) 1,289.6 ft

Problem 1.17) Station at point of curvature (PC) is given to be 4+10 and the radius of the curve is 140 ft. Find the station at PI. (Angle of intersection is 50 degrees).

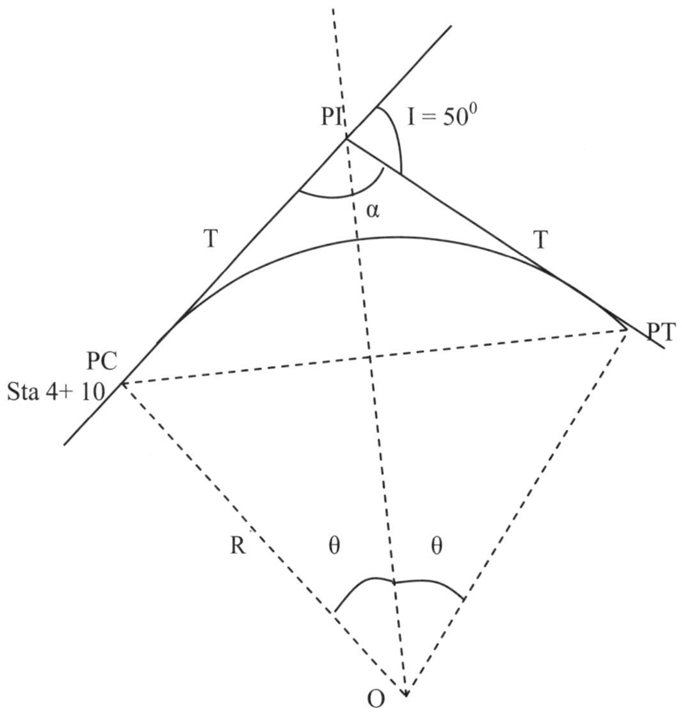

A) 4+66.4   B) 4+75.3   C) 3+93.0   D) 4+81.9

Problem 1.18): Engineers have designed a horizontal curve so that traffic at PC station will be able to see the traffic at PT station. Property owner would like to build a structure as shown. What's the closest distance to the road that the structure can be built?. (Distance AB)
Radius of the curve = 900 ft
Angle of intersection is 18^0.

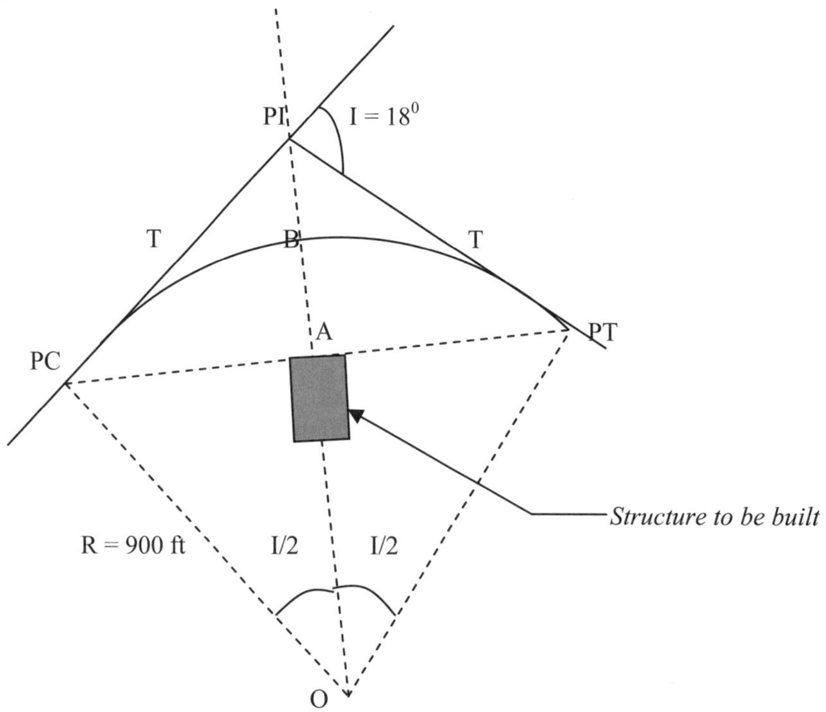

A) 13.4 ft B) 23.5 ft C) 3.4 ft D) 11.1 ft

Problem 1.19): Degree of curve of a horizontal curve (arc method) is given to be 4.3 degrees. What's the radius of the curve?

A) 1,332.4 ft B) 291.8 ft C) 1,621.1 ft D) 871.8 ft

Problem 1.20) Find the deflection at the end of the hollow steel tube shown. The outside diameter of the steel tube is 8 inches and the wall thickness is ¼ in. Young's modulus of steel is 29 x 10^6 psi

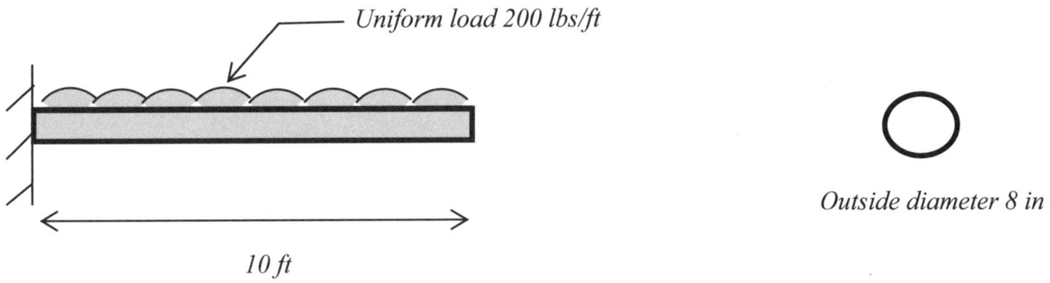

A) 3.61 in B) 0.325 in C) 12.34 in D) 5.67 in

Problem 1.21) A beam is loaded as shown. Find the maximum shear force of the beam shown.

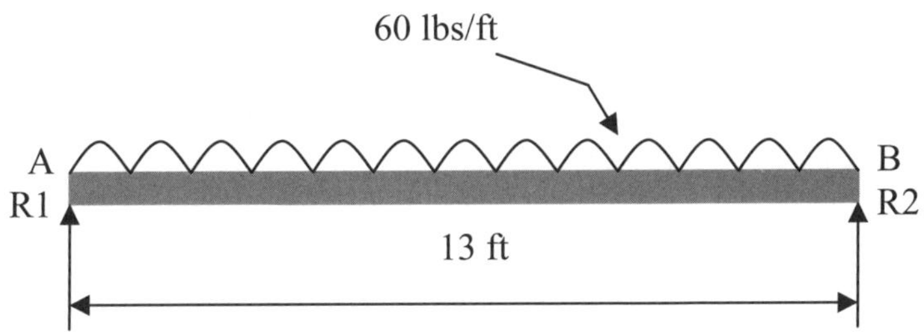

A) 450 lbs B) 390 lbs C) 780 lbs D) 810 lbs

Problem 1.22) Find the maximum bending moment for the loaded beam shown in the previous example.

A) 2,367.5 lbs. ft B) 1,345.9 lbs. ft C) 1,267.5 lbs. ft D) 2,340.8 lbs. ft

Problem 1.23) W12 x 96 column has a radius of gyration of 3.09 inches along weaker axis and 5.43 inches along stronger axis. Effective length factor (K) along both axis is equal to 1.0. Young's modulus of steel is 29 x 10^6 in^2. Area of the W section is 28.2 in^2. Unbraced length of the column is 15 ft. Find the elastic buckling load (Euler buckling load) of the column.

A) 1,245 kips B) 2,368 kips C) 1,765 kips D) 876 kips

Problem 1.24): What is true about liquefaction of soil?

A) Liquefaction mainly occur in soft clays
B) Liquefaction mainly occur in loose saturated sands
C) Liquefaction mainly occur in soft rock
D) Liquefaction mainly occur in any type of soil

Problem 1.25) What is true about natural period of buildings as related to seismic engineering?

A) Natural period of short buildings could be several seconds
B) Natural period of tall buildings could be less than one second
C) Natural period of short buildings and tall buildings are almost the same
D) Natural period of tall buildings could be several seconds

Problem 1.26): What is spectral acceleration as related to earthquakes?
A) Acceleration of soil particles during an earthquake
B) Acceleration of the building during an earthquake
C) Acceleration of the seismic wave
D) None of the above

Problem 1.27): Following successor relationships are available for a project. What is the duration of the project?

Activity ID	Duration (days)	Successors
A	5	B
B	12	C, D
C	6	D, F
D	8	E
E	12	F
F	3	

A) 22 days B) 26 days C) 46 days D) 40 days

Problem 1.28) A contractor is assigned with a project to install a 6" pipe at a depth of 5 ft. The width and length of the trench is 1 ft and 1,000 ft respectively. Contractor is planning to reuse the soil excavated to backfill the trench. Swell factor of the soil is 23% and shrinkage factor of the soil is 15%. Assume borrow pit soil also have the same swell factor and shrinkage factor. How many trucks of soil he may have to ship out (or brought in) if each truck can carry 10 CY of soil.
A) 1 B) 2 C) 3 D) 4

Problem 1.29): Contractor is trying to pull a tree stump. He believes 6 ton pull force will be enough to pull the stump. What is the pull force he needs from the pulling tractor?

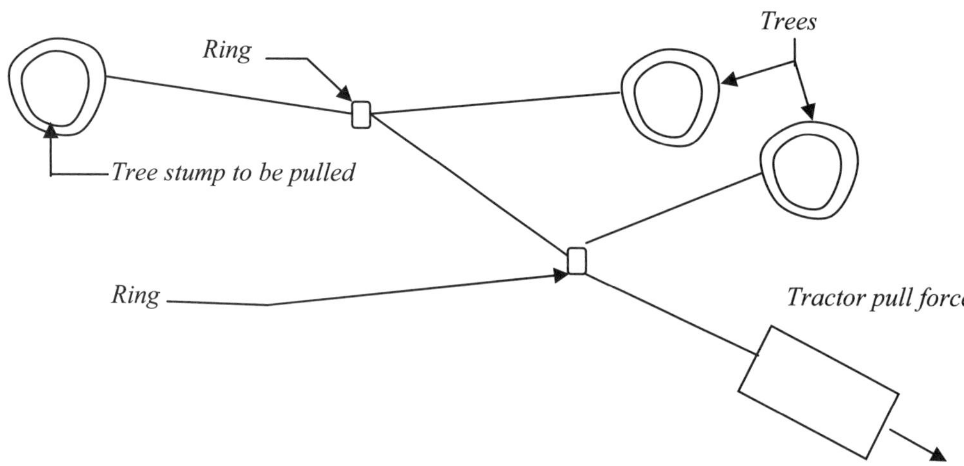

A) 6 tons B) 4 tons C) 1.5 tons D) 2 tons

Problem 1.30): What equipment is not needed for a SMAW welding setup?
A) Power source and cables B) Electrodes and electrode holders C) Plasma generator or plasma console D) AC or DC power supply

Problem 1.31) A construction crew consists of 3 laborers, 3 operators and one foreman. What is the crew hour rate and labor hour rate?

Hourly wages of workers;
Laborers - \$55
Operators - \$80
Foreman - \$90

A) 495 and 70
B) 495 and 45
C) 404 and 80
D) 430 and 67

Problem 1.32) 120 ft long, 11 ft high concrete wall needs to be constructed. The wall needs to be formed on both sides. Concrete forming crew consists of 4 carpenters, 3 laborers and 2 operators. What is the cost of labor for formwork?
Hourly wages of workers;
Carpenters - \$ 70
Laborers - \$ 55
Operators - \$ 80
Productivity of the crew = 80 SF/hour

A) \$17,672
B) \$19,230
C) \$19,965
D) \$21,108

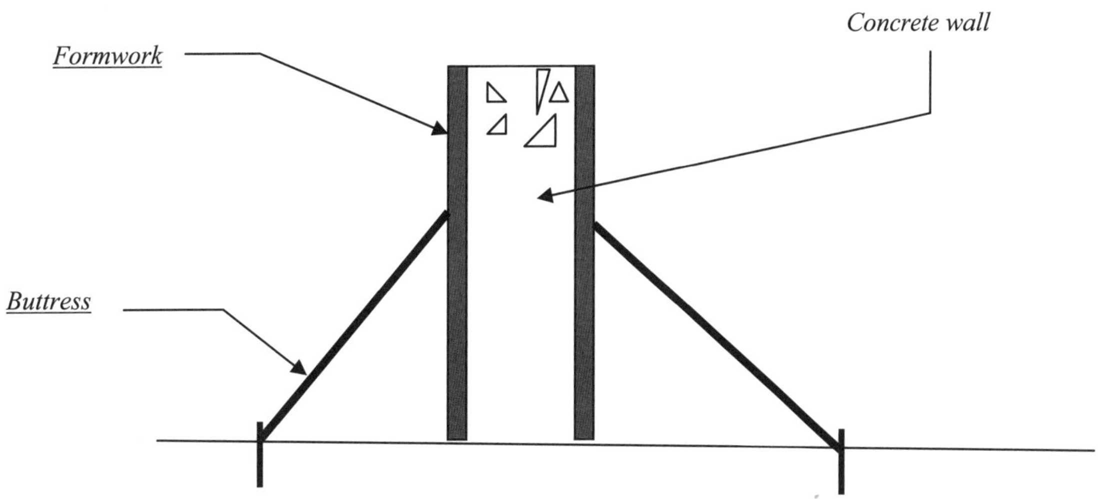

Problem 1.33): Find the allowable bearing capacity of the footing in tons.

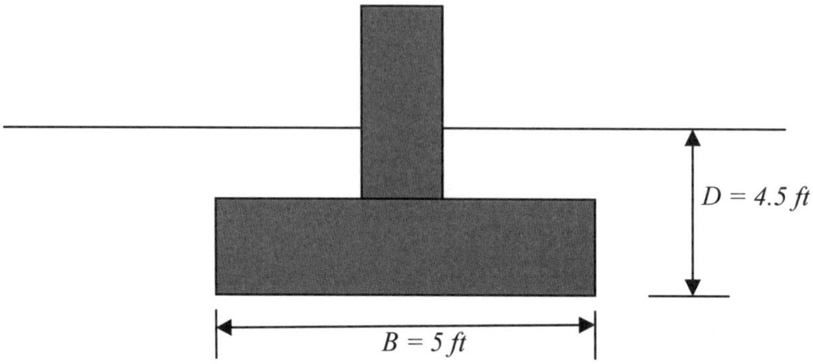

Length of the footing = 30 ft
Width of the footing = 5 ft
$N_c = 12.9$
$N_q = 4.4$
$N_\gamma = 2.5$
Total density of soil (γ) = 120 pcf
Cohesion of the soil = 300 psf
Factor of safety = 3.0
Use the Terzaghi Bearing Capacity Equation:

Ultimate Bearing Capacity in psf = $c.N_c + q.\ N_q + \frac{1}{2}\ B.\ \gamma.\ N_\gamma$
Ignore shape factors and depth factors.
Find the allowable bearing capacity of the footing in tons.

A) 134 tons B) 327 tons C) 102 tons D) 161 tons

Problem 1.34): Specific gravity of a soil sample is given to be 2.65. Moisture content and degree of saturation are 0.6 and 0.7 respectively. Find the void ratio.

A) 1.34 B) 2.27 C) 4.89 D) 0.78

Problem 1.35) Modified Proctor mould has an inner diameter of 4" and a height of 4.584 inches. Moisture content of soil is measured using a small tray that weigh 6 ounces.

Compacted soil + Mould = 15.76 lbs
The weight of the mould = 12 lbs
Water content measurement:

Prior to oven drying:

Weight of Soil + Tray = 1.21 lbs

After oven drying:

Weight of Soil + Tray = 1.15 lbs
Find the dry density of the sample.

A) 123.65 lbs/cu.ft B) 112.34 lbs/cu.ft C) 104.71 lbs/cu.ft D) 99.87 lbs/cu.ft

Problem 1.36) Find the three lateral earth pressure coefficients. (Lateral earth pressure coefficient at rest (K_0), active lateral earth pressure coefficient (K_a) and passive lateral earth pressure coefficient (K_p).

Soil friction angle (φ) = 33^0

	K_0	K_a	K_p
A)	0.53	0.34	0.23
B)	0.57	0.14	0.23
C)	0.45	0.29	3.39
D)	0.76	0.19	2.23

Problem 1.37): A Gabion wall is shown in the figure. Each gabion basket is 4 ft x 4 ft x 4 ft. Gabion baskets are filled with rock and has a density of 160 pcf. Soil backfill has a density of 110 pcf and has a friction angle of 30^0. Find the factor of safety against overturning.

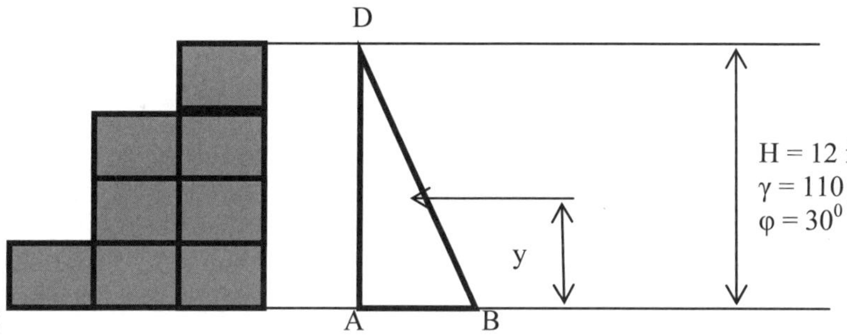

A) 3.2 B) 13.1 C) 3.7 D) 11.2

Problem 1.38) What is the new factor of safety against overturning if due to silt deposition water level has risen to 3 ft below the ground surface in the gabion wall given in the previous problem

A) 5.6 B) 8.8 C) 3.4 D) 7.9

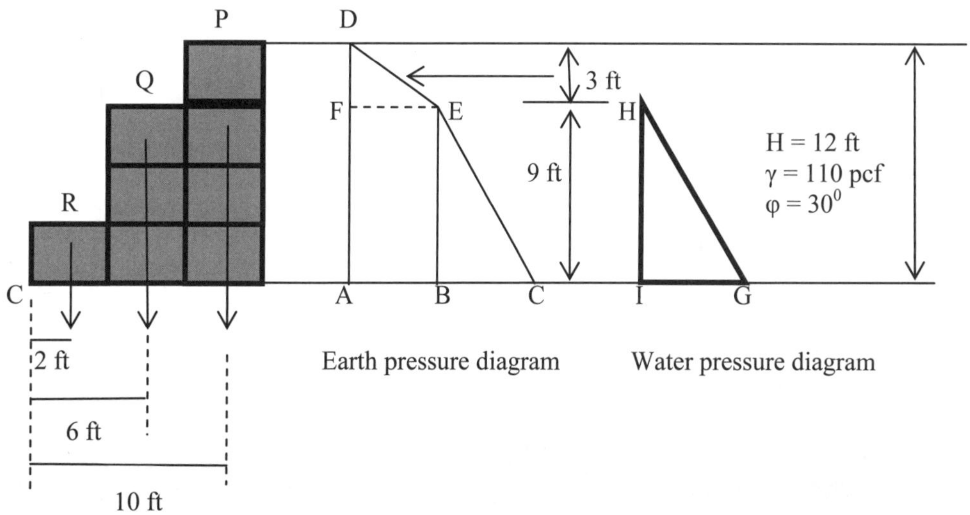

Problem 1.39): What is the correct order of processes in a wastewater plant?

A) Screens, grit chamber, primary clarifier, secondary clarifier, trickling filter, Chlorine and metal removal system, UV radiation unit
B) Screens, grit chamber, primary clarifier, trickling filter, secondary clarifier, Chlorine and metal removal system, UV radiation unit
C) Screens, grit chamber, primary clarifier, trickling filter, Chlorine and metal removal system, secondary clarifier, UV radiation unit
D) primary clarifier, trickling filter, Screens, grit chamber, Chlorine and metal removal system, secondary clarifier, UV radiation unit

Problem 1.40) Primary clarifier (Primary sedimentation tank) which has a diameter of 40 ft and a height of 8 ft receives 0.7 MGD of wastewater. What is the retention time of the clarifier?

A) 2.13 hrs   B) 2.57 hrs   C) 3.81hrs   D) 5.67 hrs

SAMPLE EXAM 1 (SOLUTIONS)

Problem 1.1): Find the wetted perimeter of the channel shown. Side slopes are 3 to 1 as shown.

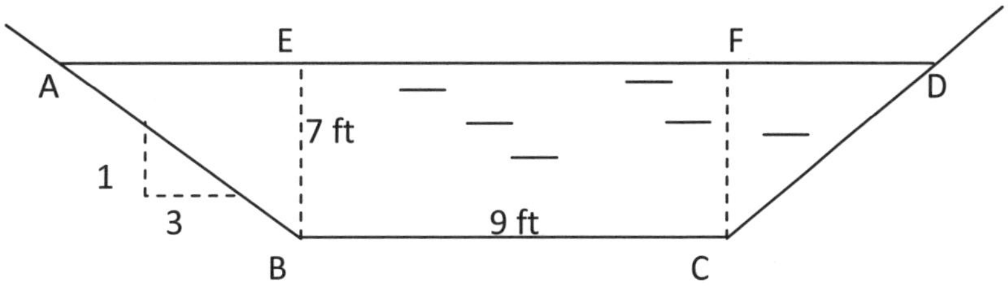

A) 23.5   B) 44.8   C) 53.2   D) 43.2

Solution 1.1):

Wetted perimeter = AB + BC + CD

If BE is 7, AE should be 21. (7 x 3).
This is due to 3:1 slope.

STEP 1: Find the length AB;

From Pythagoras theorem,

$AB^2 = AE^2 + BE^2$
$AB^2 = (21^2 + 7^2) = 490$
$AB = 22.1$ ft

AB = CD from symmetry.
Wetted perimeter = AB + BC + CD = 22.1 + 9 + 22.1 = 53.2 ft   (Ans C)

Discussion on Open Channel Flow:

Open channels are needed to transport water from one location to another. Water flow occurs due to gravity. Hence we can imagine that slope of the channel would affect the flow. It is obvious that a channel with a steep slope would have a higher flow than a channel with a gradual slope.

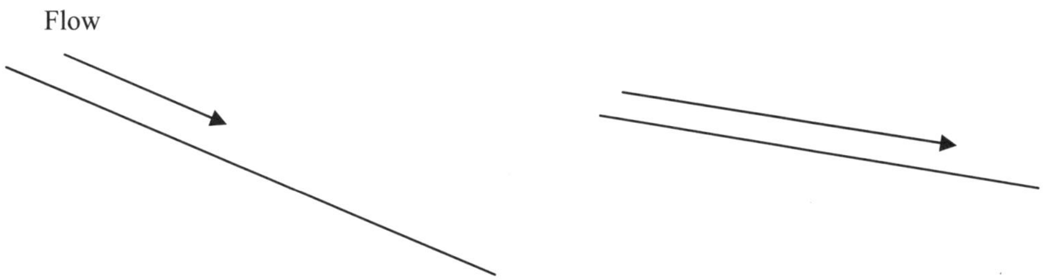

Steep slope (High velocity)   Gradual slope (Low velocity)

Other than the slope, there are other aspects of a channel that would affect the velocity of water.

Researchers have found that the shape of the channel also affects the velocity of water.

Channels that have a smaller wetted perimeter would have smaller loss of energy due to friction for a given area. Assume two channels with same slope and same area. The channel with smaller wetted perimeter would have a higher velocity.

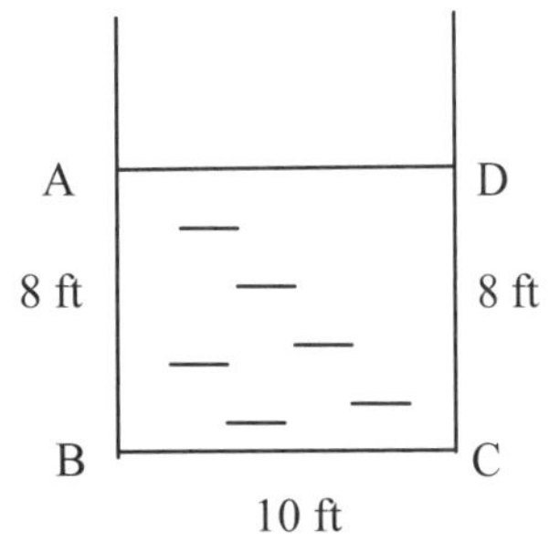

a) Higher flow velocity

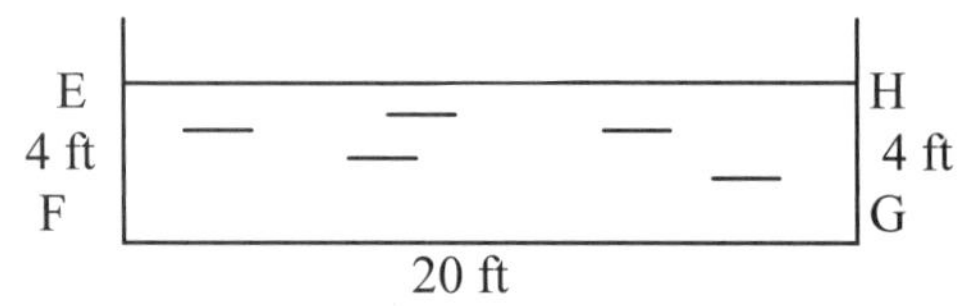

b) Low flow velocity

Areas of two channels shown above are same at 80 sq. ft.

Above figure "a" has a wetted perimeter of 26 ft (8 + 10 + 8). Which is AB + BC + CD.
Above figure "b" has a wetted perimeter of 28 ft (20 + 4 + 4). Which is EF + FG + GH.

Channel "a" would have a higher flow velocity assuming channel walls are made of same material.

<u>Hydraulic Radius (R)</u>: Hydraulic radius is defined as the ratio of area to the wetted perimeter.

R = A/P

R = Hydraulic radius;
A = Area
P = Wetted perimeter

If wetted perimeter is low, R is high and flow velocity would be high for a given area.

So far we can see that flow velocity of a channel is dependent on slope, roughness of channel walls and wetted perimeter. All these concepts were incorporated into the Manning equation.

$$v = \frac{1.49 \times R^{2/3} S^{1/2}}{n}$$

v = Velocity,
R = Hydraulic Radius = A/P
A = Area;
P = Wetted perimeter
S = Slope
n = Manning coefficient (Accounts for roughness of channel walls)

Implications of this equation are the following.

a) Higher the slope, higher the velocity
b) Higher the R, higher the velocity
c) Manning coefficient (n) is dependent upon the roughness of the wetted perimeter.

High "n" means wetter perimeter is rough. Concrete would have a lower "n" value compared to soil. Lower "n" value would result in a higher velocity.

Figure 1.1 *Concrete lined trapezoidal channel*

Problem 1.2) Water is flowing thru a rectangular channel at a flow rate of 800 cu. ft/sec. The depth of the channel is 6 ft and the width is 10 ft. What is the slope of the channel if Manning's coefficient is 0.02?

A) 0.8% B) 0.3% C) 0.15% D) 0.09%

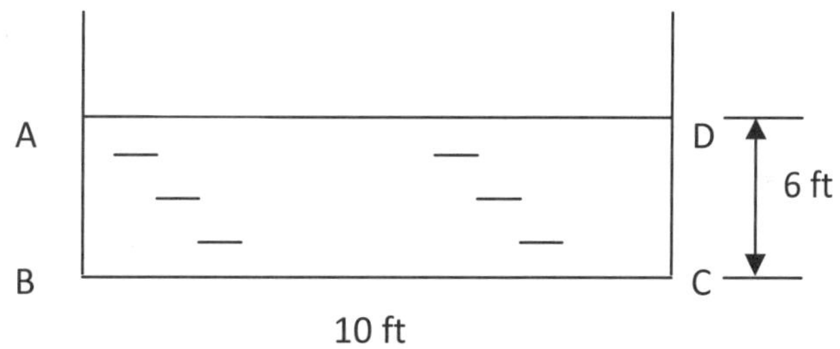

Solution 1.2)

STEP 1: Gather all the data given

In this case, flow rate (Q) and channel geometry is given.

$$v = \frac{1.49 \times R^{2/3} S^{1/2}}{n}$$

v = Velocity,
R = Hydraulic Radius = A/P
A = Area;
P = Wetted perimeter
S = slope
n = Manning's roughness coefficient

If you look at the Manning's equation, all the information has provided except the slope.

v = Velocity = Flow/Area = Q/A = 800/(6 x 10) = 13.3 ft/sec

R = A/P

P = Wetted perimeter = ABCD = 6 + 10 + 6 = 22 ft
R = 60/22 = 2.73

STEP 2: Input all the known values into Manning's equation:

$$v = \frac{1.49 \times R^{2/3} S^{1/2}}{n}$$

$$13.3 = \frac{1.49 \times (2.73)^{2/3} S^{1/2}}{0.02}$$

$13.3 = 145.5\ S^{1/2}$
$S^{1/2} = 0.091$

S = 0.008 ft/ft

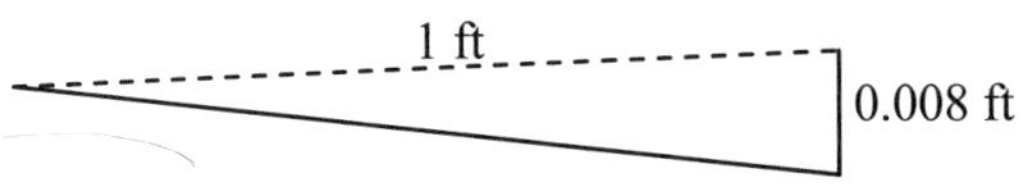

Typically, the slope is given as a percentage.
S = 0.8% (Ans A)

Finding the Froude Number of a Channel:

Problem 1.3) Find the Froude number of the channel shown if the flow is 13 cu.ft/sec. Channel width is 11 ft and channel depth is 4.3 ft.

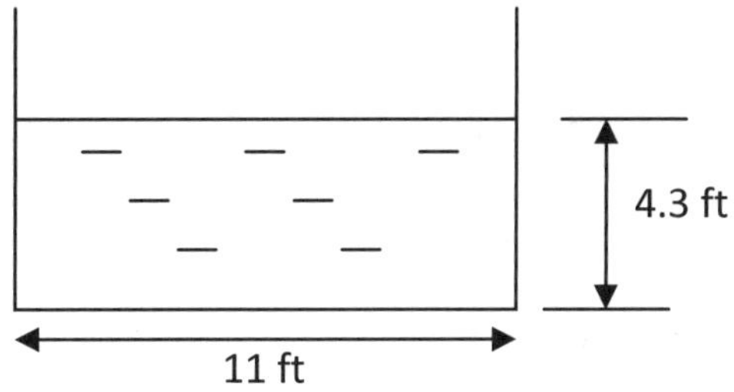

A) 0.034   B) 0.024   C) 0.765   D) 0.876

Solution 1.3):

The Froude number (F_R) is given by the following equation.

$$F_R = V/(gD)^{1/2}$$

V = Velocity;
D = Hydraulic Depth = A/T,
A = Area
T = width of the top surface.   (In this case T = 11 ft)

For rectangular channels, width of the channel at the bottom and top are the same. This is not the case for trapezoidal channels

- Note that hydraulic depth is different than the hydraulic radius.
- Hydraulic Depth = A/T;
- Hydraulic Radius = A/P

g = Gravitational constant (32.2 ft/sec^2 for English units or 9.81 m/sec^2 for metric units).

STEP 1:   Find the velocity;

Velocity (V) = Q/Area
Q = Flow
Velocity (V) = 13/(11 x 4.3) = 0.275 ft/sec

STEP 2:   Find "D"   (D = Hydraulic depth)
D = A/T
A = 11 x 4.3
T = Width of the top surface = 11 ft

D = (11 x 4.3)/11 = 4.3 ft
Hence Froude number (F_R) = $V/(gD)^{1/2}$ = $0.275/(32.2 \times 4.3)^{1/2}$ = 0.024   (Ans B)

Finding the Critical Depth of a Rectangular Channel:

Problem 1.4) Find the critical depth of the channel shown if the flow is 17.5 cu.ft/sec. Channel width is 11 ft.

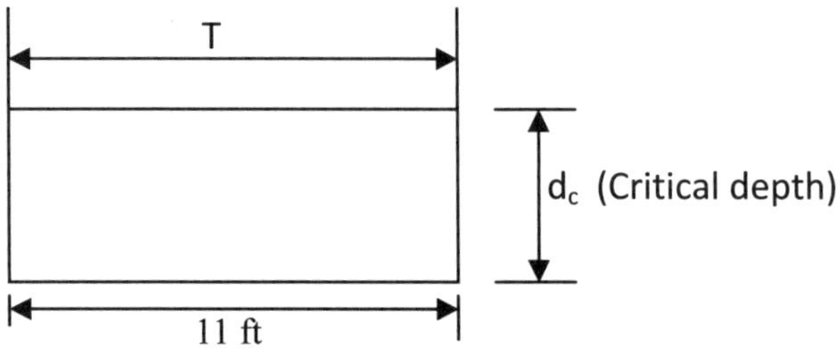

A) 0.428 ft B) 0.254 ft C) 0.654 ft D) 0.876 ft

Solution 1.4): When the flow is critical, Froude number becomes 1.0.

The Froude number (F_R) is given by the following equation.

$$F_R = V/(g.D)^{1/2}$$

V = Velocity;

D (hydraulic depth) = A/T, where A = Area and T = width of the top surface.

STEP 1: Find the velocity;

Velocity (V) = Q/Area Q = Flow
Velocity (V) = $17.5/(11 \times d_c)$

STEP 2: Find "D" (Hydraulic depth)

D = A/T
A = Area = $11 \times d_c$
T = Width of the top surface = 11 ft
D = A/T = $(11 \times d_c)/11 = d_c$ ft
Hence Froude number (F_R) = $V/(gD)^{1/2} = [17.5/(11 \times d_c)] / [(32.2 \times d_c)^{1/2}]$

STEP 3: Make Froude number equal to 1.0 to find the critical depth of the channel. At critical depth, Froude number becomes equal to 1.0.

$(F_R) = V/(gD)^{1/2} = [17.5/(11 \times d_c)] / [(32.2 \times d_c)]^{1/2} = 1.0$
$17.5/(11 \times d_c) = (32.2 \times d_c)^{1/2}$
$17.5/11 = (32.2 \times d_c)^{1/2} \times d_c$
$1.591 = 5.675 \times d_c^{1.5}$

$d_c^{1.5} = 0.2804$
Hence d_c (critical depth) = 0.428 ft (Ans A)

Discussion:

When Fr > 1.0 Flow is supercritical
When Fr < 1.0 Flow is subcritical
When Fr = 1.0 Flow is critical

Hydraulic Jump in Rectangular Channels:

Problem 1.5) A rectangular channel is undergoing a hydraulic jump. Upstream depth of the channel is 0.4 ft and velocity of the channel is 7.0 ft/sec. What is the depth after the hydraulic jump?

A) 1.23 ft B) 2.84 ft C) 0.92 ft D) 0.21 ft

Solution 1.5)

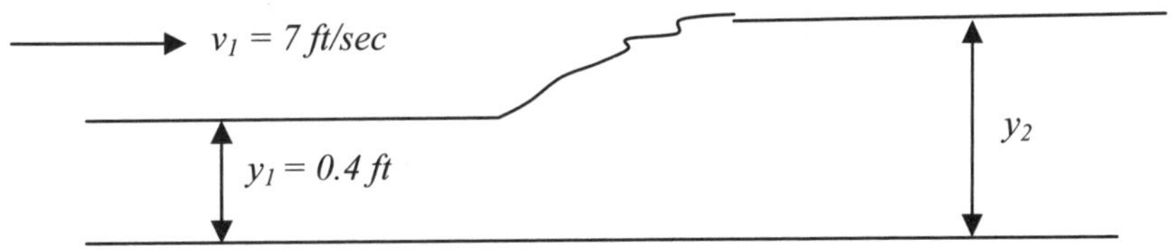

Hydraulic jump equation for a rectangular channel is shown below. Note that this equation is valid only for rectangular channels.

$$y_2/y_1 = \tfrac{1}{2}\left[(1 + 8\,v_1^2/gy_1)^{1/2} - 1\right]$$

y_2 = Depth after the jump;
y_1= Depth before the jump
v_1 = Velocity of the channel before the jump

y_1= 0.4 ft; v_1 = 7.0 ft

Using above equation; y_2 = 0.92 ft (Ans C)

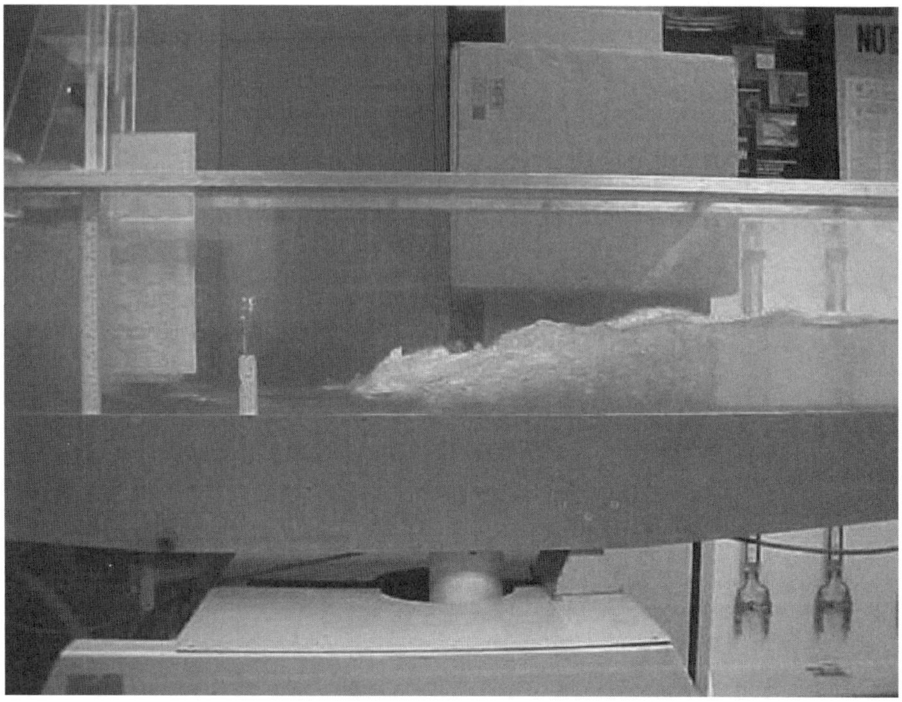

Hydraulic Jump *- Left side of the channel shows that water is flowing at very high speed. The depth is smaller before the jump. After the jump, the depth is increased. It can be clearly seen that the depth of the water on right end of the photo is much higher than the left end of the photo. The transition occurs thru a jump. Energy is dissipated during a hydraulic jump.*

Pipe Flow:

Problem 1.6): Find the velocity at point B in the pipe shown. Ignore the head loss due to friction.

A) 32.8 ft/sec   B) 28.8 ft/sec   C) 13.9 ft/sec   D) 65.8 ft/sec

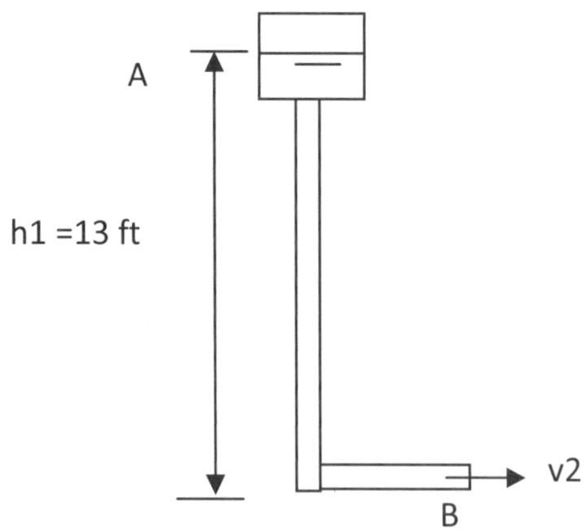

Solution 1.6):

This problem can be solved using the energy equation. Energy equation is also known as Bernoulli's equation.

STEP 1: Write the energy equation;

$$h_1 + v_1^2/2g + p_1/\gamma = h_2 + v_2^2/2g + p_2/\gamma$$

h_1 = Energy due to gravity (Datum head);
$v_1^2/2g$ = Kinetic energy
p_1/γ = Pressure head

Energy at point A = $h_1 + v_1^2/2g + p_1/\gamma$
Energy at point B = $h_2 + v_2^2/2g + p_2/\gamma$

At point A:
$h_1 = 13$ ft
$v_1 = 0$;
(Velocity of the reservoir surface is negligible).
$p_1 = 0$;
Reservoir surface is exposed to the atmosphere. Hence pressure is zero.

At point B:
$h_2 = 0$ ft
$v_2 = ?$;
$p_2 = 0$;
Water at point B is exposed to the atmosphere. Hence pressure is zero.

STEP 2: Apply the Energy equation;

$h_1 + v_1^2/2g + p_1/\gamma = h_2 + v_2^2/2g + p_2/\gamma$

$13 + 0 + 0 = 0 + v_2^2/2g + 0$

$v_2 = 28.8$ ft/sec (Ans B)

Discussion:

Energy Equation (Bernoulli's Equation):

Fluids contain energy in the form of pressure, velocity and datum. Energy cannot be created nor destroyed. Energy can dissipate from the system as heat due to friction.

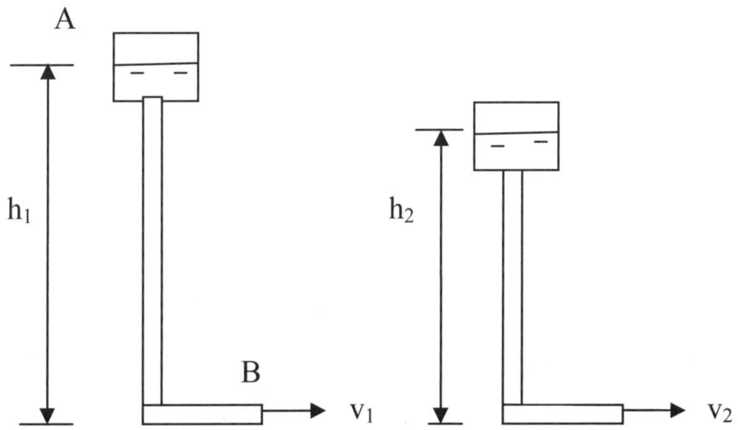

It is obvious that $v_1 > v_2$

It was known for many centuries that higher the "h", higher the velocity. The connection between these two parameters was not known until Bernoulli found the equation.

Energy at point A = Energy at point B

This is valid only when the head loss due to friction is negligible.

Energy in a pipe line can be in three different forms,

- Pressure energy
- Kinetic energy
- Energy due to elevation (Datum head)

Pressure Energy: Pressure is also a form of energy. In typical pipes, water flows under pressure. Hence pressure energy also has to be considered.

Pressure head is given by the following equation

$$\boxed{\text{Pressure head} = P/\gamma}$$

P = Pressure in psf and γ is the density of water in pcf.

Pressure head units = P (psf)/γ (pcf) = ft

Kinetic Energy (Velocity Head): Kinetic energy is due to velocity. Kinetic head is given by the following equation.

$$\boxed{\text{Kinetic head} = v^2/2g}$$

v = velocity in ft/sec.
g = Gravitational constant. (32.2 in US units)

Energy due to High Elevation (Datum Head):

Datum head is given by simply ft.

Total Energy inside a Pipe = Pressure head + Velocity Head + Datum Head
$= P/\gamma + V^2/2g + Z$

$$\boxed{\text{Total Energy} = P/\gamma + V^2/2g + Z}$$

Problem 1.7): Pressure inside a pipe is 1 psi. What is the pressure head in ft of water?

A) 3.8 ft B) 2.3 ft C) 5.8 ft D) 9.7 ft

Solution 1.7):

Pressure Conduits:

Pressure can be converted to velocity. Pressure in psi has an equivalent head. Normally all forms of energy is converted to head (ft of water).

Pressure is given in psi. This has to be converted to psf.
1 psi = 1 x 144 psf
Density of water = 62.4 pcf

Pressure head = p/γ = 144/62.4 = 2.3 ft

γ = Density of water

1 psi is equal to 2.3 ft of pressure head. (Ans B)

HYDROLOGY:

Problem 1.8): Net precipitation (after deducting for infiltration) of a watershed area was found to be 2.45 inches. The main culvert located at the most downstream end of the watershed area had 40,210 cu. ft of water passing thru it during the storm. Find the total acreage of the watershed area.

A) 2.34 Acres B) 12.76 Acres C) 4.52 Acres D) 8.76 Acres

Solution 1.8):

Following equation can be used to find the area of the drainage basin;

Runoff = Area of the drainage basin x Net precipitation

STEP 1: Find the total acreage of the watershed area:

During a rainfall event, portion of water that falls into ground would be infiltrated. What's left would be the runoff. In this case runoff is given to be 40,210 cu. ft.

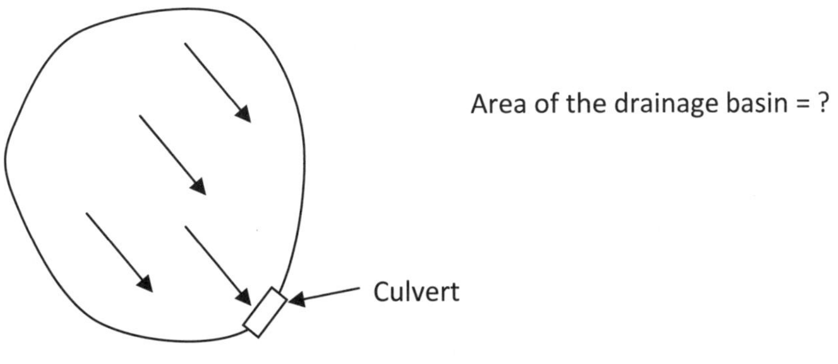

Area of the drainage basin (watershed area) = Need to be found

Total discharge thru the culvert (Runoff) = 40,210 cu. ft
Net precipitation (after deducting the infiltration) = 2.45 inches = 2.45/12 ft

Runoff = Area of the watershed basin x Net precipitation

40,210 cu. ft = Area of the watershed basin x 2.45/12 ft

Area of the watershed basin = 40,210 x 12/2.45
= 196,947 sq. ft = 196,947/43,560 acres
= 4.52 acres (Ans C)

Note 1: 1 Acre = 43,560 sq. ft
Note 2: Watershed basin is also known as drainage basin and catchment area.

Water flows from two sides into the stream at the center. The stream then carries water to a river.

Number of streams are seen in the Canyon. Each stream would be recharged during a rain event. All streams would feed a major river. (See below).

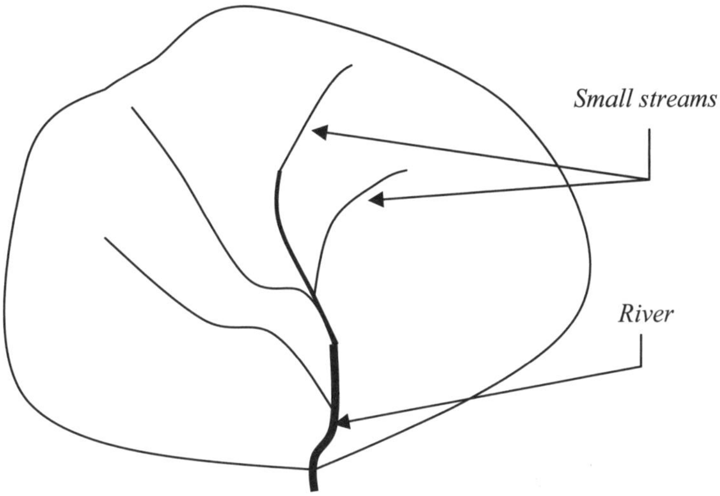

Problem 1.9): 20 year and 50 year storms are shown in the figure.

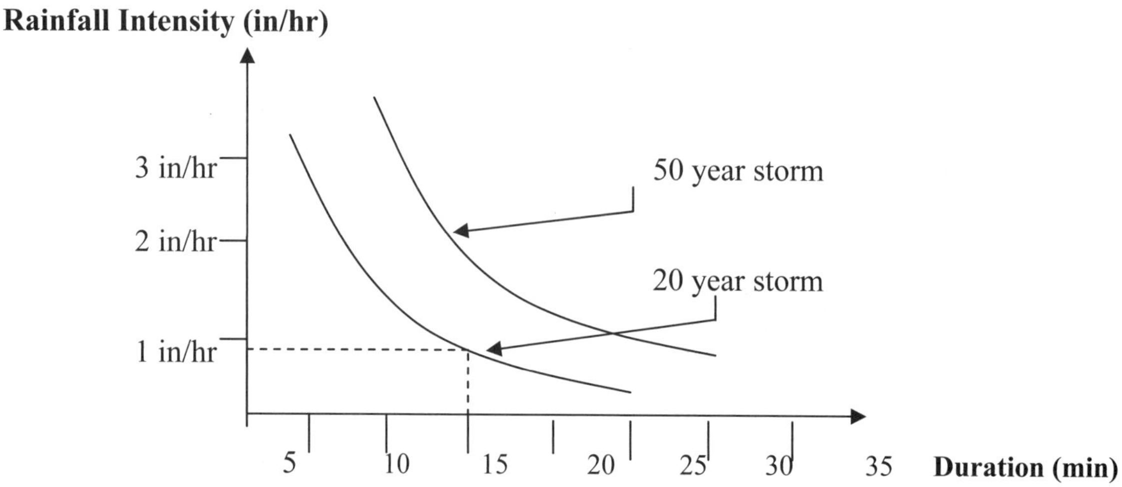

Drainage basin has an area of 13 acres. Time of concentration for the drainage basin was found to be 15 min. What is the design flow if the decision was made to design for a 20 year rainfall? Runoff coefficient (C) of the drainage area is 0.73.

A) 4.56 cu.ft/sec B) 8.61 cu.ft/sec C) 6.90 cu.ft/sec D) 12.78 cu.ft/sec

Solution 1.9)

STEP 1: This problem needs to be solved using the rational formula; Q = CIA

$$Q = CIA$$

Q = Design flow (Acre. in/hr)
C = Runoff coefficient (No units)
I = Design rainfall intensity (in/hr)
A = Area of the drainage basin (Acres)

STEP 2: Find the rainfall Intensity "I":

"I" can be obtained by using the graph for 20 year rainfall. Time of concentration is given to be 15 minutes. Time of concentration is the time taken by rainwater from the *hydrologically most distant point to reach the culvert.*

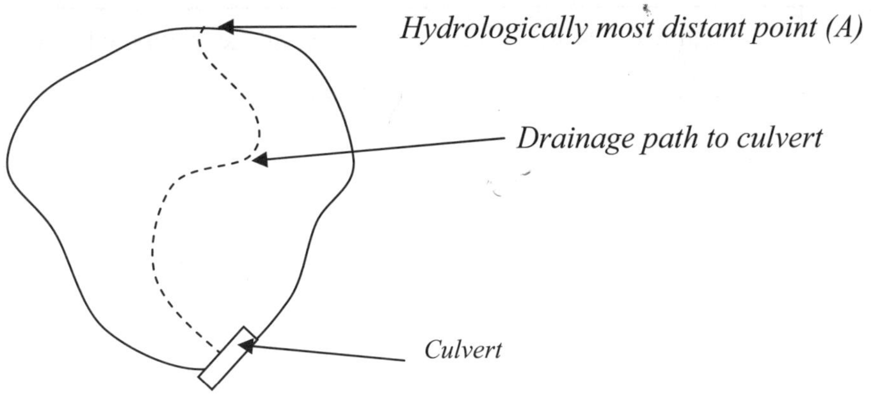

The idea is to find the maximum flow at the culvert. Hydrologically most distant point may not be the most distant point from the culvert. Water runoff from a given point to the culvert is not a straight line. If the duration of the rainfall is less than 15 minutes, rain would stop before water from the most distant point reach the culvert. Hence one would not get the maximum flow. To obtain the maximum flow, one needs to wait till the flow from all parts of the drainage basin has reached the culvert.

Q = C.I.A

C = 0.73 (Given);
I = 0.9 in/hr (For a time of concentration of 15 min "I" was obtained from the graph given).
A = 13 Acres; (Given)

Q = 0.73 x 0.9 x 13 = 8.541 Acre in/hr
1 Acre = 43,560 Sq. ft
Q = 8.541 x 43,560/12 = 31,003 cu.ft/hr = 8.61 cu.ft/sec (Ans B)

Problem 1.10): Hydrograph of a drain pipe is shown below. Find the maximum flow in the drain pipe (cu.ft/sec) if the net precipitation is 3.5 inches. Area of the drainage basin is 4 acres.

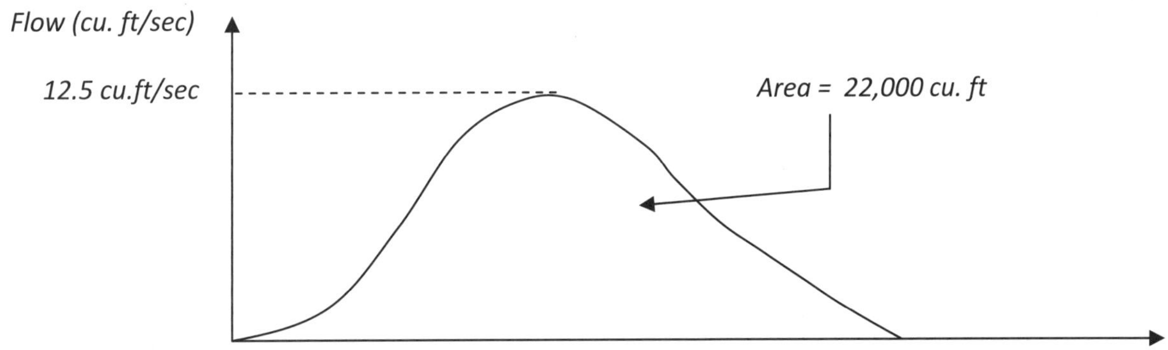

A) 12.6 cu.ft/sec   B) 3.6 cu.ft/sec   C) 28.9 cu.ft/sec   D) 23.7 cu.ft/sec

Solution 1.10): Following equations are necessary to solve hydrograph problems.

Runoff = Area of the Hydrograph

Runoff = Net Precipitation x Area of the drainage basin

STEP 1: Find the runoff:

Runoff volume is given by the area of the hydrograph.
Runoff = 22,000 cu.ft.

STEP 2: Find the net precipitation;

Runoff = Net precipitation x Area of the drainage basin

Area of the drainage basin = 4 acres;

1 acre = 43,560 sq. ft
Runoff = Net precipitation x Area of the drainage basin
22,000 = Net precipitation x (4 x 43560)

Net precipitation = 0.1263 ft
= 1.515 inches

STEP 3: Find the maximum flow due to 1 inch rainfall;

According to the hydrograph given, maximum flow of 12.5 cu.ft/sec is achieved due to a net precipitation of 1.515 inches.

Maximum flow due to 1.0 inch = 12.5/1.515 = 8.25 cu.ft/sec

STEP 4: Find the maximum flow due to 3.5 inch rainfall;

Maximum flow due to 1 inch rainfall = 8.25 cu.ft/sec
Maximum flow due to 3.5 inch rainfall = 8.25 x 3.5 cu.ft/sec = 28.9 cu.ft/sec (Ans C)

Discussion:

Following three equations are needed for hydrographs problems.

Runoff = Area of the hydrograph

Runoff = Net precipitation x Area of the drainage basin

Runoff in a unit hydrograph = 1" x Area of the drainage basin

TRANSPORTATION:

Problem 1.11) A truck was left on top of a hill without brakes. The truck rolls down the hill for 500 ft. The friction coefficient between tires and the road is 0.02. The grade of the hill is 2.5%. What is the speed of the vehicle after rolling down 500 ft.

A) 21.3 mph   B) 12.7 mph   C) 34.5 mph   D) 19.2 mph

Solution 1.11):

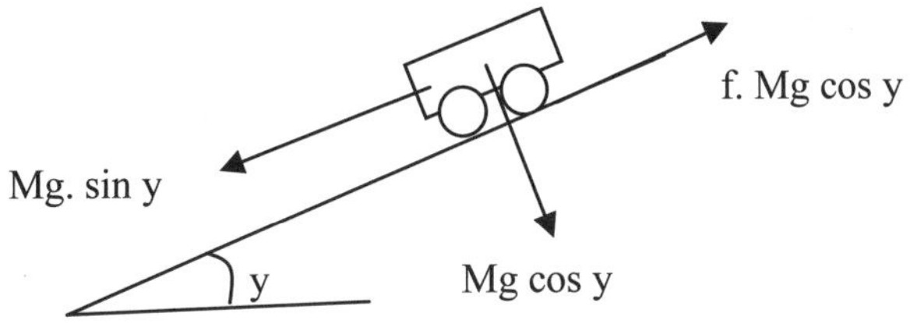

f = friction coefficient

STEP 1: Write down necessary parameters;

First let us look at the forces acting on the truck. Gravity is acting vertically downward. Gravitational force can be resolved into two components. One force is parallel to the slope and the other perpendicular to the slope.
Gravitational force parallel to the slope = mg. sin y
Gravitational force perpendicular to the slope = mg. cos y

Then there is friction acting opposite to the movement of the truck.
The grade of the slope is given to be 2.5%. Grade is the tangent of the slope angle.
tan y = Grade = 0.025
Hence; $y = \tan^{-1} 0.025 = 1.43$ degrees

STEP 2: Resolve forces parallel to the slope;

Force parallel to the slope = mg sin y – f. mg. cos y
f = Friction coefficient

STEP 3: Apply Newton's equation; P = m. a

$$P = m. a$$

P = Force;	*m = mass;*	*a = acceleration*

Force = Force parallel to the slope = mg sin y – f. mg. cos y
Apply the equation; P = m. a
P = mg sin y – f. mg. cos y = m. a
Divide the whole equation by "m"
g sin y – f. g. cos y = a
32.2 sin 1.43 - 0.02 x 32.2 cos 1.43 = a
$a = 0.7923$ ft/sec^2.

STEP 4: Find the velocity of the vehicle after rolling down 500 ft.

We just found the acceleration of the vehicle to be 0.7923 ft/sec^2. We need to find the velocity after travelling 500 ft. Let us look at three dynamic equations;
Three equations of dynamics;

$v = u + at$ ---------------------------------------(1)
$s = ut + 1/2at^2$---------------------------------(2)
$v^2 = u^2 + 2as$ ---------------------------------(3)

u = Initial velocity;
v = Final velocity;
s = Distance;
t = time;
a = acceleration

What we know so far;
Initial velocity (u) = 0.0
Acceleration (a) = 0.7923 ft/sec^2
Travel distance (s) = 500 ft (given)
Need to find final velocity (v);
Use above equation (3).

$v^2 = u^2 + 2a.s$
$v^2 = 0^2 + 2 \times 0.7923 \times 500$
v = 28.14 ft/sec = 19.2 mph Ans D

Problem 1.12) A car is travelling downhill in a 3% grade and collides with a truck. Impact speed was estimated to be 35 mph. Brake marks were visible for 210 feet. Coefficient of friction is 0.35. What is the speed of the car before applying brakes?

A) 67.9 mph B) 56.9 mph C) 45.6 mph D) 32.1 mph

Solution 1.12:
STEP 1: Draw the force diagram;

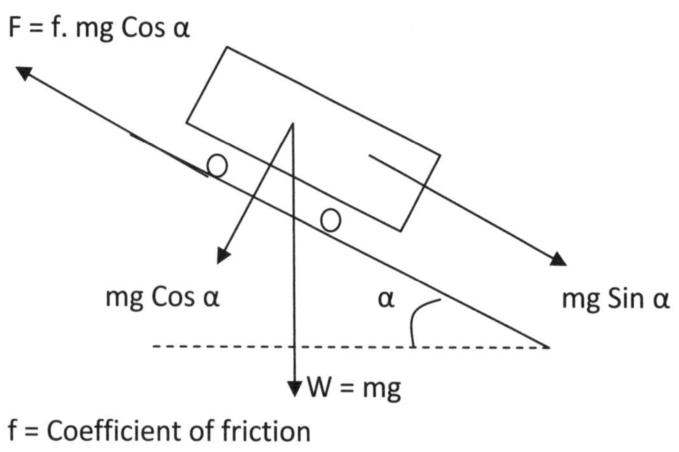

Weight of the vehicle is mg. This weight is acting vertically downward.
mg sin α = Component of the weight acting downward along the slope
mg cos α = Component of the weight acting perpendicular to the slope
f. mg cos α = Force due to friction. Always acts against the movement

STEP 2: Resolve forces along the slope;

$$\text{Acceleration force (P)} = mg \sin\alpha - f.\, mg \cos\alpha$$

STEP 3: Find α;

Grade is given to be 3%. (3% = 0.03)
Grade = tan α = 0.03
Hence α = 1.72^0

STEP 4: Find the acceleration:

Apply Newton's second law (P = mf) to find the acceleration of the vehicle.

$$P = m \times a$$

P = Acceleration force (lbf);
a = acceleration (ft/sec^2)

$mg \sin\alpha - f\, mg \cos\alpha = m.\, a$
$g \sin 1.72^0 - f.\, g \cos 1.72^0 = a$ (Divide by m)
g = 32.2 and f = 0.35
Hence a = -10.3 ft/sec^2.

Negative acceleration means it is a *deceleration*. Since the brakes are applied, the vehicle had started to decelerate.

STEP 5: Find the initial velocity;

Three equations of dynamics;

$$v = u + at \text{ ---------------------------------------(1)}$$
$$s = ut + \tfrac{1}{2}.\, a.t^2 \text{ ---------------------------------(2)}$$
$$v^2 = u^2 + 2.a.s \text{ ---------------------------------(3)}$$

u = Initial velocity;
v = Final velocity;
s = Distance;
t = time;
a = acceleration

In this problem, final velocity or the impact velocity is given. (v = 35 mph).
Travel distance is given to be 210 ft. Hence, s = 210 ft.
Initial velocity "u" can be found using the above third equation.

$v^2 = u^2 + 2.a.s$

v = 35 mph = 51.3 ft/sec;
a = -10.3 ft/sec^2. (Found earlier)
s = 210 ft

$51.3^2 = u^2 + 2 \times (-10.3) \times 210$
$u^2 = 6958$;
u = Initial velocity = 83.4 ft/sec = 56.9 mph (Ans B)

Problem 1.13) A total station is used to find the height of a building as shown in the figure. Find the height of the building (AB) from the information given. Angle ACD = 82^0 13' 11", Angle BCD = 65^0 45' 12"
Length DC = 165.9 ft

A) 654.4 ft   B) 274.9 ft   C) 845.9 ft   D) 454,9 ft

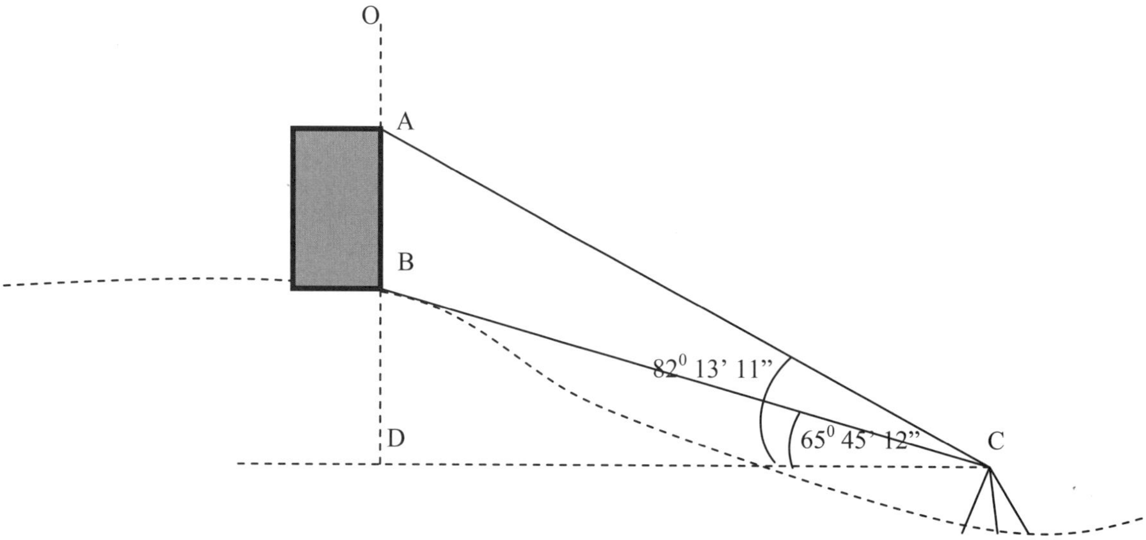

Solution 1.13:
Angle ACD = 82^0 13' 11" = 82.2197^0
Length AD = DC tan 82.2197^0
DC distance is given to be 165.9 ft
Hence length AD = 165.9 x tan 82.2197^0 = 165.9 x 7.319 = 1,214.2 ft
Angle BCD = 65^0 45' 12" = 65.7533^0
Length BD = DC x tan 65.7533^0 = 165.9 x 2.22 = 368.3 ft
Building height = AB = AD – BD = 1,214.2 – 368.3 = 845.9 ft
Ans C

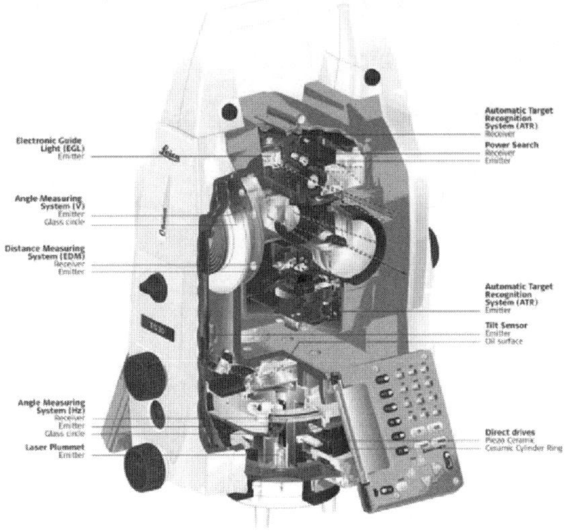

Total station

Total station can be used to measure distances, horizontal angles and vertical angles. Electro magnetic waves are used to measure distances.

Problem 1.14) Intersection angle of a horizontal curve is 144^0 12' 14". Radius of the curve is 531.3 ft. Find the area of the hatched section.

A) 1.544 acres B) 3.549 acres C) 8.155 acres D) 9.534 acres

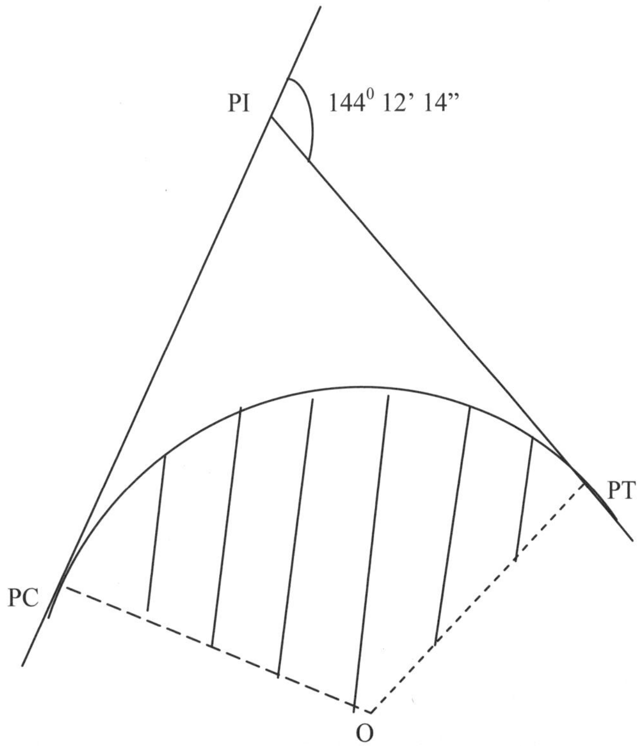

Solution 1.14):

STEP 1: Find the angle PC O PT

Angle PC O PT is same as the intersection angle.

Angle PC O PT = 144^0 12' 14" = 144.2039^0

STEP 2: Find the area;

Area of a full circle = $\pi . R^2$
Full circle has 360 degrees.

Area inside one degree arc = $\pi . R^2/360$
Area inside 144.2039 degree arc = $\pi . R^2/360 \times 144.2039$

R is given to be 531.3

Hence area of the hatched section = $\pi . (531.3)^2/360 \times 144.2039$ = 355,225.4 sq. ft
= 355,225.4/43,560 acres = 8.155 acres
Note that one acres is equal to 43,560 sq. ft.
Ans C

Problem 1.15) Passenger car equivalent service flow rate (v_p) in a freeway is 1,500 veh/hr/ln. Free flow speed of the freeway is 55 mph. What is the level of service of the freeway.

Level of Service (LOS)

	A	B	C	D	E	F
Density (vehicles/mile per lane)	Less than or equal to 11	Greater than 11, less than or equal to 18	Greater than 18, less than or equal to 26	Greater than 26, less than or equal to 35	Greater than 35, less than or equal to 45	Greater than 45

HCM– Highway Capacity Manual

A) LOS A B) LOS B C) LOS C D) LOS D

Solution 1.15):

If the density is less than or equal to 11, then the level of service is considered to be A, which is the best level of service. Level of service is considered to be "B" when density is between 11 and 18.

In this problem, flow is given to be 1,500 veh/hr/ln and speed is given to be 55 mph.

Flow (Vehicles/hr) = Density x Speed
1,500 = Density x 55
Density = 27.3 vehicles/mile.

Since the density is higher than 26, this road does not qualify to be considered as LOS C. Hence this road is LOS D.

Discussion:

General description for level of service:

Level of service A: Traffic is moving approximately 90% of the free flow speed (FFS). Drivers do not get delayed due to other vehicles in the road. Vehicles are far apart.

Level of service B: Traffic is moving approximately 70% of the free flow speed (FFS). Drivers get slightly delayed due to other vehicles in the road. Vehicles are far apart.

Level of service C: More congested than LOS B. But the traffic is moving steadily.

Level of service D: Traffic is moving approximately 40% of the free flow speed (FFS). More congested than LOS C.

Level of service E: Traffic is moving approximately 33% of the free flow speed (FFS). Extensive delays are experienced at some intersections.

Level of service F: Traffic is moving approximately 25% to 33% of the free flow speed (FFS). Extensive delays are experienced. Traffic is moving at very low speed.

Level of service F

Level of service probably B or C

Level of service A

LOS A or B on left side and LOS D or E on right side

Problem 1.16): Annual traffic of a freeway was found to be 15.8 million vehicles. "K" factor was found to be 0.3 for the freeways in the region. What is the design hourly volume (DHV) in vehicles/hr.?

A) 65,879 B) 12,345 C) 12,986 D) 34,500

Solution 1.16):

STEP 1: Find AADT;
AADT = Average annual daily traffic.
Total traffic during a year is measured and the average per day is obtained.
Annual traffic = 15.8 million = 15,800,000
Divide annual traffic by 365 to obtain AADT. (Average annual daily traffic).

Note: There is a difference between ADT and AADT.
ADT - (Annual daily traffic) can be obtained by measuring traffic flow for few months and obtaining the daily traffic.
AADT- (Average annual daily traffic) – To obtain AADT one has to measure traffic flow for one full year.

AADT = 15,800,000/365 = 43,287 veh/day

K factor is defined as follows.

$$K = DHV/AADT$$

DHV = Design Hourly Volume
Design hourly volume is found by finding the flow rate for each hour within a year. Then sort the flow rates in descending order. 30th highest flow rate is considered to be the DHV.

DHV = K. AADT = 0.3 x 43,287 = 12,986 vehicles/hr (Ans C)
Note that AADT and DHV are not in same units.

Problem 1.17) Station at point of curvature (PC) is given to be 4+10 and the radius of the curve is 140 ft. Find the station at PI. (Angle of intersection is 50 degrees).

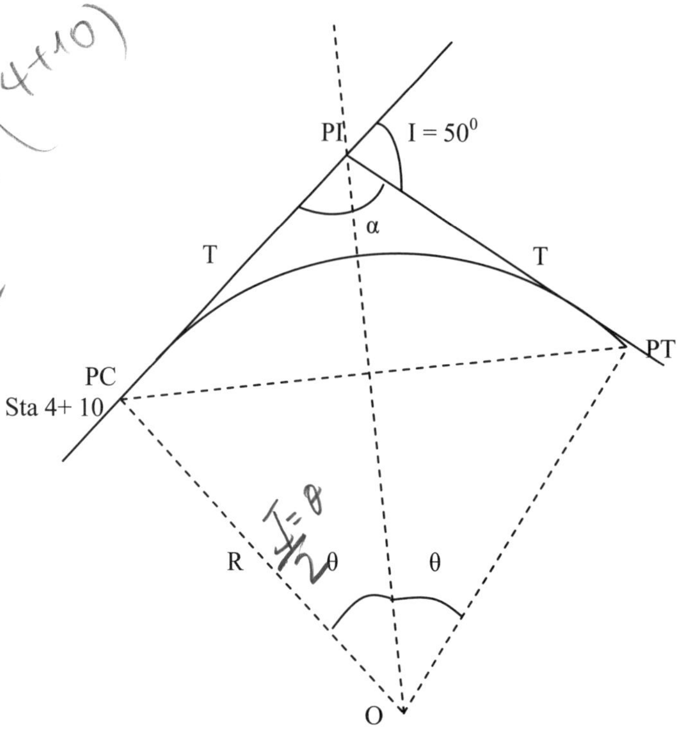

A) 4+66.4 B) 4+75.3 C) 3+93.0 D) 4+81.9

Solution 1.17):

Following nomenclature is used for horizontal curves.

PC = Start of the curve is known as point of curvature (also known as BC, Beginning of curve)
PT = End of the curve is known as point of tangent (also known as EC, end of curve).
PI (Point of intersection) = Intersection of two tangents at PC and PT. Intersection angle at PI is generally represented with "I". In this problem it is given to be 50^0.
R = Radius of the curve
Length between PC and PI is normally represented by "T".

In this problem, station at PC, radius (R) and intersection angle are given.

STEP 1: Find the distance T

From trigonometry, T = R tan θ

STEP 1: Find θ

Consider the angle at PT PI PC.
Angle at PT PI PC = α = 180 - I

Hence angle O PI PC should be half of that.
Angle O PI PC = α /2 = (180 - I)/2
= 90 - I/2

Now consider the triangle PI PC O.

Angle $\theta = 90 - \alpha/2$

Since $\alpha = 180 - I$

$\theta = 90 - (180 - I)/2 = 90 - (90 - I/2) = I/2$

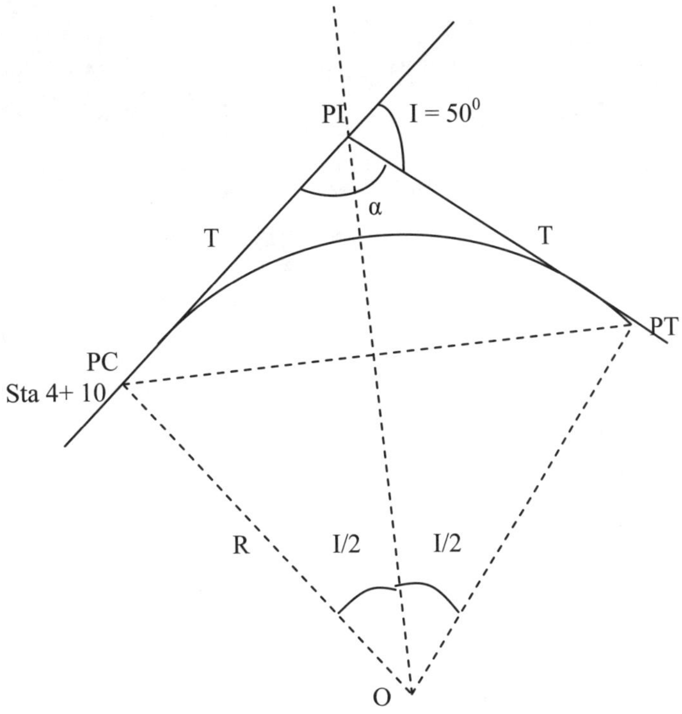

$T = R \times \tan I/2$

$T = R \times \tan(50/2) = 140 \times \tan 25 = 65.3$ ft

Station at PI = 410 + 65.3 = 475.3 = 4 + 75.30 Ans B

Discussion:

Distance is represented with stations. Some examples are shown below.

100 ft = 1+ 00 Station

410 ft = 4 + 10 station

17,500 ft = 175 + 00 station

Horizontal curve in a highway

Horizontal curve of a railway (PI station of this curve would be in the hill to the left.

Problem 1.18): Engineers have designed a horizontal curve so that traffic at PC station will be able to see the traffic at PT station. Property owner would like to build a structure as shown. What's the closest distance to the road that the structure can be built?. (Distance AB)
Radius of the curve = 900 ft
Angle of intersection is 18^0.

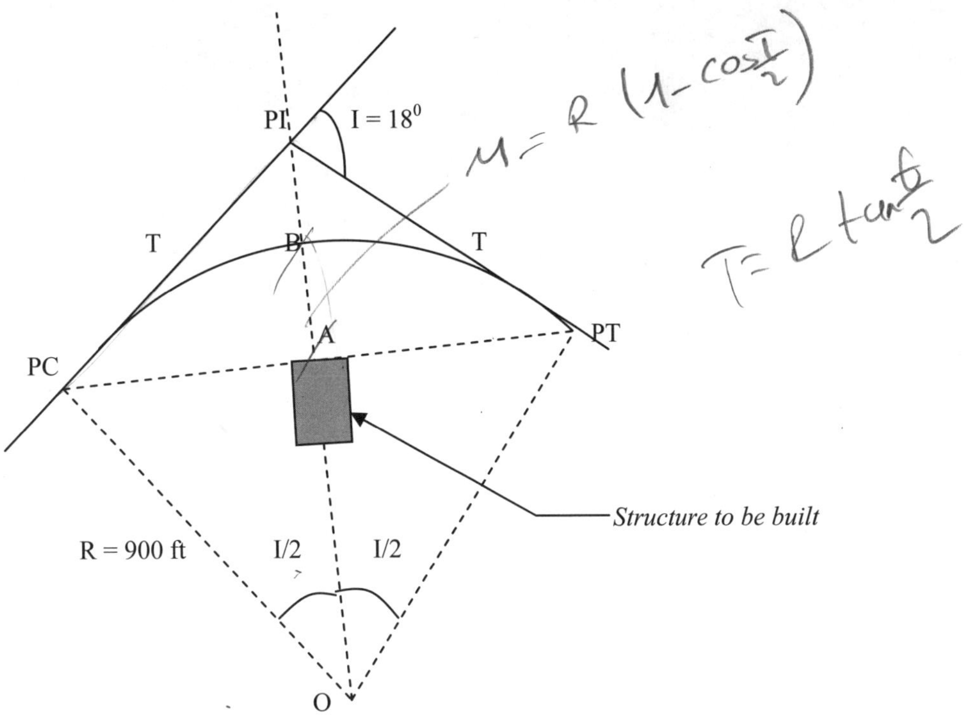

A) 13.4 ft   B) 23.5 ft   C) 3.4 ft   D) 11.1 ft

Solution 1.18):

OB = Radius of the curve = 900 ft
AB = OB – OA

OA = R Cos I/2

Intersection angle (I) is given to be 18^0.

OA = 900 Cos (18/2) = 888.9 ft

AB = OB – OA = 900 - 888.9 = 11.1 ft   (Ans D)

Problem 1.19): Degree of curve of a horizontal curve (arc method) is given to be 4.3 degrees. What's the radius of the curve?

A) 1,332.4 ft   B) 291.8 ft   C) 1,621.1 ft   D) 871.8 ft

Solution 1.19): There are two methods used in highway design.

- Degree of curve (arc method)
- Degree of curve (chord method)

Degree of curve (arc method): In this method, angle subtended at the center of a circle with an arc of 100 ft is known as degree of curve (D).

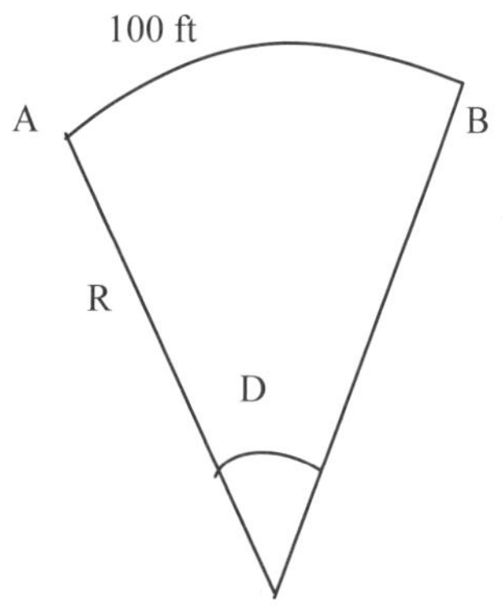

Arc Method (Arc AB = 100 ft)

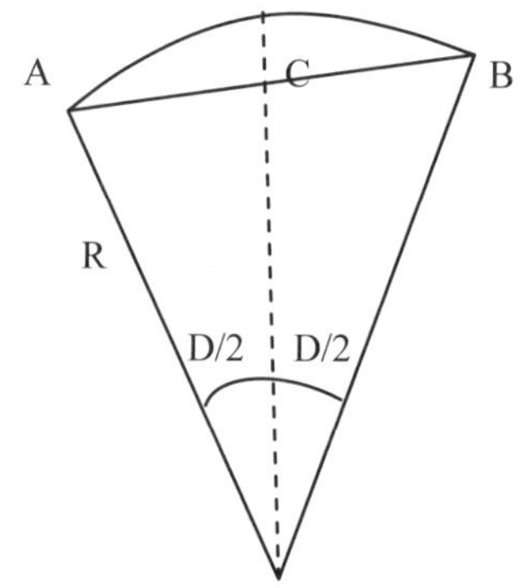

Chord Method (Chord AB = 100 ft)

Arc Method;

As per definition of degree of curve, D degrees will produce an arc of 100 ft.

Full circle is 360 degrees. Full circle produces $2\pi R$.

360 degrees ----→ $2\pi R$

1 degree --------→ $2\pi R/360$

D degrees ------→ $2\pi R.D/360$

As per definition, D degrees would produce an arc of 100 ft.
D degrees ------→ $2\pi R.D/360 = 100$ ft

$$D = \frac{100 \times 360}{2\pi R} = \frac{5{,}729.5}{R}$$

Degree of curve (arc method)

$$R = \frac{5{,}729.5}{D} \quad \text{----------------------}(1)$$

If "D" degrees is known, R can be computed.

Chord Method:

Chord AB (Straight line) = 100 ft

$AC = R \sin D/2$
$AB = 2\,AC = 2\,R \sin D/2$

In chord method 100 ft chord is used.
Hence

$$100 = 2\,R \sin D/2$$

$$R = 50/\sin D/2 \quad \text{--------------------------------------}(2)$$

Degree of curve (chord method)

R = 50/Sin D/2---------------------(2)

Use equation 1 for arc method and equation 2 for chord method to find the radius of a horizontal curve.

Note: Instead of giving the radius of a circle, the problem may give the degree of curve. Use above equations 1 and 2 to find the radius.

Degree of curve in this problem = 4.3 degrees.
Since it is arc method, use equation 1.

$$R = \frac{5{,}729.5}{D} = \frac{5{,}729.5}{4.3}$$

R = 1,332.4 ft   (Ans A)

STRUCTURAL:

Problem 1.20) Find the deflection at the end of the hollow steel tube shown. The outside diameter of the steel tube is 8 inches and the wall thickness is ¼ in. Young's modulus of steel is 29×10^6 psi

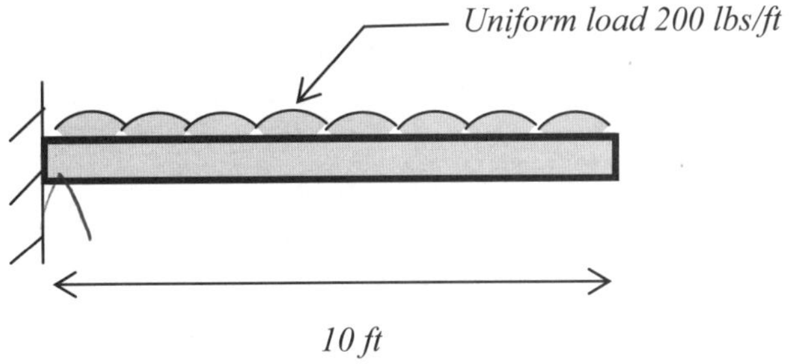

A) 3.61 in   B) 0.325 in   C) 12.34 in   D) 5.67 in

Solution 1.20:

Deflection (y) at the end of a cantilever section with a uniform load is given by

$$y = \frac{w L^4}{8 E I}$$

w = uniform load = 200 lbs/ft
I = Moment of inertia

Moment of inertia of a hollow tube is given by $(I) = \pi \frac{(r_2^4 - r_1^4)}{4} = \frac{\pi (4^4 - 3.75^4)}{4} = 45.75 \text{ in}^4$

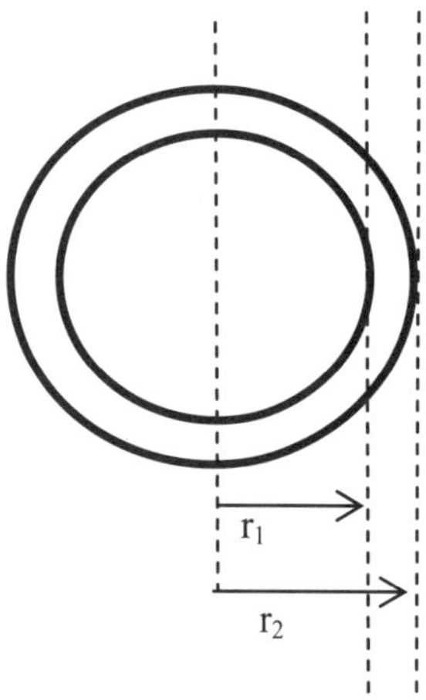

E = 29 x 10^6 psi

w = Uniform load = 200 lbs/ft = 200/12 lbs/in = 16.67 lbs/in
L = 10 ft = 120 in

$$y = \frac{w L^4}{8 E I} = \frac{16.67 \times (120)^4}{8 \times 29 \times 10^6 \times 45.75} = 0.325 \text{ in}$$

Note: Moment of inertia values of different sections are given in steel tables in AISC.

Hollow sections used for structural work is shown above. (Left: Box girders; Middle and left: Hollow tubes)

Problem 1.21) A beam is loaded as shown. Find the maximum shear force of the beam

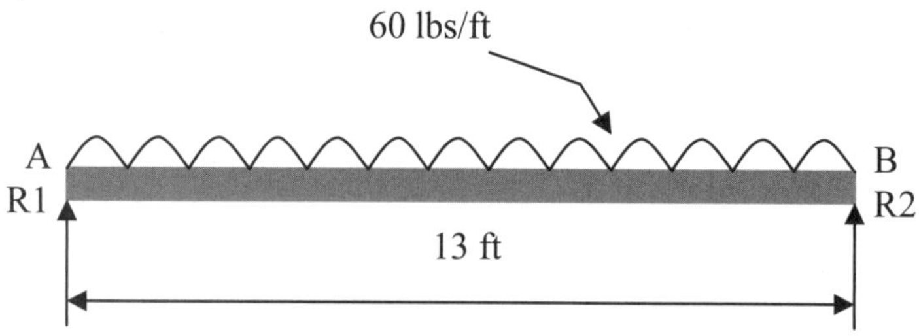

A) 450 lbs   B) 390 lbs   C) 780 lbs   D) 810 lbs

Solution 1.21):

STEP 1: Find the support reactions R1 and R2.

$$R1 + R2 = 60 \times 13 \text{ lbs} = 780 \text{ lbs}$$

From symmetry; R1 = R2 = 390 lbs

STEP 2: Draw the shear force diagram

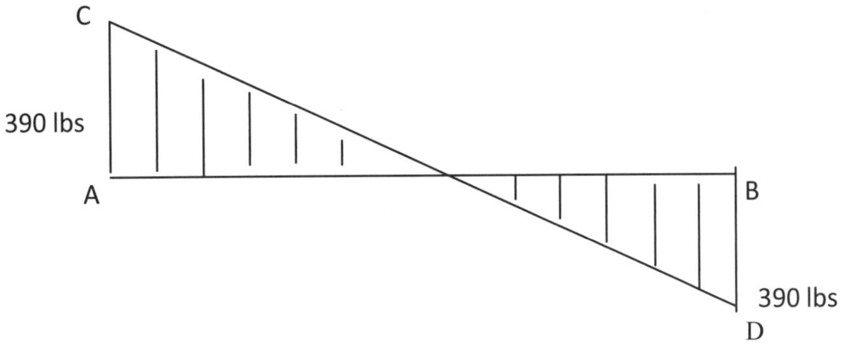

How to draw the shear force diagram?
Assume upward shear is positive and downward shear is negative.
At support A, reaction R1 is upward. Hence draw a line 390 lbs upwards. (Line AC)
Then due to distributed load, line drops down.
Shear force drops down at a rate of 60 lbs/ft.
Shear force diagram starts at point A and then go up 390 lbs. Then comes down to point D due to distributed load. Then goes up another 390 lbs to account for R2.

Maximum shear force occurs at points A and B. (390 lbs or -390 lbs). In the case of shear, sign does not matter. (Ans B)

Problem 1.22) Find the maximum bending moment for the loaded beam shown in the previous example.

A) 2,367.5 lbs. ft B) 1,345.9 lbs. ft C) 1,267.5 lbs. ft D) 2,340.8 lbs. ft

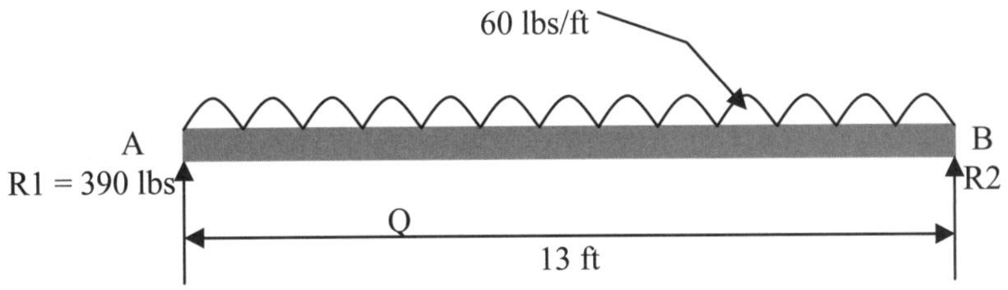

Solution 1.22):

Maximum bending moment occurs when shear force is zero

For the beam given, shear force becomes zero at the midpoint of the beam. (See the previous problem). R1 and R2 were found in the previous problem.

Find the bending moment at the midpoint of the beam.

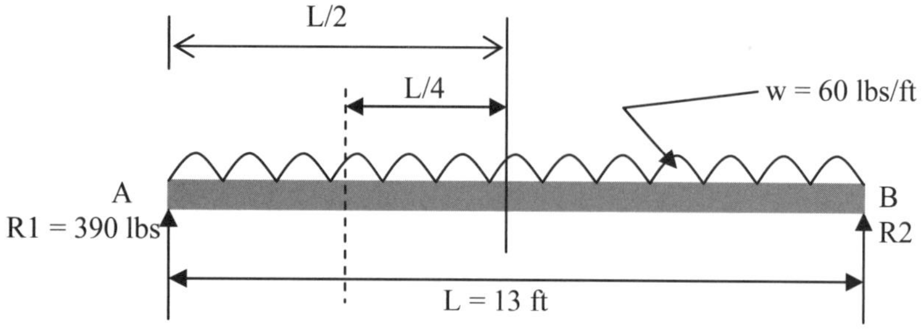

Cut a section at midpoint and take moments:

$M_{midpoint}$ = R1. L/2 – (w. L/2). L/4

w = Distributed load;
L = Length of the beam
w.L/2 term indicates the total load in half of the span.
L/4 indicates the distance to center of gravity of the distributed load from the midpoint of the beam.

M = 390 x 13/2 - (60 x 13/2) x 13/4 = 1,267.5 lbs.ft (Ans C)

Discussion:

How to draw the bending moment diagram?

Though not necessary for this problem, below analysis shows how to find the bending moment at any point of a beam.

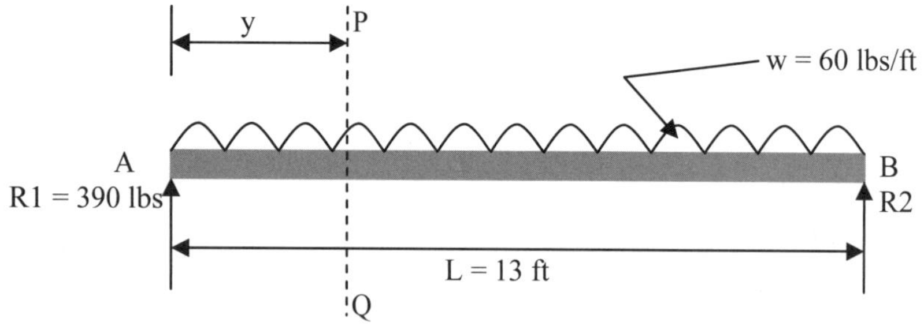

Cut the beam at a section, distance "y" from R1 as shown above.

Obtain moments from line PQ.

Note that R1 was found to be 390 lbs in the previous example.

M = R1. y - w.y x (y/2)

M = R1. y - 60y x (y/2)

R1.y term is the moment due to support reaction.
60y is the total load from A to PQ.

y/2 is the distance to the center of gravity of the distributed load.

When “y” is zero, M is zero. Hence moment at point A is zero.

M at mid-point, y = 6.5 ft.

M at mid-point = 390 x 6.5 - 60 x 6.5 x (6.5/2) = 1,267.5 lbs. ft

M at point B, y = 13 ft

M at point B = 390 x 13 - 60 x 13 x (13/2) = 0

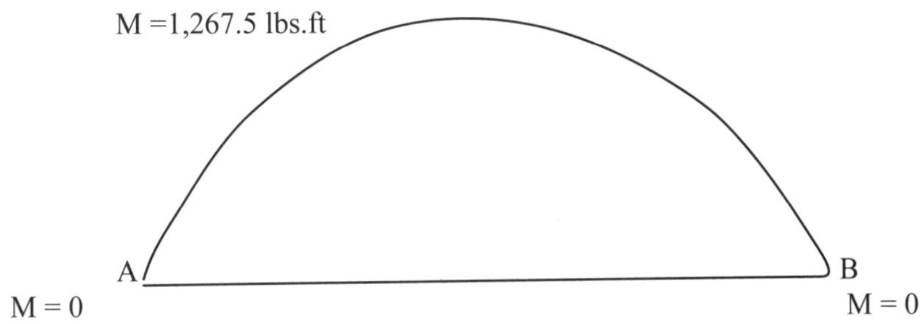

Failure of a beam due to flexure (bending moment). Beam failed due to tension in bottom fibers

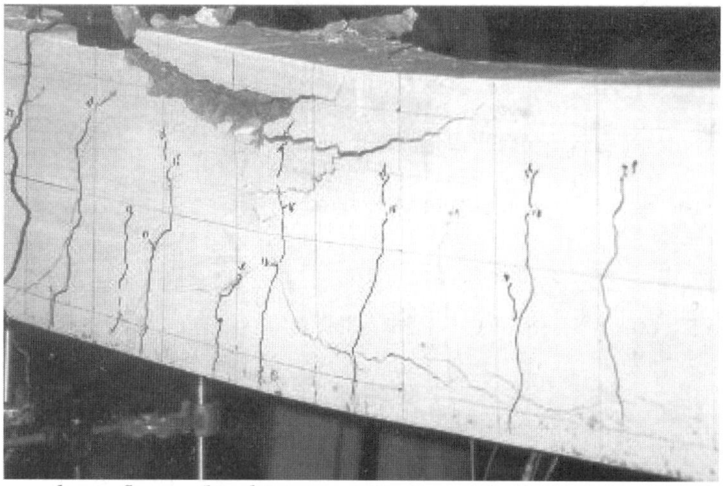

Failure of a beam due to flexure (bending moment). Beam failed due to compression in top fibers

Failure of a beam due to shear.

Failure of a beam due to shear at the corner

Problem 1.23) W12 x 96 column has a radius of gyration of 3.09 inches along weaker axis and 5.43 inches along stronger axis. Effective length factor (K) along both axis is equal to 1.0. Young's modulus of steel is 29×10^6 in^2. Area of the W section is 28.2 in^2. Unbraced length of the column is 15 ft. Find the elastic buckling load (Euler buckling load) of the column.

A) 1,245 kips   B) 2,368 kips   C) 1,765 kips   D) 876 kips

Solution 1.23):

Elastic buckling stress or the Euler buckling stress is given by the following equation.
Euler Equation for Elastic Buckling

$$F_e = \frac{\pi^2 E . A}{(K.L/r)^2}$$

F_e = Euler buckling load (lbs)
E = Young's modulus of steel *(psi)*.
Young's modulus of structural steel is 29×10^6 psi. (Given in this problem).
K = Effective length factor *(No units)*
L = Unbraced length *(inches)*
r = Radius of gyration along the weaker axis. *(inches)*
A = Area (*sq. inches*)

In this problem, radius of gyration along both axis are given. To solve the problem, you should use the smaller value of r.
Convert all units to inches and apply the equation.

E = 29×10^6 psi;   L = 15 x 12 inches;   K = 1.0;   r = 3.09 inches

Area of the section = 28.2 in^2.

$$F_e = \frac{\pi^2 \text{ x } 29 \text{ x } 10^6 \text{ x } 28.2}{(1.0 \text{ x } 15 \text{ x } 12/3.09)^2} \text{ lbs} = 0.084 \text{ x } 10^6 \text{ x } 28.2 \text{ lbs}$$

$$= 2{,}368 \text{ kips}$$

Note that this is the elastic buckling load. In reality, the column would fail way before this load due to inelastic buckling.

Discussion on Elastic Buckling and Inelastic Buckling:

Leonard Euler in 17th century developed the equation we use today for elastic buckling. This equation is valid as long as the column is made without any impurities using pure steel. In reality, due to impurities and uneven heating and cooling during the manufacturing process, some parts of the column would be weaker than the rest. These weaker areas would yield prior to other areas. Hence real buckling load would be smaller than the value obtained using the Euler equation. AISC table 4.1 provides a capacity of 990 kips for this column.

Problem 1.24): What is true about liquefaction of soil?

A) Liquefaction mainly occur in soft clays
B) Liquefaction mainly occur in loose saturated sands
C) Liquefaction mainly occur in soft rock
D) Liquefaction mainly occur in any type of soil

Solution 1.24): Liquefaction is the process of loose sandy soils converting into a liquid like paste during an earthquake. Non-technical term for liquefaction is "quick sand" condition. Liquefaction occurs in saturated loose sandy soils. Liquefaction does not occur in dry soil even though the soil is loose.

Ans B

Discussion:

Consider five jars filled with sand.

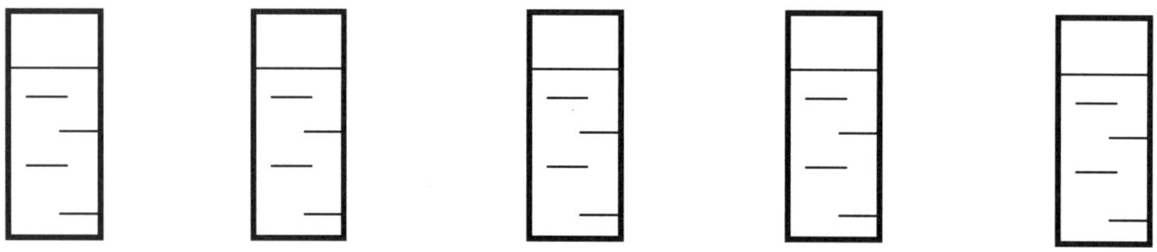

Dense dry sand *Dense wet sand* *Loose dry sand* *Loose wet sand* *Clay*

If the five jars were violently shaken, jar with loose wet sand would liquefy.

Saturated loose sandy soils turning into a semi liquid paste

Soil liquefaction and formation of a sink hole

Condition for liquefaction: Sand should be loose and wet.

Problem 1.25) What is true about natural period of buildings as related to seismic engineering?

A) Natural period of short buildings could be several seconds
B) Natural period of tall buildings could be less than one second
C) Natural period of short buildings and tall buildings are almost the same
D) Natural period of tall buildings could be several seconds

Solution 1.25): Take a ruler and hold tightly from one end. Then give a small blow to the ruler with your finger.

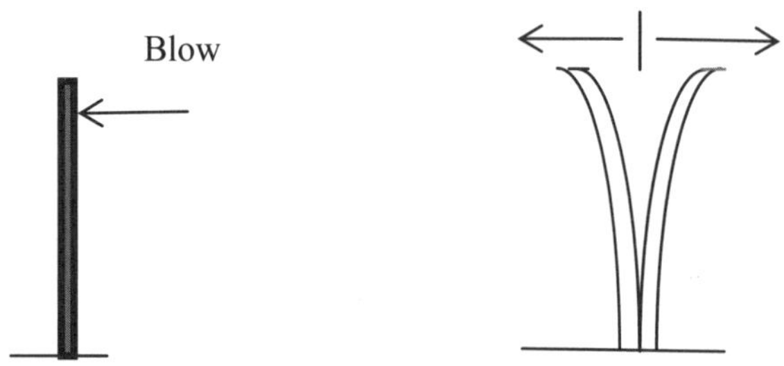

Rod or ruler is cantilevered and blow is given to the free end. The rod would start to vibrate as shown. Period is the time taken for the rod to move from a given point and come back to the same point again.

Longer the rod, longer the period. Shorter rods vibrate fast and hence have a shorter period.

During an earthquake event, the building is jolted from the base.

Force due to earthquake

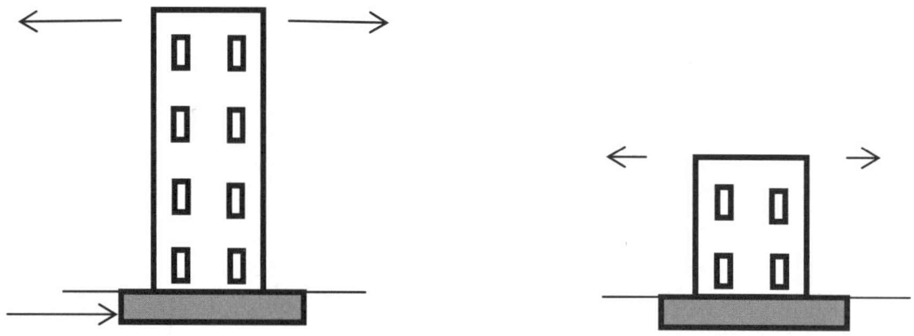

Tall high rise buildings would have a longer natural period, typically in the range of few seconds.
Short buildings would have a shorter natural period, typically in the range of 0.2 to 1.0 seconds. (Ans D).

Problem 1.26) What is spectral acceleration as related to earthquakes?

A) Acceleration of soil particles during an earthquake
B) Acceleration of the building during an earthquake
C) Acceleration of the seismic wave
D) None of the above

Solution 1.26):

During an earthquake, buildings tend to oscillate like a pendulum. The acceleration of the building during oscillations is known as spectral acceleration.

Shorter buildings tend to undergo high accelerations compared to taller buildings for the same earthquake. Also note that short buildings have a short natural period compared to tall buildings.

Spectral acceleration vs natural period is shown below.

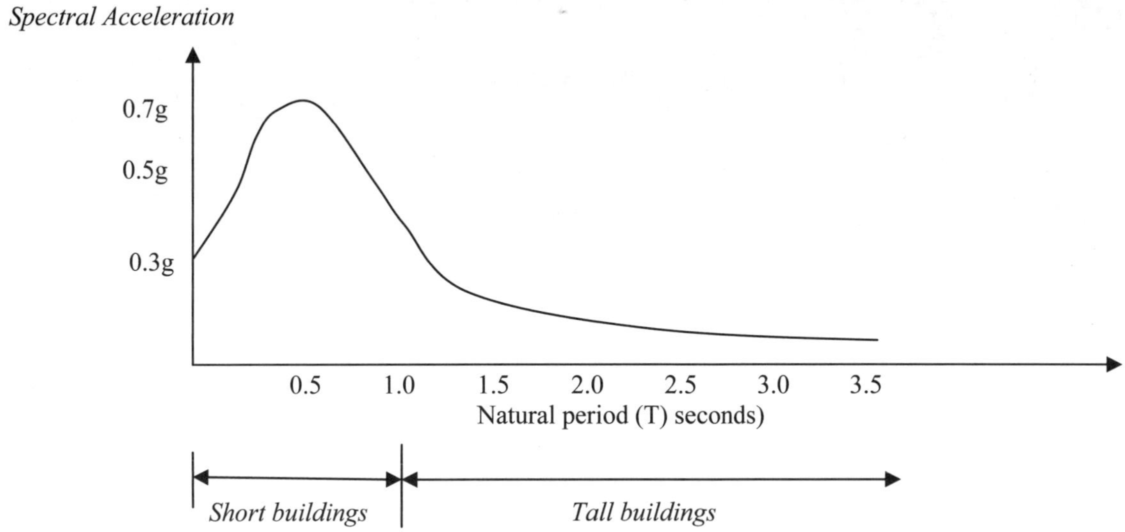

Above graph shows acceleration of buildings for a given earthquake.
Natural period is shown in the X axis and acceleration of the building is shown in the Y-axis. Tall buildings would have a high natural period and a low acceleration. Short buildings would have a short natural period and a high acceleration. (Ans B).

CONSTRUCTION:

Problem 1.27): Following successor relationships are available for a project. What is the duration of the project?

Activity ID	Duration (days)	Successors
A	5	B
B	12	C, D
C	6	D, F
D	8	E
E	12	F
F	3	

A) 22 days B) 26 days C) 46 days D) 40 days

Solution 1.27):

STEP 1: Draw an activity on node diagram for the data given.

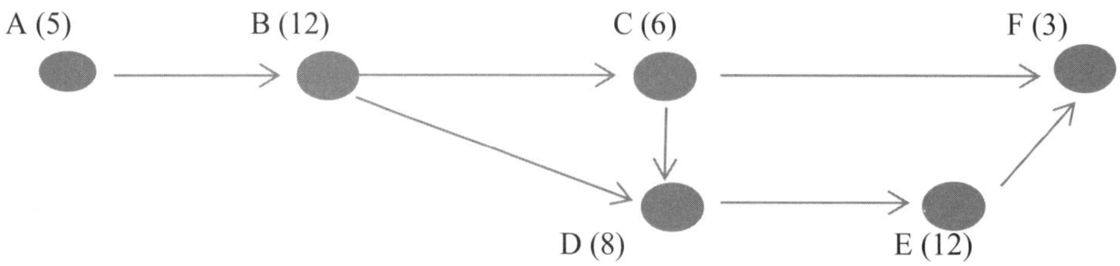

Activity A is the first activity. B is the successor of A.
C and D are successors of B.

STEP 2: Find the maximum duration;

Path A, B, C, F = Total duration = 5 + 12 + 6 + 3 = 26
Path A, B, C, D, E, F = Total duration = 5 + 12 + 6 + 8 + 12 + 3 = 46
Path A, B, D, E, F = Total duration = 5 + 12 + 8 + 12 + 3 = 40

To complete the project, ABCDEF path need to be completed.

Maximum duration = 46 days (Ans C)

Problem 1.28) A contractor is assigned with a project to install a 6" pipe at a depth of 5 ft. The width and length of the trench is 1 ft and 1,000 ft respectively. Contractor is planning to reuse the soil excavated to backfill the trench. Swell factor of the soil is 23% and shrinkage factor of the soil is 15%. Assume borrow pit soil also have the same swell factor and shrinkage factor. How many trucks of soil he may have to ship out (or brought in) if each truck can carry 10 CY of soil.

A) 1 B) 2 C) 3 D) 4

Solution 1.28)

Following nomenclature is important to solve this problem.

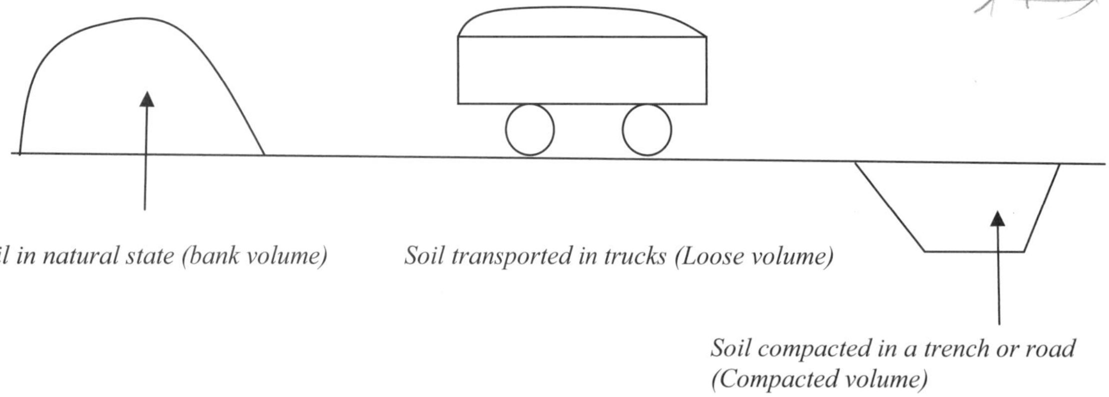

When the soil is in natural state, the volume is known as bank volume.
When the soil is excavated and filled in trucks for transportation, the soil volume is known as loose volume. Loose volume is always higher than the bank volume.

When the soil is compacted in a trench or road, soil volume is known as compacted volume. Compacted volume is always less than loose volume. In most cases compacted volume is less than bank volume.

STEP 1: Find the volume of soil to be excavated (Bank volume)

Length of the trench = 1,000 ft,
Width of the trench = 1 ft
Depth of the trench = 5 ft.

Volume of the soil to be excavated (Bank Volume) = 1,000 x 5 x 1 = 5,000 cu.ft

STEP 2: Find the volume of soil needed (Compacted basis);

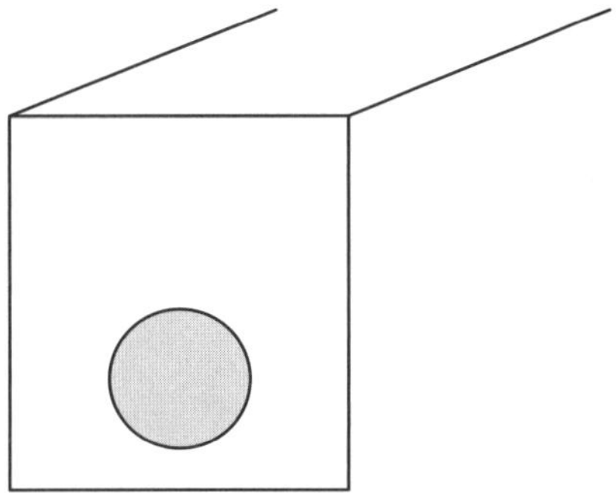

The volume of the trench is 5,000 cu.ft. Once the pipe is installed, the pipe would take up some volume.
Volume of the pipe = $\pi \times D^2/4 \times$ Length = $\pi \times 0.5^2/4 \times 1{,}000$ = 196.35 cu.ft

Volume of soil needed (Compacted basis) = Volume of the trench – Volume of the pipe
= 5,000 – 196.35 = 4,803 cu.ft

STEP 3: Volume of soil (Compacted basis):

In step 1, we found that the volume of soil excavated (bank basis) to be 5,000 cu. ft.
This need to be converted to compacted basis. This can be done using the shrinkage factor.

$$\text{Shrinkage Factor} = \frac{\text{Bank Volume} - \text{Compacted Volume}}{\text{Bank Volume}}$$

Shrinkage factor is given to be 15%.

$$0.15 = \frac{5{,}000 - \text{Compacted Volume}}{5{,}000}$$

Compacted Volume = 5,000- 5,000 x 0.15

Compacted volume of soil = 4,250 cu.ft

Excavated soil when compacted will yield 4,250 cu. ft.

Volume of soil needed (Compacted basis) = 4,803 cu.ft (Found in step 2)

As you could see more soil is needed than what was excavated.

STEP 4: Find the volume of soil need to be brought in (Loose Volume):
When the soil is brought in or shipped out, they have to be in trucks. When soil is loaded into trucks, the soil is in loose state. Hence loose volume need to be computed.

Amount of soil needed (Compacted basis) = 4,803 – 4250 = 553 cu.ft
This volume needs to be converted to loose volume.

Compacted Volume = Loose Volume x $(1 - S_H)/(1 + S_w)$

S_W = Swell factor; S_H = Shrinkage factor

553 = Loose Volume x (1 – 0.15)/(1 + 0.23)
Compacted volume is 553 cu. ft, shrinkage factor is 0.15 and swell factor is 0.23.

Loose Volume = 800.22 cu. ft = 800.22/27 CY
= 29.63 CY

Number of trucks needed = 3.
Each truck can carry 10 CY.

Problem 1.29): Contractor is trying to pull a tree stump. He believes 6 ton pull force will be enough to pull the stump. What is the pull force he needs from the pulling tractor?

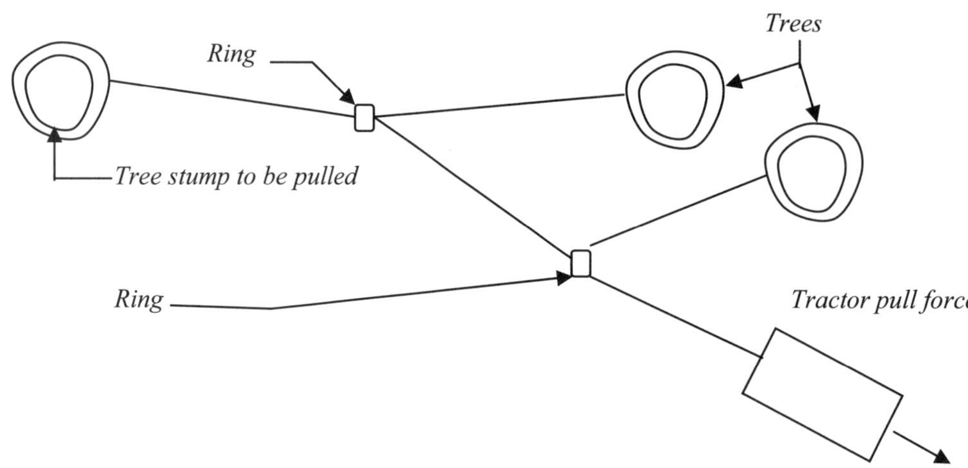

A) 6 tons B) 4 tons C) 1.5 tons D) 2 tons

Solution 1.29): Ans C

Many loggers use existing tees to increase the pull force.

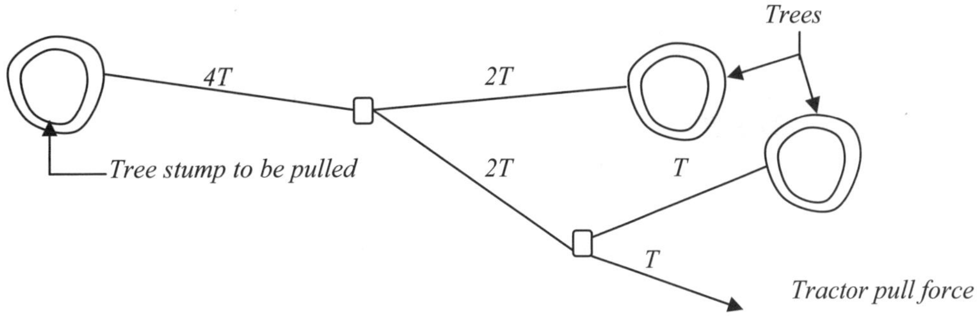

If the tractor pull force is T, force distribution would be as shown. If the pull force from the tractor is T, then one would get a force of 4T at the stump from the shown arrangement.
T = 1.5 tons (Ans C)

Problem 1.30): What equipment is not needed for a SMAW welding setup?
A) Power source and cables B) Electrodes and electrode holders C) Plasma generator or plasma console D) AC or DC power supply

Solution 1.30): SMAW stands for "Shielded metal arc welding". Before we discuss SMAW, we need to ask "What is an arc?" We all have seen electric arcs. Whenever there is a voltage difference, an electric arc would be generated. Biggest arcs are generated in rain clouds and we call them "lightening".

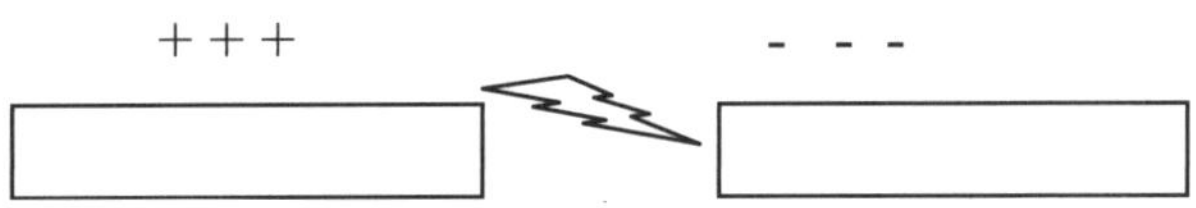

Voltage difference creates an arc

If an arc can be sustained for a longer period of time, one can generate heat. Heat is the main ingredient for welding. In Oxyacetylene welding, heat was provided to the metal by burning of Acetylene.
Equipment required for shielded metal arc welding:

- Power source: (Power source can be AC current or DC current)
- Electrodes – Electrodes are made of metals. Electrodes have two purposes. One is to create an arc. The other purpose is to act as a filler rod. Metal electrodes are coated with chemicals that act as a flux.
- Cables and ground wires

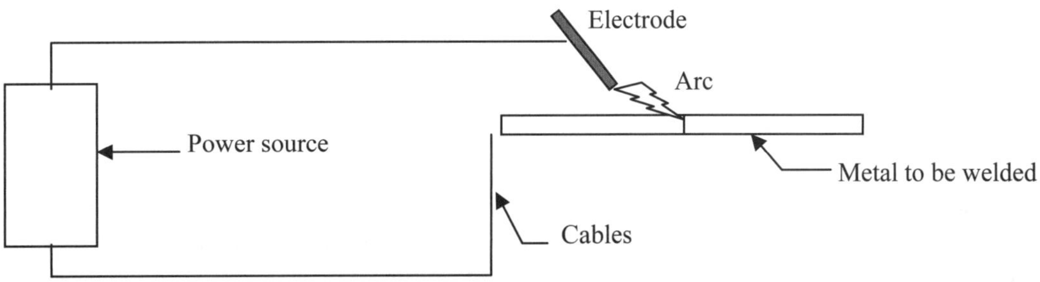

Note: Plasma welding is a different process than arc welding. (Ans C)

Correct and safe electric welding station

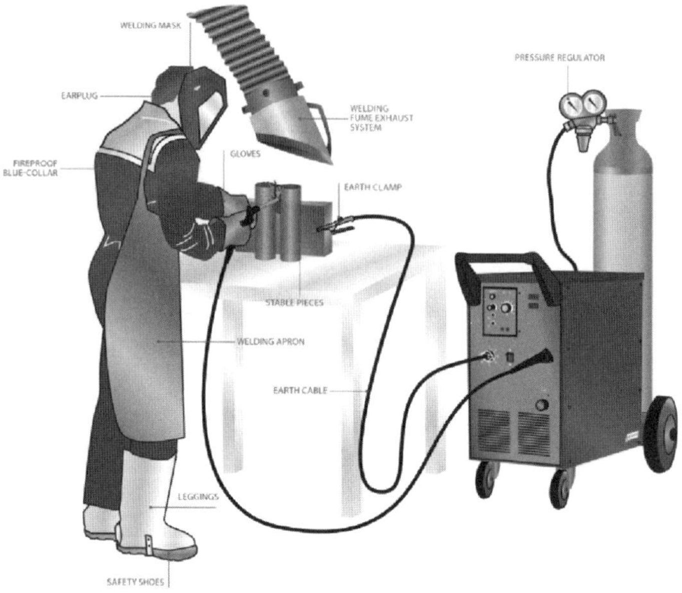

Welding machine on the right produces a voltage difference in two cables. Voltage difference would produce an arc

Problem 1.31) A construction crew consists of 3 laborers, 3 operators and one foreman. What is the crew hour rate and labor hour rate?

Hourly wages of workers;

Laborers - \$55
Operators - \$80
Foreman - \$90

A) 495 and 70
B) 495 and 45
C) 404 and 80
D) 430 and 67

Solution 1.31)

Crew hour rate is the amount of money that need to be spent on the crew per hour.

Crew hour rate (CH) = (3 x 55) + 3 x(80) + 1 x (90) = \$ 495

Labor hour rate (LH) is defined as the crew hour rate divided by number of workers.

Labor Hour Rate (LH) = Crew Hour Rate (CH)/Number of workers

Labor Hour Rate (LH) = CH/Number of workers

Labor Hour Rate (LH) = Crew Hour Rate (CH)/Number of workers

Labor Hour Rate (LH) = 495/ (3 + 3 + 1) = $70.7

(Ans A)

Problem 1.32) 120 ft long, 11 ft high concrete wall needs to be constructed. The wall needs to be formed on both sides. Concrete forming crew consists of 4 carpenters, 3 laborers and 2 operators. What is the cost of labor for formwork?

Hourly wages of workers;

Carpenters - $ 70
Laborers - $ 55
Operators - $ 80

Productivity of the crew = 80 SF/hour

A) $17,672
B) $19,230
C) $19,965
D) $21,108

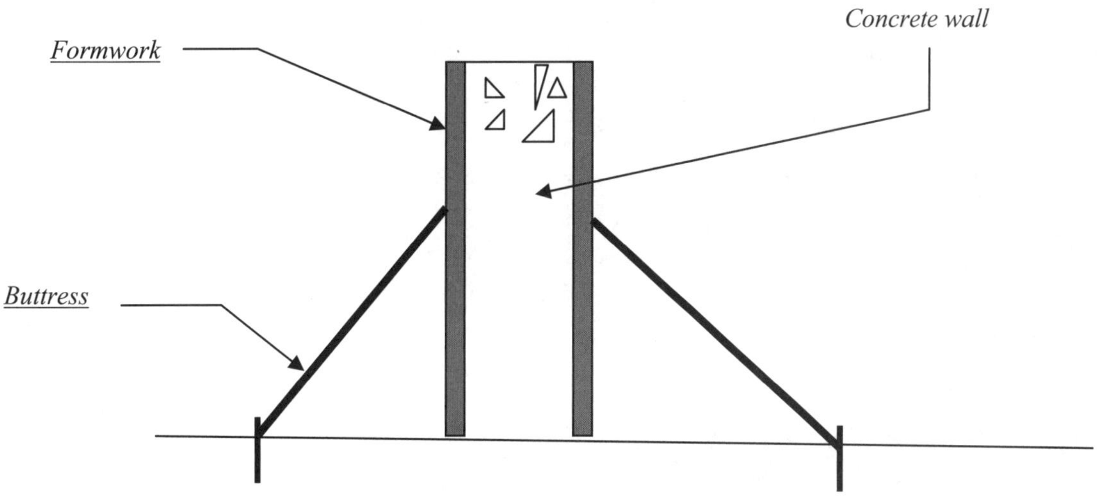

Solution 1.32)

STEP 1: Find the crew hour rate:

Crew hour rate (CH) = (4 x 70) + 3 x(55) + 2 x (80) = $ 605

STEP 2: Find the work hours needed to complete the formwork for the wall;

Total square footage of the formwork = (120 x 11) x 2 = 2,640 SF

Note that the wall needs to be formed on both sides.

Productivity of the crew = 80 SF/hour
Number of work hours needed to complete the wall = 2,640/80 = 33 hours

STEP 3: Find Cost of labor for the formwork:

Cost of labor for formwork = Number of crew hours x Crew hour rate
= 33 x 605 = $ 19,965 (Ans C)

Wall formwork is shown in this figure. The formwork consists of steel frame and buttresses.

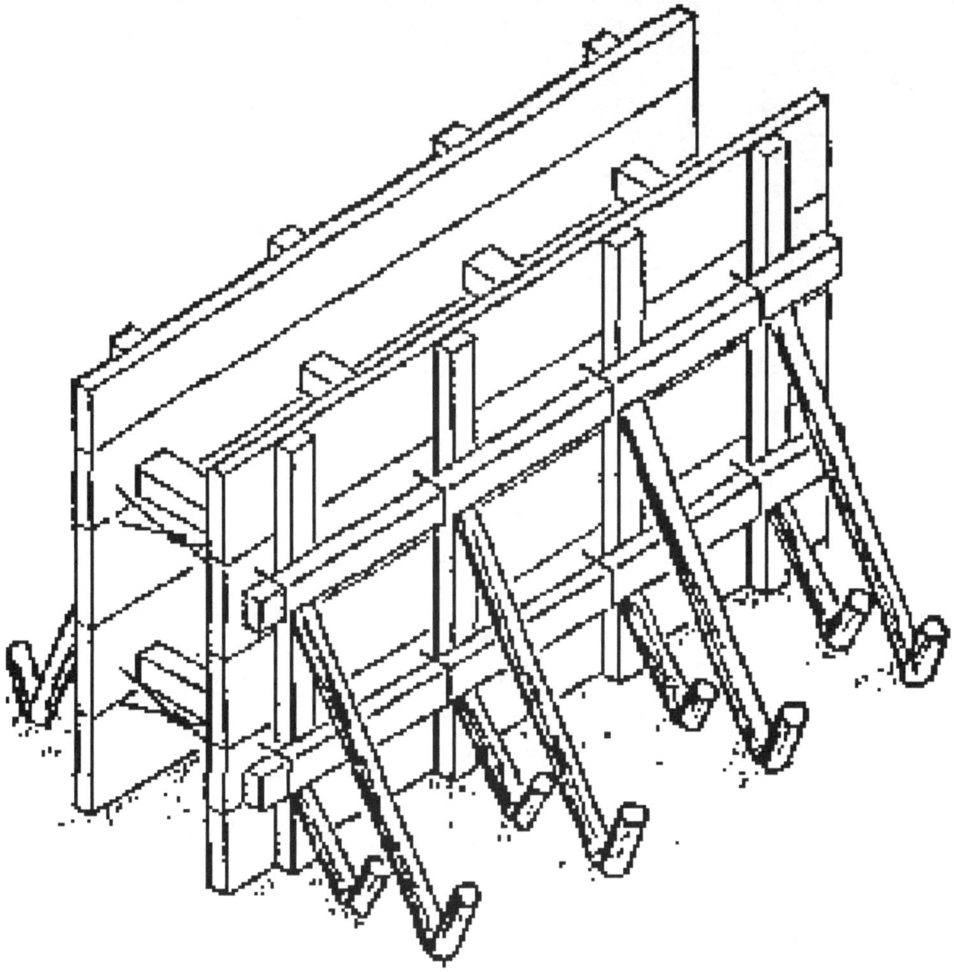

Wooden formwork for a wall

Geotechnical:

Problem 1.33): Find the net allowable bearing capacity of the footing in tons.

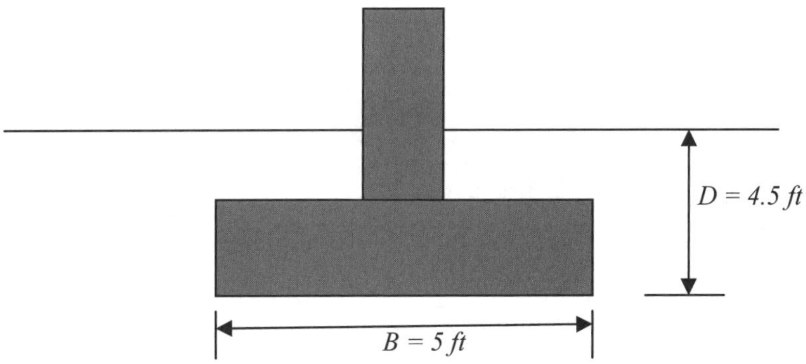

Length of the footing = 30 ft
Width of the footing = 5 ft
N_c = 12.9
N_q = 4.4

$N_\gamma = 2.5$
Total density of soil (γ) = 120 pcf
Cohesion of the soil = 300 psf
Factor of safety = 3.0

Use the Terzaghi Bearing Capacity Equation:

Ultimate Bearing Capacity in psf = $c.N_c + q.\ N_q + \frac{1}{2}\ B.\ \gamma.\ N_\gamma$
Ignore shape factors and depth factors.
Find the allowable bearing capacity of the footing in tons.

A) 134 tons B) 327 tons C) 102 tons D) 161 tons

Solution 1.33):

STEP 1: Write down the Terzaghi Bearing Capacity Equation;

Ultimate Bearing Capacity in psf (q_{ult}) = $c.N_c + q.\ N_q + \frac{1}{2}\ B.\ \gamma.\ N_\gamma$

This is the basic bearing capacity equation. Shape factors, inclination factors and depth factors are not included in this equation.

This is the basic equation for bearing capacity. There are other factors that normally use to account for the shape of the footing, depth, inclination of load etc. The problem states to ignore these factors.

q_{ult} = Ultimate bearing capacity
c = Cohesion of soil (psf)
N_c = Bearing capacity factor for cohesion. (N_c = 12.9)
Tables are provided by various researchers to find this factor. In this problem N_c is given to be 12.9
$q = \gamma.\ D$
"q" is known as overburden.
γ = Density of soil. In this problem γ is given to be 120 pcf.
D = Distance from top of ground to bottom of footing. In this problem given to be 4.5 ft
N_q = Terzaghi bearing capacity factor for overburden. In this problem, given to be 4.4
B = Width of the footing. In this problem B is given to be 5 ft.
N_γ = Terzaghi bearing capacity factor for soil density. In this problem, N_γ is given to be 2.5

STEP 2: Apply the Terzaghi bearing capacity equation;

Ultimate Bearing Capacity in psf (q_{ult}) = $c.N_c + q.\ N_q + \frac{1}{2}\ B.\ \gamma.\ N_\gamma$

$q = \gamma.D$ = 120 x 4.5 = 540 psf

Ultimate Bearing Capacity in psf (q_{ult}) = (300 x 12.9) + (540 x 4.4) + (½. 5 x 120 x 2.5)
Ultimate Bearing Capacity in psf (q_{ult}) = 3,870 + 2,376 + 750
Ultimate Bearing Capacity in psf (q_{ult}) = 6,996 psf

STEP 3: Find q_{net};

q_{net} is the **additional** pressure that can be placed on the soil at bottom of footing. I have emphasized the word "additional".

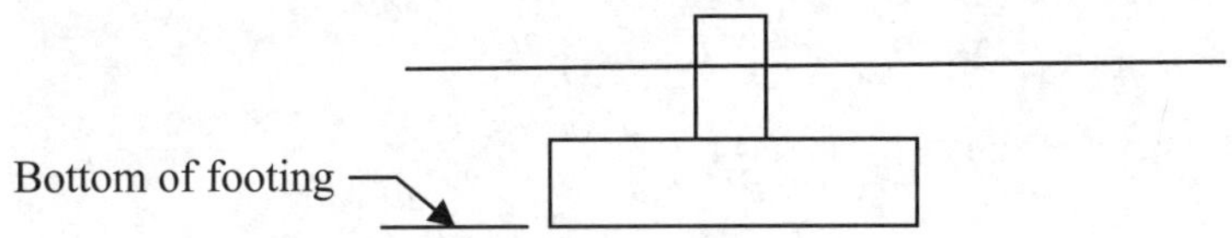

We have found that when you place 6,996 psf (q_{ult}), the soil would fail. This is the total pressure that can be placed at bottom of footing.
What was the pressure there before?
The pressure at the bottom of footing prior to excavation of the footing = $\gamma.D$
Hence **additional** pressure that can be placed at bottom of footing level = $q_{ult} - \gamma.D$

$q_{net} = q_{ult} - \gamma.D$
$q_{net} = 6{,}996 - 120 \times 4.5 = 6{,}456$ psf
q_{net} is the pressure that can be placed at the bottom of the footing in excess of what was there before. q_{net} value is important for settlement.

STEP 4: Find the total bearing capacity of the footing;

Total Bearing capacity of the footing (tons) = Net bearing capacity (psf) x Area/2,000

Length of the footing = 30 ft
Width of the footing = 5 ft

Total Bearing capacity of the footing (tons) = 6,456 x (30 x 5)/2,000
= 484.2 tons

STEP 4: Find the allowable bearing capacity ($q_{allowable}$) of the footing;

$$\text{Allowable bearing capacity} = \frac{\text{Total Bearing capacity}}{\text{Factor of safety}}$$

$$\text{Allowable bearing capacity} = \frac{484.2}{3.0} = 161.4 \text{ tons}$$ (Ans D)

Note to candidates: Though bearing capacity equation looks long and difficult, once you complete few problems you would feel comfortable.

Procedure to find the allowable bearing capacity:

STEP 1: Use the Terzaghi bearing capacity equation and obtain the ultimate bearing capacity (q_{ult})
STEP 2: Find the net bearing capacity (q_{net})
STEP 3: Multiply the net bearing capacity by area of the footing and obtain the total bearing capacity
STEP 4: Divide the total bearing capacity by a factor of safety to obtain the allowable bearing capacity

Excavation and formwork for a shallow foundation

Discussion of q_{net}: Let us imagine a point in the ground 3 ft below the surface. At this point, at the present time existing pressure is γ.D, where D = 3 ft. If the soil density is 120 pcf, pressure at a point 3 ft below the surface = 3 x 120 = 360 psf.

The settlement due to this pressure is zero. We assume that soil has settled and attained stability.

In order to construct the footing we will have to remove 3 ft of soil.

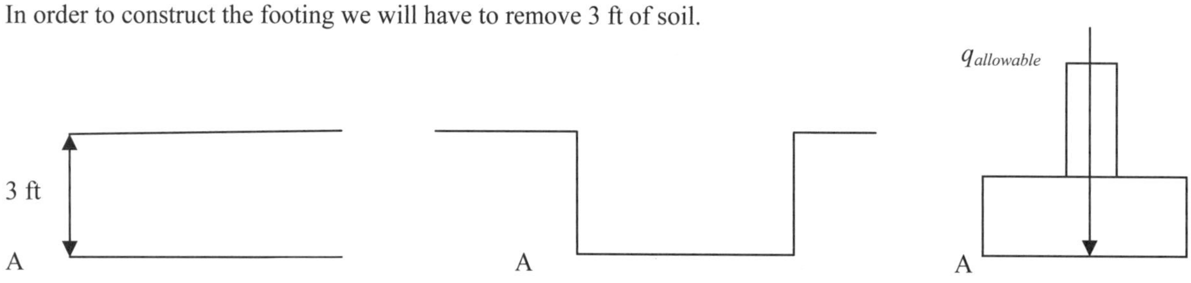

Fig 1   Fig 2   Fig 3

Figure 1 shows a soil profile. Figure 2 shows after excavation for the footing. Figure 3 shows after construction of the footing.

Fig. 1: The pressure at point A = γ.D   (No settlement. Soil is stable).

Fig 2: The pressure at point A = 0   (Soil relief or slight upheaval)

Fig 3: The pressure at point A = Load due to footing and column = $q_{allowable}$

If $q_{allowable}$ = γ.D, there will not be any settlement.

q_{net} is defined as the pressure at bottom of footing level **in excess** of what was there before.

Hence $q_{net} = q_{allowable} - \gamma.D$

Now let us try to see the significance of q_{net}.
Now let us assume $q_{allowable} = \gamma.D$.
What is the settlement if $q_{allowable} = \gamma.D$?
If $q_{allowable} = \gamma.D$, then settlement would be zero. This is due to the fact that there was no settlement in the soil before the excavation. Prior to excavation, the soil was subjected to a pressure of γ.D and the soil had attained stability. If you remove a pressure of γ.D and put back γ.D, then soil will not undergo any additional settlement. Settlement occurs only if the pressure due to footing exceeds γ.D.
The pressure in excess of γ.D will cause settlement. The pressure in excess of γ.D is known as q_{net}. This value is important for structural engineers since they are interested in soil settlement.

Concreting of a shallow foundation

Problem 1.34): Specific gravity of a soil sample is given to be 2.65. Moisture content and degree of saturation are 0.6 and 0.7 respectively. Find the void ratio.

A) 1.34   B) 2.27   C) 4.89   D) 0.78

Solution 1.34):

This is a soil phase relationship problem. See the appendix for complete discussion of soil phase relationships.

In this problem, specific gravity of soil, moisture content and degree of saturation is given. You are required to find the void ratio.

Specific gravity = G = 2.65
Moisture content = w = 0.6
Degree of saturation = S = 0.7
Void ratio = e

$$S.e = G.w$$

0.7 x e = 2.65 x 0.6
e = 2.27 (Ans B)

Problem 1.35) Modified Proctor mould has an inner diameter of 4" and a height of 4.584 inches. Moisture content of soil is measured using a small tray that weigh 6 ounces.

Compacted soil + Mould = 15.76 lbs
The weight of the mould = 12 lbs
Water content measurement:

Prior to oven drying:

Weight of Soil + Tray = 1.21 lbs

After oven drying:

Weight of Soil + Tray = 1.15 lbs
Find the dry density of the sample.

A) 123.65 lbs/cu.ft B) 112.34 lbs/cu.ft C) 104.71 lbs/cu.ft D) 99.87 lbs/cu.ft

Solution 1.35): Modified Proctor tests are conducted to obtain the dry density of a soil at a pre determined level of compaction.

Step by step procedure of conducting a Modified Proctor test;

STEP 1: Weigh the mould

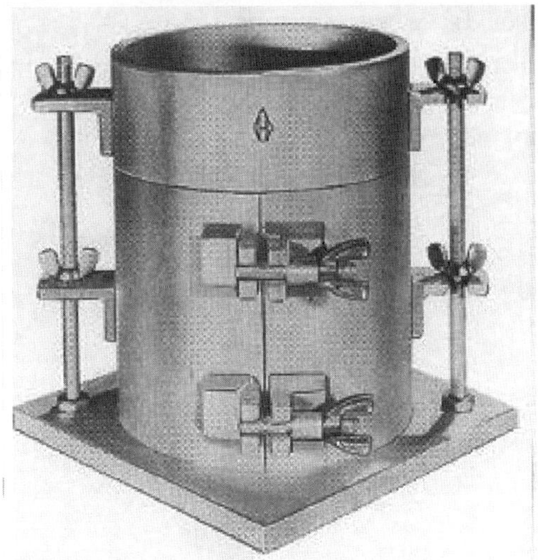

Modified Proctor mould (volume is 1/30 cu.ft as per ASTM D 698).

STEP 2: Fill 1/5th of the mould with soil and compact with a ram. Ram needs to be dropped 25 times.

STEP 3: Fill with another layer and compact with the ram. Compact the soil in 5 layers with each layer getting 25 blows.

STEP 4: Weigh the mould + soil

STEP 5: Obtain the total density of soil.

$$\text{Total Density} = \frac{\text{Weight of soil}}{\text{Volume of soil}}$$

Weight of soil = (Weight of soil + mould) – Weight of mould;

Since weight of mould is given, weight of soil can be computed.
Volume of soil is equal to the volume of the mould.

STEP 6: Moisture content measurement;

Remove small portion of soil from the mould and place in a tray and weigh. Then oven dry the soil and weigh again.

STEP 7: Obtain the moisture content.

Dry Density (γ_d) = $\gamma/(1 + w)$

γ_d = Dry density;
γ = Total density (also known as wet density);
w = Moisture content
γ = Soil weight/Volume = M/V
w = Moisture content = M_w/ M_s
M_w = Weight of water
M_s = Weight of solids

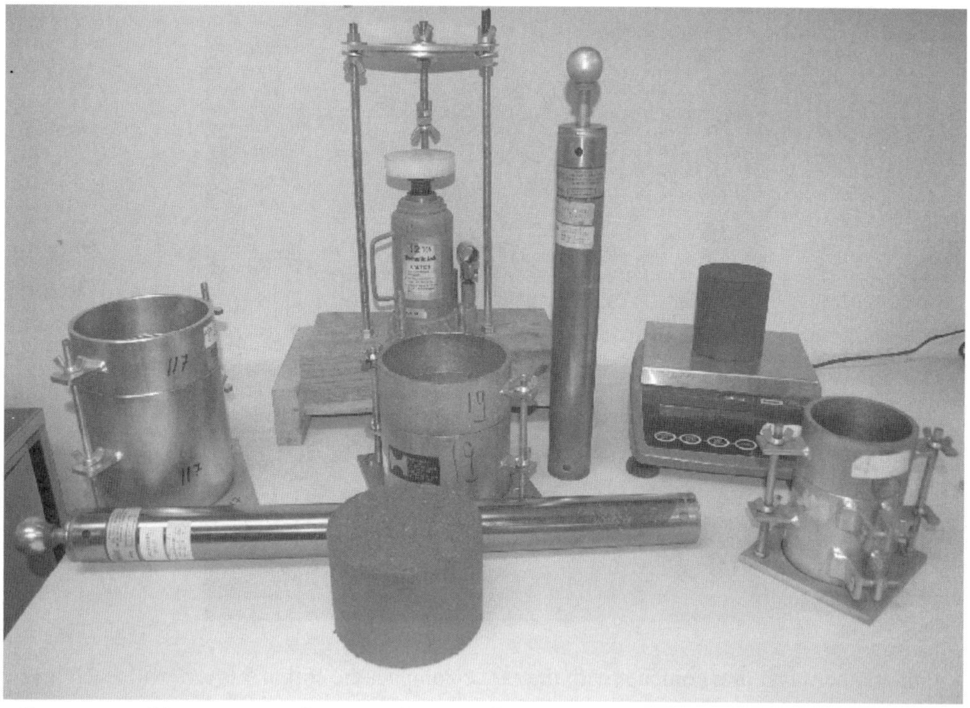

Proctor moulds, rams, weighing machines and compacted soil sample shown in the figure

Compacted soil + Mould = 15.76 lbs
Weight of the mould = 12 lbs
Weight of compacted soil = 3.76 lbs

Volume of the mould = π x $D^2/4$ x Height = π x $4^2/4$ x 4.584 = 57.60 cu. in = 0.0333 cu. ft

Diameter of the mould is given to be 4" and height is given to be 4.584 inches.

Total Density of soil (γ) = Mass /Volume = 3.76/0.0333 = 112.81 lbs/cu.ft ---------------------(1)

Water content measurement:

Small soil sample from the mould is obtained to measure the water content.

Prior to oven drying:

Weight of Soil + Tray = 1.21 lbs

After oven drying:
Weight of Soil + Tray = 1.15 lbs

Wt. of water = Wt. of soil before oven drying – Wt. of soil after oven drying

Weight of water = M_w = 1.21 - 1.15 = 0.06 lbs
Weight of solids M_s = Weight of oven dried soil - weight of tray = 1.15 - 6/16 = 0.775 lbs
Weight of tray is given to be 6 ounces. (6 ounces = 6/16 lbs).

Moisture content = M_w/M_s = 0.0.06/0.775 = 0.0774

Dry Density (γ_d) = $\gamma/(1 + w)$

γ = Total density (Also known as wet density)
w = Moisture content

Dry Density (γ_d) = 112.81/(1 + 0.0774) = 104.71 lbs/cu.ft (Ans C)

Problem 1.36) Find the three lateral earth pressure coefficients. (Lateral earth pressure coefficient at rest (K_0), active lateral earth pressure coefficient (K_a) and passive lateral earth pressure coefficient (K_p).

Soil friction angle (φ) = 33^0

	K_0	K_a	K_p
A)	0.53	0.34	0.23
B)	0.57	0.14	0.23
C)	0.45	0.29	3.39
D)	0.76	0.19	2.23

Solution 1.36):

Three lateral earth pressure coefficients are given by following equations.

Lateral earth pressure coefficient at rest (K_0) = $1 - \sin \varphi$
= $1 - \sin 33$ = 0.455

Active lateral earth pressure coefficient (K_a) = $\tan^2(45 - \varphi/2)$

$= \tan^2(45 - 33/2) = 0.295$

Passive lateral earth pressure coefficient (K_p) $= \tan^2(45 + \varphi/2)$
$= \tan^2(45 + 33/2) = 3.39$

Ans (C)

Discussion;

In the case of water, pressure is same in all directions.

Water. $(p_H = p_V)$.

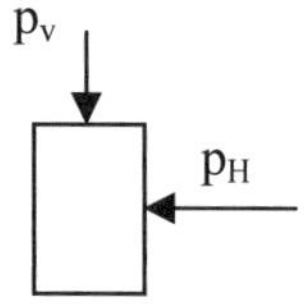

In the case of water, pressure acting on an object is same in all directions. ($p_H = p_V$). But this is not the case with soils.

Soil. $(p_H \neq p_V)$.

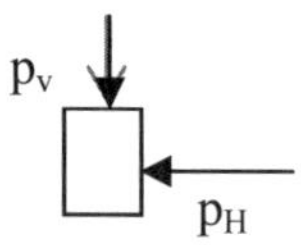

In the case of soil, lateral earth pressure is given by following equation.

$p_H = K.\ p_v$

K = Lateral earth pressure coefficient
p_H = Horizontal earth pressure;
p_v = Vertical pressure or vertical effective stress
Lateral earth pressure coefficient (K) can have three different values.

Three possible values of K are,

1) K_0 = At rest condition
2) K_a = Active condition
3) K_p = Passive condition

At rest condition applies for this problem.
Active condition applies when soil is relaxed (*ex: back of a retaining wall*). Passive condition applies when the soil is stressed. (*ex: front of a retaining wall*).

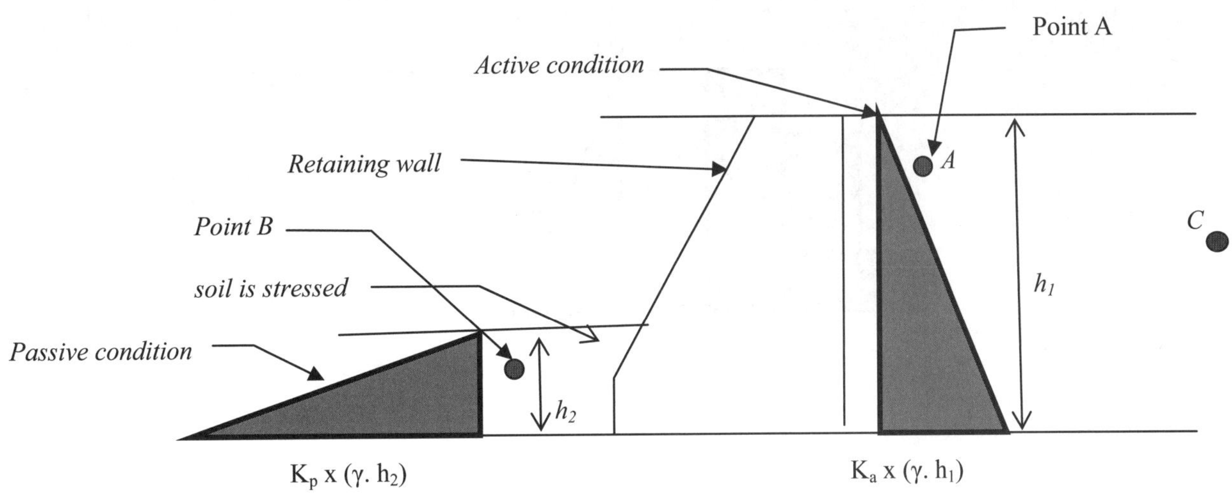

Point A (Active state) : Soil at point A is relaxed due to movement of the retaining wall. The soil in this region is in active state. K_a should be used to find the horizontal stress.

Point B (Passive state): Soil in this region is stressed due to the movement of the retaining wall. The horizontal stress triangle is much more elongated in this region. Soil in this region is in passive state. K_p should be used to find the horizontal stress.

Point C (At rest state): Soil at point C is far away from the retaining wall and effect of the movement of the retaining wall is not felt. Hence soil at point C is at rest. (K_0)

Point A = Active condition
Point B = Passive condition
Point C = At rest condition

$K_p > K_0 > K_a$

$K_0 = 1 - \sin\varphi = 1 \sin 33 = 0.455$
$K_a = \text{Tan}^2 (45 - \varphi/2) = \text{Tan}^2 (45 - 33/2) = 0.294$
$K_p = \text{Tan}^2 (45 + \varphi/2) = \text{Tan}^2 (45 + 33/2) = 3.39$

(Ans C)

Problem 1.37): A Gabion wall is shown in the figure. Each gabion basket is 4 ft x 4 ft x 4 ft. Gabion baskets are filled with rock and has a density of 160 pcf. Soil backfill has a density of 110 pcf and has a friction angle of 30^0. Find the factor of safety against overturning.

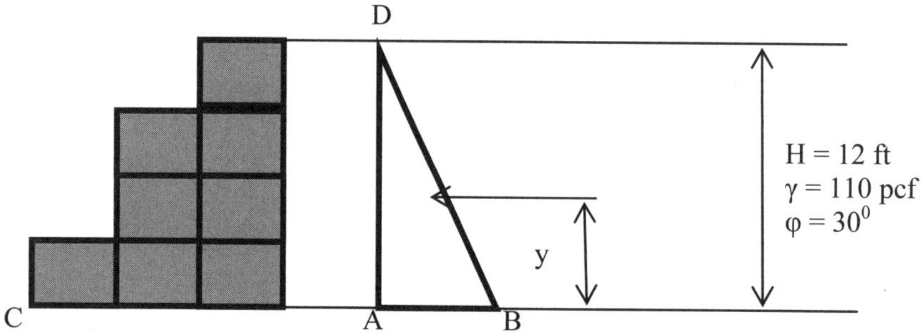

A) 3.2 B) 13.1 C) 3.7 D) 11.2

Solution 1.37):

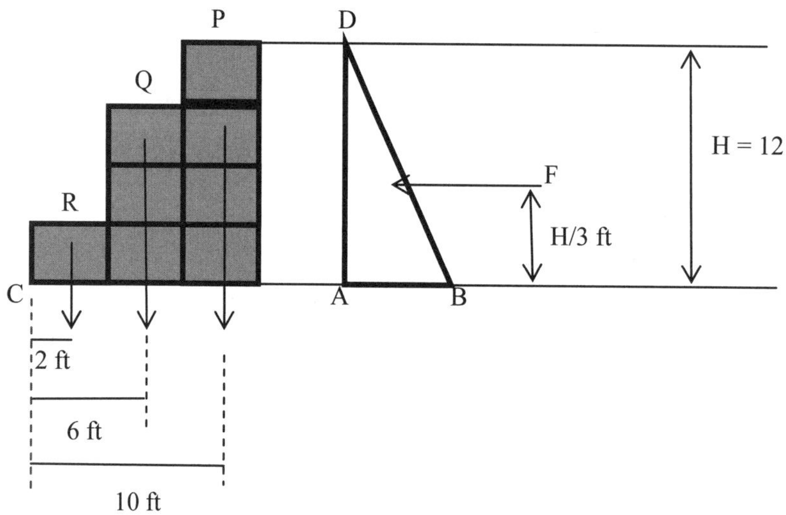

STEP 1: Find the coefficient of earth pressure

Soil in right hand side of the gabion wall is in active state .

Earth pressure is always equal to K. γ. h. You need to find correct K, γ and h.

Following rules should be used for this purpose.

Rule 1: If the soil is relaxed due to movement of the retaining wall, then use active earth pressure coefficient (K_a)
Rule 2: If the soil is stressed due to movement of the retaining wall, then use passive earth pressure coefficient (K_p)
Rule 3: Above groundwater use total density of the soil for γ.
Rule 4: Below groundwater use buoyant density of the soil.

Buoyant density = $\gamma - \gamma_w$
γ = Density of soil
γ_w = Density of water

Active side:
$K_a = \tan^2(45 - \varphi/2) = \tan^2(45\text{–}30/2)$
$K_a = \tan^2(30) = 0.333$

Passive side:
$K_p = \tan^2(45 + \varphi/2)$
Note: There is no soil in the passive side.

STEP 2: Find Earth Pressure
$AB = K_a\,\gamma.\,H = 0.333 \times 110 \times 12 = 439.6 \text{ lbs/ft}^2$.

Total force due to earth pressure is given by the area of the earth pressure diagram.

Total force (F) = Area of triangle ABD = AB x H/2
= 439.6 x 12/2 = 2,637.6 lbs/linear ft

STEP 3: Find Overturning moment:

Force due to earth pressure acts at the center of gravity of the earth pressure diagram.
Center of gravity of the earth pressure diagram is at H/3 distance from the bottom.
Overturning moment = F x H/3 = 2,637.6 x 12/3 = 10,550.4 lbs.ft/linear ft

STEP 4: Find the Resisting Moment:

Resisting moment comes from the weight of gabion baskets.
Gabion baskets are 4 ft x 4 ft x 4 ft. But the force is computed for 1 (one) linear foot of wall.

Weight of one gabion basket/LF = 4 x 4 x 1 x 160 lbs/linear ft of the retaining wall.
= 2,560 lbs/linear ft

Gabion baskets are 4 ft x 4 ft x 4 ft. But the force is computed for 1 LF of the wall.

Note that we are finding the weight per linear ft of the wall. Hence length is taken to be 1 ft.

The gabion wall will rotate around point C. Hence take moments around point C.
Resisting moment due to 4 baskets in line P = 4 x (2,560) x 10 = 102,400 lbs. ft/linear ft

Note: 4 x 2,560 is the weight of 4 baskets per linear foot.
10 ft is the length to the center of gravity of four baskets from point C, where overturning occur. (see the figure).
Resisting moment due to 2 baskets in line Q/LF =2 x (2,560) x 6 = 30,720 lbs. ft/linear ft
Resisting moment due to 1 basket in line R =1 x (2,560) x 2 = 5,120 lbs. ft/linear ft

Total resisting moment = 102,400 + 30,720 + 5,120 = 138,240 lbs. ft/linear ft

STEP 5: Find the factor of safety against overturning:

Factor of safety against overturning = Resisting moment/Overturning moment
=138,240/10,550.4 = 13.1 Ans (B)

Gabion Baskets

Gabion Wall

Problem 1.38) What is the new factor of safety against overturning if due to silt deposition water level has risen to 3 ft below the ground surface in the gabion wall given in the previous problem

A) 5.6 B) 8.8 C) 3.4 D) 7.9

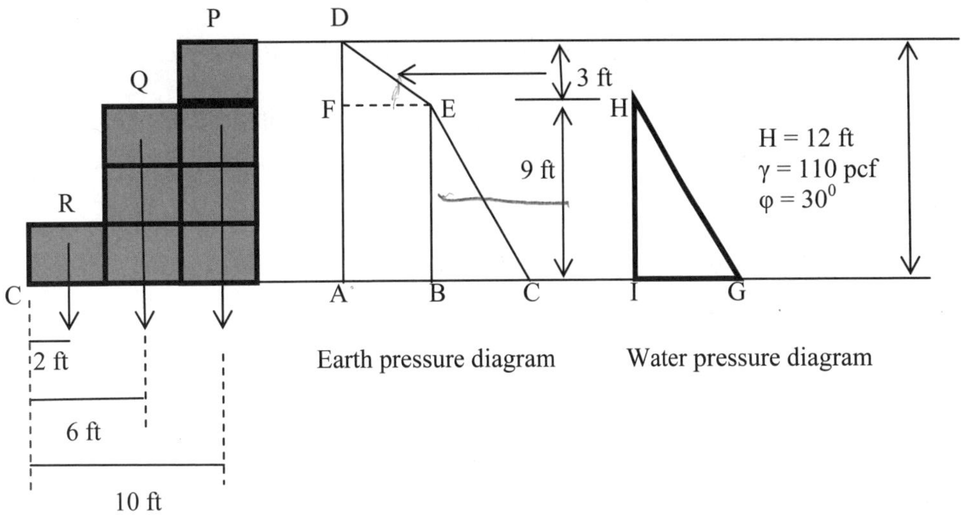

Solution 1.38)

Active side:

STEP 1: Find the dimensions of pressure triangles:

$K_a = \tan^2(45 - \varphi/2) = \tan^2(45–30/2)$

$K_a = \tan^2(30) = 0.333$

$FE = AB = K_a\, \gamma .\, 3 = 0.33 \times 110 \times 3 = 110$ lbs/linear ft of wall

Buoyant density is used below groundwater.

Buoyant density = $(\gamma - \gamma_w)$

$BC = K_a\, (\gamma - \gamma_w).\, h_2$

$= 0.33 \times (110 - 62.4) \times 9 = 141.4$ lbs/linear ft of wall

Pressure diagram due to water:

$IG = \gamma_w \times 9 = 62.4 \times 9 = 561.6$ lbs/linear ft of wall

STEP 2: Find the total force:

Total force = Triangle DFE + Rectangle FEAB + Triangle EBC + Triangle HIG -------(2)

Triangle DFE = FE x 3/2 = 110 x 3/2 = 165 lbs/linear ft of wall

Rectangle FEAB = FE x 9 = 110 x 9 = 990 lbs/linear ft of wall

Triangle EBC = BC x 9/2 = 141.4 x 9/2 = 636.3 lbs/linear ft of wall

Triangle HIG = IG x 9/2 = 561.6 x 9/2 = 2,527.2 lbs

Total force = 165 + 990 + 636.3 + 2,527.2
= 4,318.5 linear ft of wall

STEP 3: Find the Overturning Moment:

Take moments around point C;

Overturning Moment due to DFE = 165 x (9 + 3/3) = 1,650 lbs. ft/linear ft of wall

Force of triangle DFE act at the center of gravity of the triangle DFE. Center of gravity of triangle DFE is acting 3/3 ft above FE line. Moment arm is measured from the bottom (ABC line). Hence moment arm is (9 + 3/3).

Moment due to Rectangle FEAB = 990 x 9/2 = 4,455 lbs ft/linear ft of wall
Moment due to Triangle EBC = 636.3 x 9/3 = 1,909.9 lbs ft/linear ft of wall
Moment due to Triangle HIG = 2,527.2 x 9/3 = 7,581.6 lbs.ft//linear ft of wall

Total overturning moment = 1,650 + 4,455+ 1,908.9 + 7,581.6 = 15,595.5 lbs.ft/linear ft of wall

STEP 5: Find the factor of safety against overturning:

Factor of safety against overturning= Resisting moment/Overturning moment
=138,240/15,595.5 = 8.8 Ans (B)

Resisting moment was found in the previous problem.

Note that factor of safety was reduced from 13.1 to 8.8 due to silt deposition.

WASTEWATER: (WATER RESOURCES)

Problem 1.39): What is the correct order of processes in a wastewater plant?

A) Screens, grit chamber, primary clarifier, secondary clarifier, trickling filter, Chlorine and metal removal system, UV radiation unit
B) Screens, grit chamber, primary clarifier, trickling filter, secondary clarifier, Chlorine and metal removal system, UV radiation unit
C) Screens, grit chamber, primary clarifier, trickling filter, Chlorine and metal removal system, secondary clarifier, UV radiation unit
D) primary clarifier, trickling filter, Screens, grit chamber, Chlorine and metal removal system, secondary clarifier, UV radiation unit

Solution 1.39): Ans B

Screens:

Screens are used to remove wood, sticks, plants, plastic bottles etc.

Bar screens – Remove wood, sticks, plastics and other large objects

Grit Chamber:

Grit chamber is used to remove sand, gravel and other heavy particles. Basically a large tank is used to remove grit or small particles such as sand. The wastewater is passed thru a grit chamber.

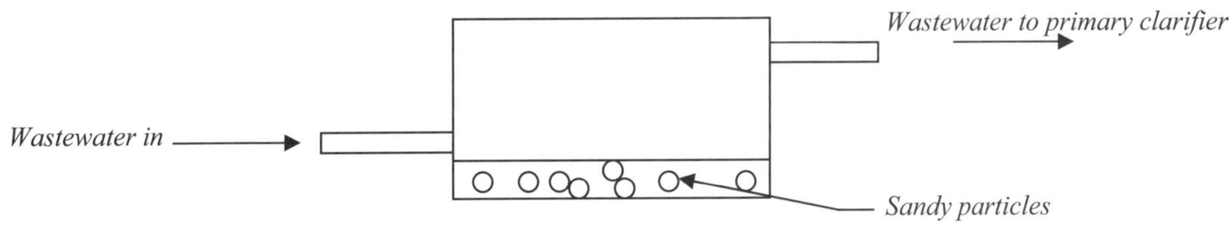

Primary Clarifiers (Sedimentation tanks):

Primary clarifiers are used to remove solids. One of the most common steps is a primary clarifier. Clarifiers are also known as sedimentation tanks.

Assume you take a wastewater sample and put it in a jar and left alone for a day. After one day, you would see some solids settled at the bottom. This is known as sludge. At the same time you would see some solids floating on top. That is known as scum.

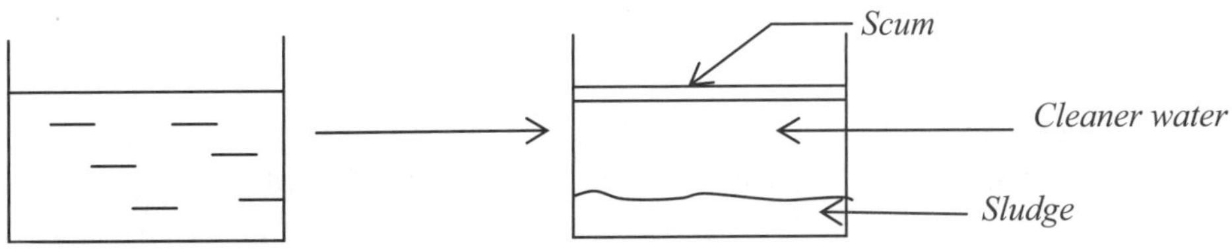

After sludge and scum is removed, one would be left with much cleaner wastewater. Primary clarifiers are one of the most cheapest way to clean wastewater.

Unfortunately in real life, one do not have the luxury to leave wastewater in a tank and left alone since wastewater keeps coming without any break. All what one can do is to make sure that the retention time is sufficient enough for the sludge and scum to form.

Primary Clarifiers (or sedimentation tanks) shown in the figure above. Note the peripheral weirs and the scum removing mechanism . Also note the bridge for technicians to travel to the center of the tank.

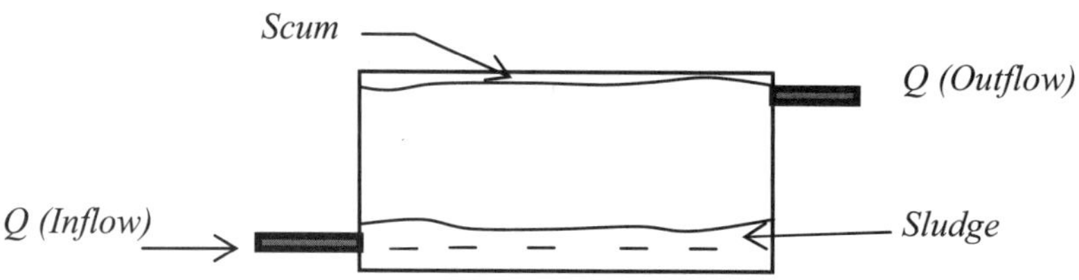

Retention time is defined as the volume of the tank divided by the wastewater flow.

$$\text{Retention Time} = \frac{\text{Volume of the Tank}}{\text{Wastewater Flow}}$$

- If the retention time is high, then more sludge and scum would form. Once the sludge and scum is removed one would left with a cleaner wastewater that needs less processing.
- If the retention time is low, then less sludge and scum would form. Hence the water left would need more secondary processing.

Wastewater flow of a city cannot be changed. To achieve the required retention time, one may have to increase the size of the tank or have more tanks. Typically 2 to 3 hr retention time is desired.

<u>Trickling Filters:</u>

Trickling filter is a system with gravel. Water is allowed to pass through gravel. During the process wastewater would be aerated and bacteria would grow. Bacteria would breakdown organic matter into CO_2 and O_2.

Aerate (Add Oxygen)

Organic matter + Bacteria → Organic matter + More Bacteria → Organic matter will be used as food by Bacteria. Since more Bacteria have grown due to adding Oxygen, ortganic matter would be broken down soon into CO_2 and O_2.

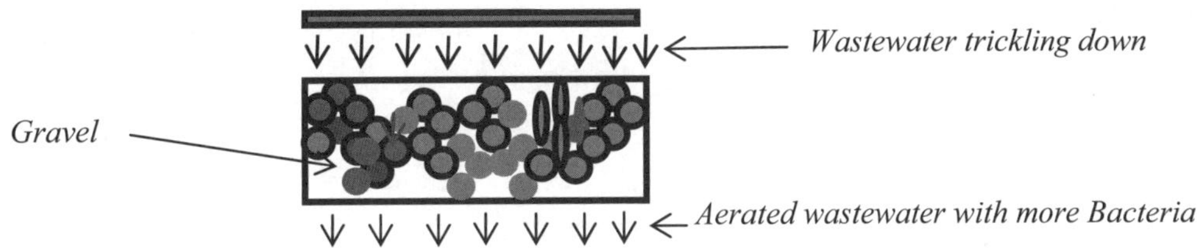

Bacteria use organic matter as food and convert organic matter to CO_2 and O_2. Increasing the number of Bacteria would accelerate the process.

Trickling filters consists of a rock or plastic media and a trickling device. Bacteria culture is introduced to the rock or plastic media. Bacteria culture living in the rock or plastic media is known as bio-film.

<u>Modern trickling filter (water trickle thru the gravel bed. Bacteria and other biological agents use organic matter as food during the process</u>

Trickling bed used in the past

UV Radiation Units: These are used to kill the Bacteria. First Bacteria is multiplied to use them to convert organic matter into CO_2 and O_2. Then Bacteria is killed using UV radiation. (Ans B).

Problem 1.40) Primary clarifier (Primary sedimentation tank) which has a diameter of 40 ft and a height of 8 ft receives 0.7 MGD of wastewater. What is the retention time of the clarifier?

A) 2.13 hrs B) 2.57 hrs C) 3.81hrs D) 5.67 hrs

Solution 1.40):

For primary clarifiers, retention time is obtained by dividing the volume of the tank by flow.

Retention time = Volume of the Tank/Flow

Retention time is also known as detention time.

MGD = Million gallons per day

Volume of the tank = $\pi \times D^2/4 \times$ Height = $\pi \times 40^2/4 \times 8$ = 10,053 cu.ft

1 cu. ft = 7.48 gallons

Hence Volume = 10,053 x 7.48 gallons = 75,196 gallons

Wastewater flow = 0.7 MGD (million gallons per day)
This has to be converted to gallons per hour.
Wastewater flow = 0.7 MGD = 0.7 x 10^6 gallons per day = 0.7 x 10^6/24 gallons per hr
= 29,166.7 gallons per hour

Retention time = Volume/Wastewater flow = 75,196 gallons/29,166.7 gallons per hr = 2.57 hrs (Ans B).

SAMPLE EXAM 2

Water Resources:

Problem 2.1) Find the hydraulic radius of the channel shown.

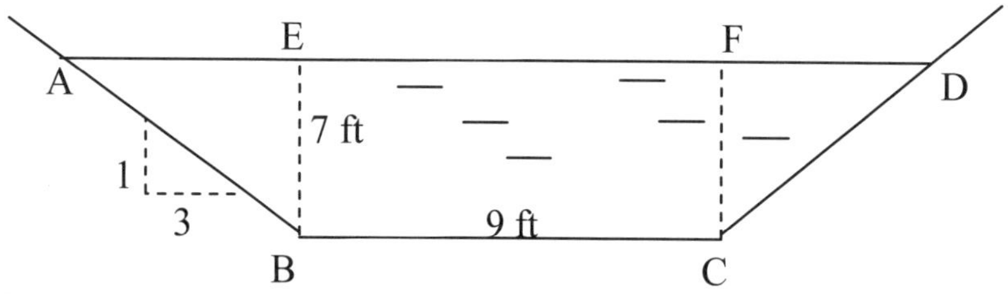

A) 3.95 ft B) 6.54 ft C) 5.31 ft D) 2.10 ft

Problem 2.2) Water is flowing thru a trapezoidal channel at a depth of 6 ft. The slope of the channel is given to be 0.4% and Manning roughness coefficient is given to be 0.002. What is the flow rate (cu. ft/sec) of the channel?

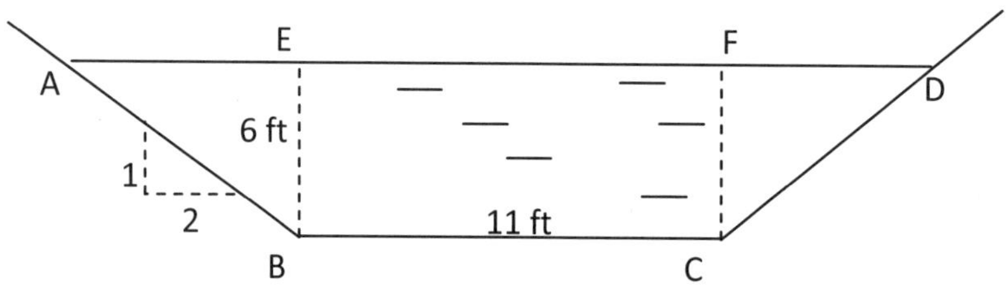

A) 12,340 B) 15,414 C) 19,089 D) 16,679

Problem 2.3) During a hydraulic jump, upstream depth found to be 2.5 ft and downstream depth found to be 8.4 ft. What is the energy dissipation due to the jump?

A) 2.445 ft B) 1.887 ft C) 5.617 ft D) 9.238 ft

Problem 2.4): Find the pressure at point B in the pipe shown below. Assume there is no energy loss due to friction.

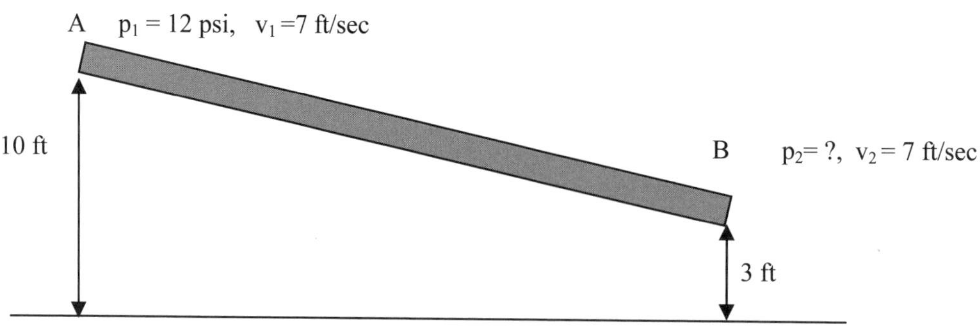

A) 15.03 psi B) 13.03 psi C) 2.89 psi D) 17.98 psi

Problem 2.5: A pipe is shown below. Find the pressure at point B.
(Use Hazen Williams formula to calculate the head loss due to friction).

Hazen Williams coefficient = 120
Diameter of the pipe = 6 in (0.5 ft)
Length of the pipe = 35 ft

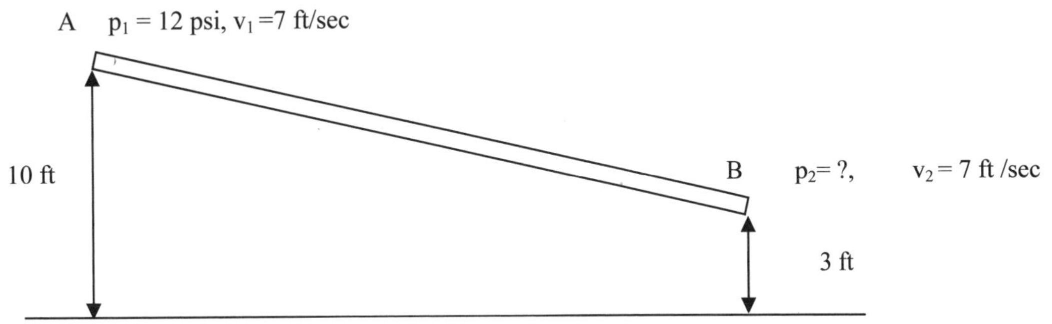

A) 18.2 psi B) 14.5 psi C) 1.2 psi D) 5.1 psi

Problem 2.6: Pitot tube is shown in the figure. Find the flow velocity.

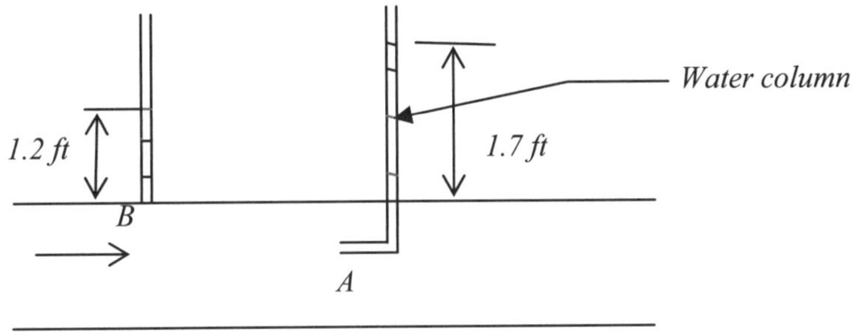

A) 12.3 ft/sec B) 5.67 ft/sec C) 3.45 ft/sec D) 9.4 ft/sec

Problem 2.7): Hydrologists have developed the following equation for a given region for 25 yr rainfall intensity.

I = 105/Duration;

I = Rainfall intensity (in/hr);
Duration (min)

Time of concentration of the drainage basin was found to be 35 minutes. Hydrologists believe that water will not start to runoff during the first 10 minutes of the rainfall due to percolation into the ground. Runoff coefficient of the drainage basin is 0.67. The area of the drainage basin is 7.5 acres. Find the design flow of the drainage basin.

A) 12.34 Acre. in/hr B) 32.67 Acre in/hr C) 23.12 Acre. in/hr D) 11.71 Acre. in/hr

Problem 2.8) Two areas are draining into a drain pipe as shown. Information of the two areas are given below.
Area 1 = Area is 12 acres, time of concentration 22 minutes, runoff coefficient 0.76
Area 2 = Area is 17 acres, time of concentration 27 minutes, runoff coefficient 0.68

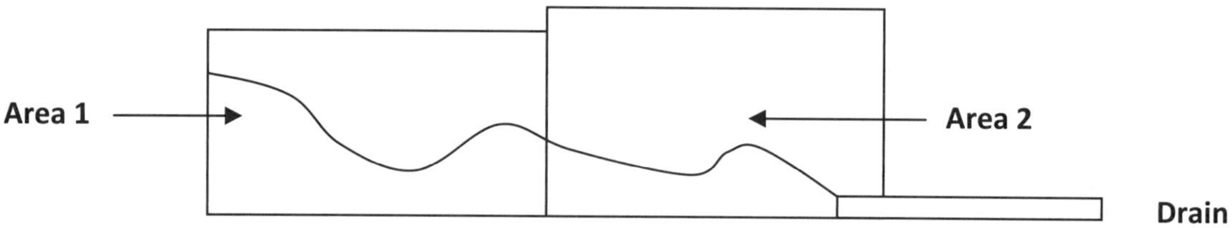

Maximum Intensity of 30 year, 49 minute rainfall is 2.3 in/hr.
Find the maximum flow for a 30 year rainfall.

A) 47.8 cu.ft/sec B) 65.0 cu.ft/sec C) 21.4 cu.ft/sec D) 34.5 cu.ft/sec

Problem 2.9) Mountain region drains into a road as shown. Engineers have designed a trench 2,000 ft long and move the water thru a pipe across the road. Time of concentration for each point of the trench is as shown in the figure. Area of the drainage basin is 5 acres and the runoff coefficient is 0.55.
Velocity of flow in the trench is 8 ft/sec.

Rainfall intensity of the region is given by the following equation.

$$I = 100/(t_c + 15)$$

I = Rainfall intensity (in/hr);
t_c = Time of concentration

Time of concentration for three points in the trench is shown.
Find the maximum flow thru the culvert.

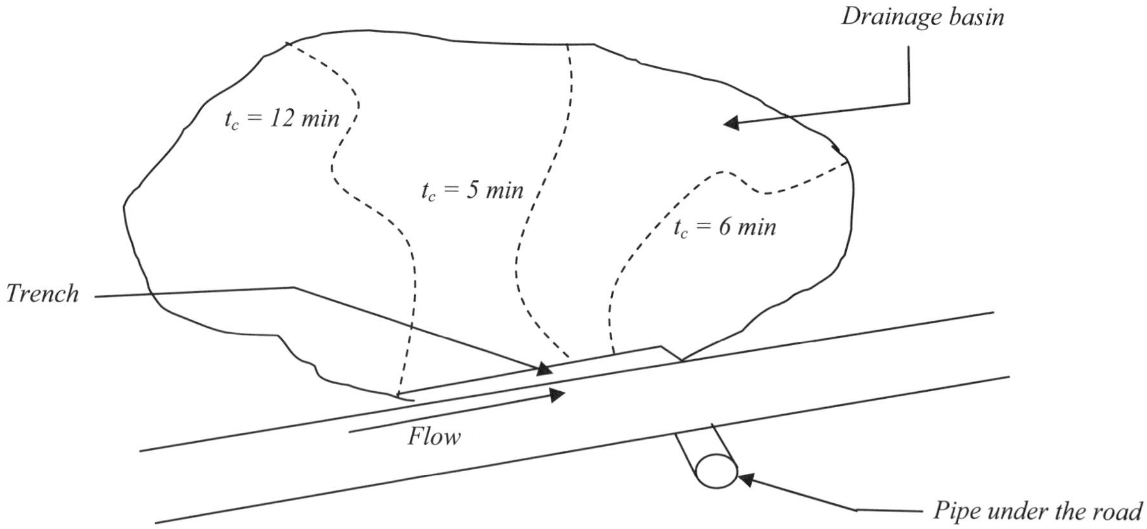

A) 12.35 cu.ft/sec B) 8.87 cu.ft/sec C) 3.45 cu.ft/sec D) 4.98 cu.ft/sec

Problem 2.10) Find the maximum flow in the drain pipe (cu.ft/sec) for a unit hydrograph. Area of the drainage basin is 4 acres.

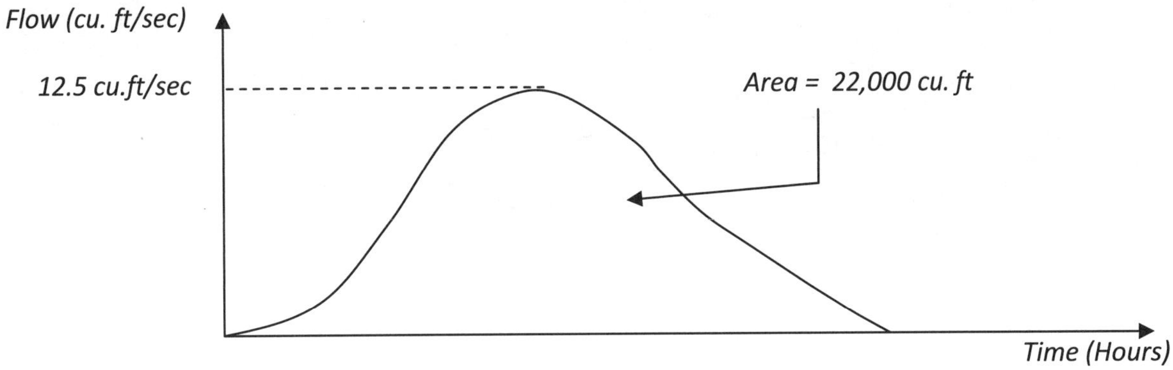

A) 7.86 cu.ft/sec B) 8.27 cu.ft/sec; C) 2.78 cu.ft/sec; D) 7.97 cu.ft/sec

Problem 2.11) A car was travelling on an asphalt road and ran out of the road and collides with a car parked away from the road in a sandy area. Impact speed was found to be 30 mph. Friction coefficient (μ) of asphalt is 0.6 and friction coefficient (μ) of sand is 0.5. Length of travel in asphalt was 200 ft and length of travel in sand was 120 ft. What was the speed of the vehicle just before applying brakes?

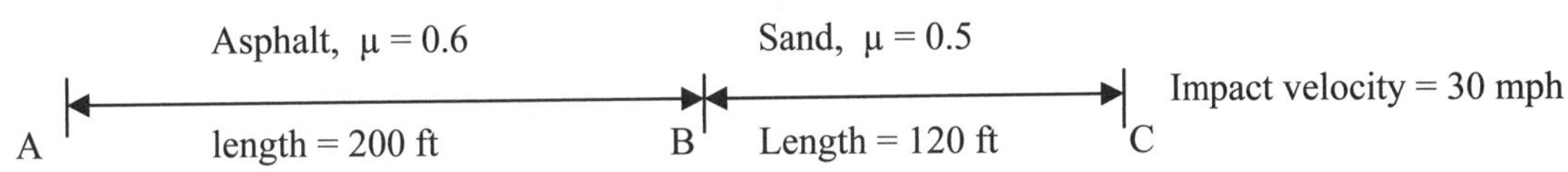

A) 116.3 ft/sec B) 231.2 ft/sec C) 78.9 ft/sec D) 43.9 ft/sec

Problem 2.12) Stopping sight distance of a road is 460 ft. Radius of the horizontal curve is 600 ft. Find the horizontal sight offset required for the horizontal curve.

A) 123.56 ft B) 25.89 ft C) 91.00 ft D) 43.55 ft

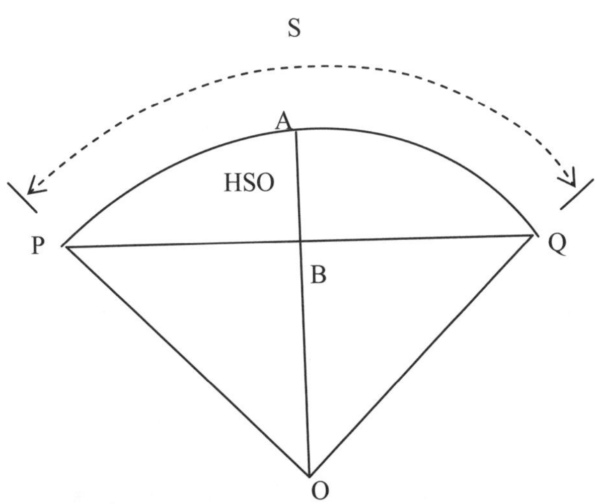

AB = HSO (Horizontal sight offset)

Problem 2.13): A traverse is done as shown. Internal angles of the traverse is as given. What is the total error of internal angles?

A) 4.662^0 B) 3.632^0 C) 4.169^0 D) 6.167^0

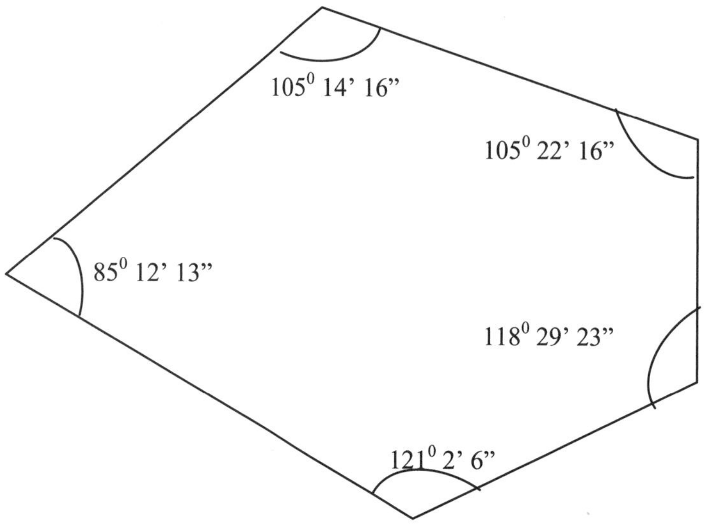

Problem 2.14) Two vehicles are travelling in the same direction in a highway next to each other. First vehicle is travelling at a speed of 50 mph and the second vehicle is travelling at a speed of 40 mph. Both drivers apply brakes at the same time. First vehicle decelerates at 12 ft/sec^2 and the second vehicle decelerates at 8 ft/sec^2. When both vehicles come to a complete stop, what's the distance between the two vehicles?

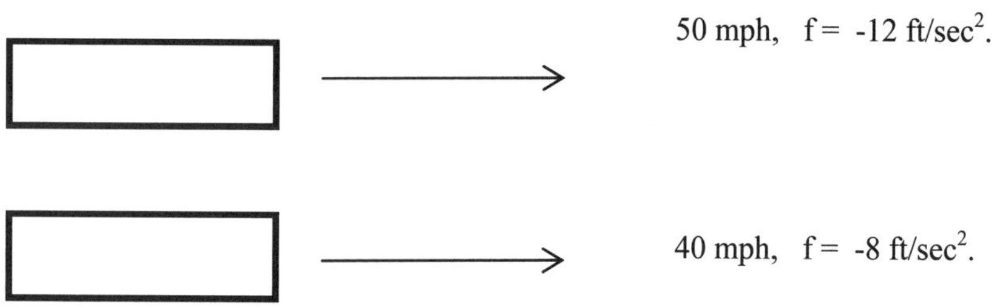

A) 12.34 ft B) 8.97 ft C) 32.45 ft D) 4.56 ft

Problem 2.15): Station at point of intersection (PI) is given to be 14+70 and the radius of the curve is 1,400 ft. Find the station at PT. (Angle of intersection is 53 degrees).

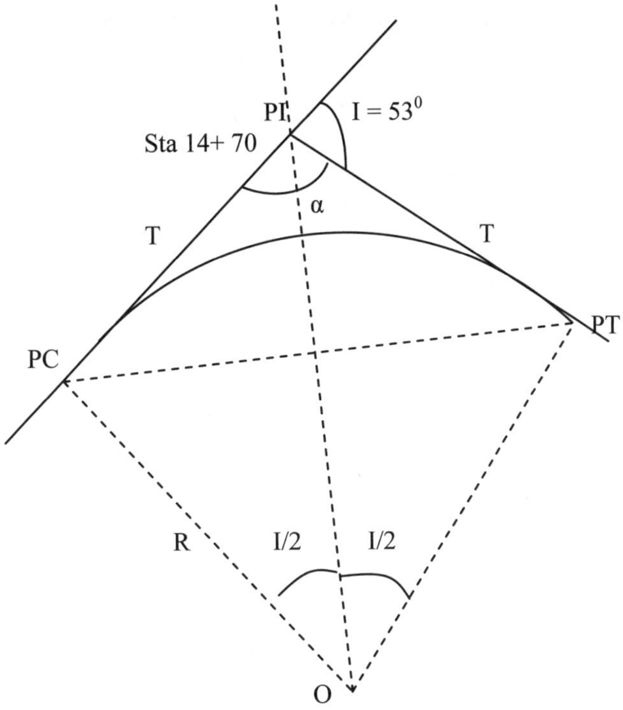

A) 14+66 B) 34+75 C) 20+67 D) 24+81

Problem 2.16): Vertical curve is shown in the figure. Upward gradient is 1.8% and downward gradient is 3.2% as shown. BVC elevation is 132.71 and BVC station is 11+53. Station of point P is given to be 20 +00. EVC station is 22+48. Find the elevation at point P.

A) 105.94 B) 113.87 C) 106.79 D) 121.97

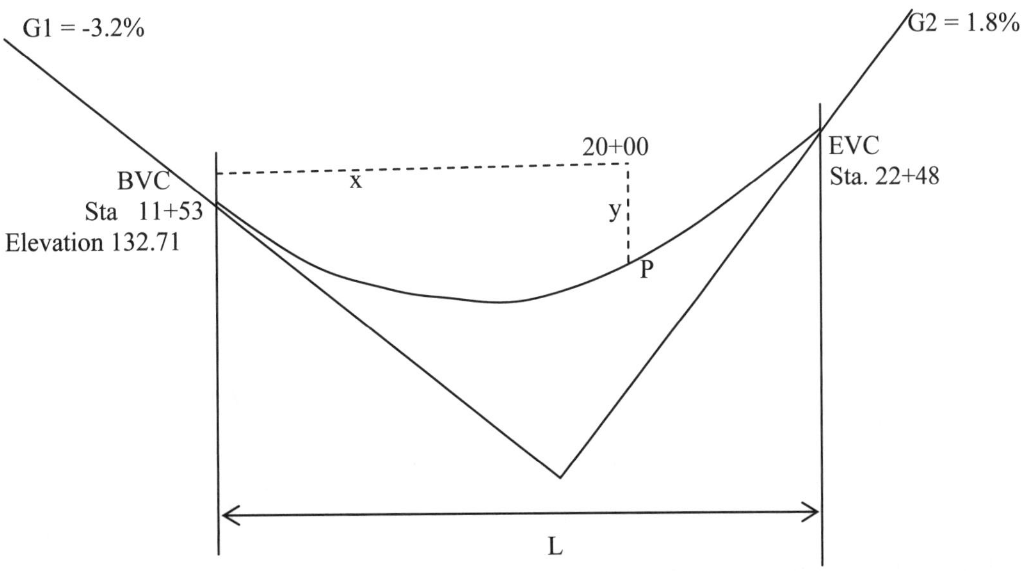

BVC = Beginning of vertical curve
EVC = End of vertical curve

Problem 2.17) Engineers are designing a roadway horizontal curve. Side friction factor of the roadway is 0.2 and the speed of vehicles is 60 mph. Radius of the horizontal curve is 700 ft. What is the superelevation required at the curve to maintain the speed.

A) 12.3% B) 14.3% C) 3.2% D) 7.9%

Problem 2.18) Horizontal curve is designed with a superlevation for a highway. The road has a normal crown and after the superelevated horizontal curve it switches back to the normal crown. What is the correct process of changing a normal crown to superelevation and then back to a normal crown?

A) Normal crown, Tangent runout, superelevation runoff, full superelevation, superelevation runoff, Tangent runout, Normal crown

B) Normal crown, superelevation runoff , Tangent runout, full superelevation, superelevation runoff, Tangent runout, Normal crown

C) Normal crown, superelevation runoff , full superelevation, superelevation runoff, Tangent runout, Normal crown

D) superelevation runoff, superelevation runout, Tangent runout, Normal crown, superelevation, superelevation runoff, Tangent runout, Normal crown

Problem 2.19) Find the deflection at the end of the hollow rectangular section shown. Outside dimensions of the section is 12 x 12 inches and the thickness is 0.25 inches. The Young's modulus of steel is 29 x 10^6 psi.

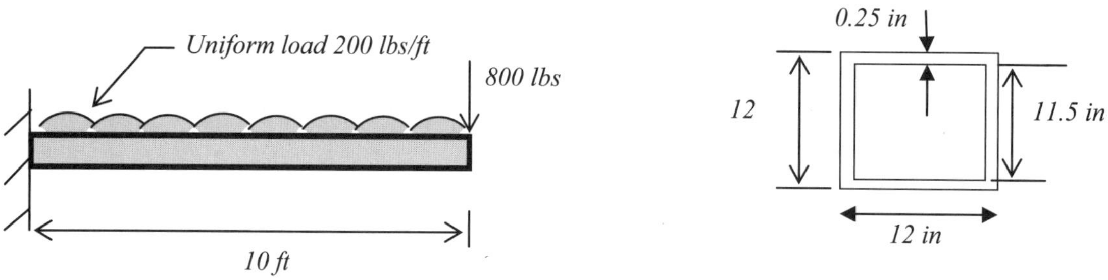

A) 0.149 in *B) 1.23 in* *C) 0.453 in* *D) 1.67 in*

Problem 2.20): Seismic design categories are developed
A) Based on soil type and building type
B) Based on risk and importance of the structure with regard to loss of life
C) Based on soil type only
D) Based on the probability of an earthquake in the region

Problem 2.21): Generalized spectral response curves in seismic engineering prepared by USGS (adopted by building codes in USA) S_1 and S_s represents;
A) S_1 represents the ground motion for one second earthquake and S_s represents the ground motion for short period earthquakes.
B) S_1 represents the spectral response acceleration for one second earthquake and S_s represents the spectral response acceleration for short period earthquakes.
C) S_1 represents the spectral response acceleration for buildings with one second natural period and S_s represents the spectral response acceleration for short buildings with natural period less than 1 second.
D) None of the above

Problem 2.22) What is not true about moment frame buildings?
A) Need less steel compared to braced frame structures
B) More labor intensive than braced frame structures

C) Larger deflections can be expected compared to braced frame structures

D) Easy to design doorways and windows.

Problem 2:23): Find the nominal moment of the beam shown.

f_y= 60,000 psi
Concrete compression strength (f_c') = 4,000 psi
A_s = 4.3 sq. in

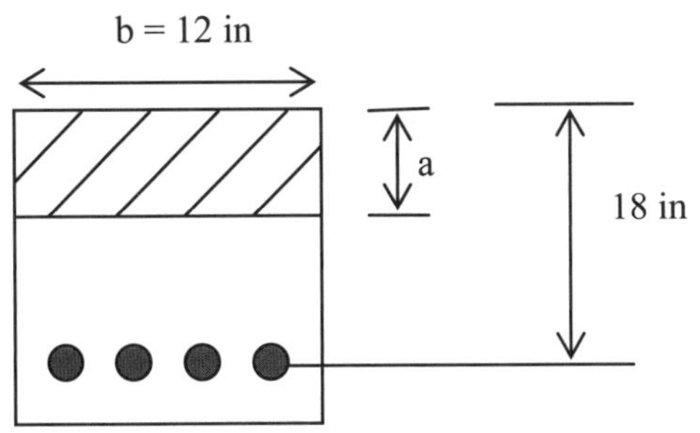

A) 213 kip. ft B) 319 kip. ft C) 154 kip. ft D) 298 kip.ft

Problem 2:24) What is the maximum steel allowed by ACI 318 for the above beam.

As per ACI 318, steel strain allowed = 0.004
Concrete crushing strain = 0.003

f_y= 60,000 psi
Concrete compression strength (f_c') = 4,000 psi

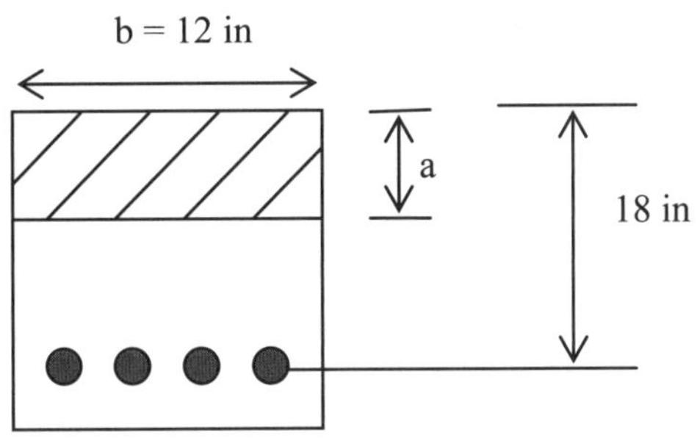

A) 3.45 sq. in B) 4.46 sq. in C) 5.93 sq. in D) 6.12 sq. in

Problem 2:25) A beam is loaded with a uniform load of 5 kips/ft and a concentrated load of 30 kips. Find the shear force at point P.

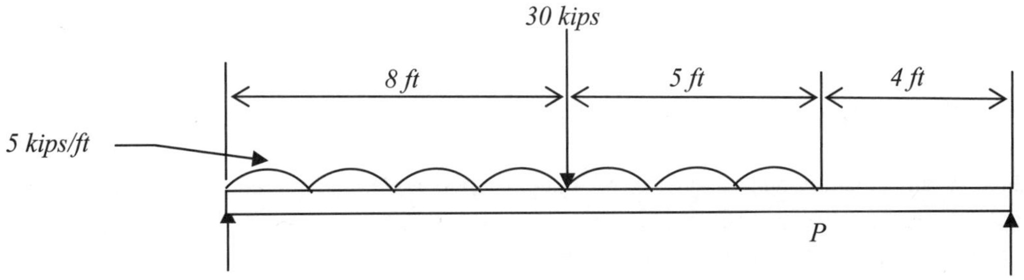

A) 34 kips B) 46 kips C) 39 kips D) 62 kips

Problem 2:26) Find the maximum bending moment of the beam given in the previous problem.

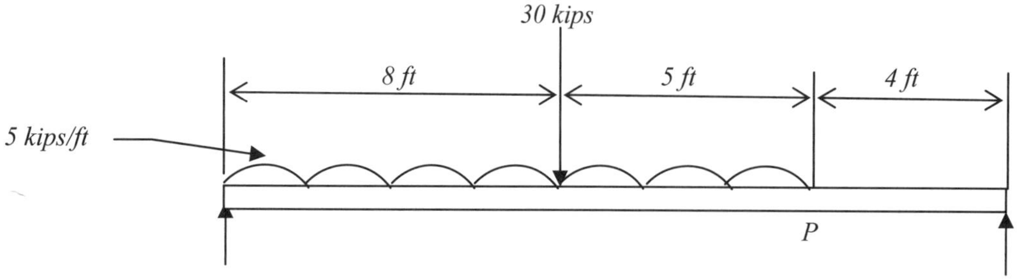

A) 288 kip.ft B) 146 kip.ft C) 239 kip.ft D) 621 kip.ft

Problem 2.27): A solution has 12 grams of dissolved Oxygen. How many moles of Oxygen are dissolved in the solution?

A) 0.375 moles B) 0.618 moles C) 0.750 moles D) 1.204 moles

Problem 2.28): 100 ml wastewater sample was diluted with 800 ml of water. Initial dissolved Oxygen level (DO_i) was found to be 14.8 mg/L. After 5 days of incubation, final dissolved Oxygen level (DO_f) was found to be 3.9 mg/L. What is the 5 day BOD of the sample?

A) 211.3 mg/L B) 93.4 mg/L C) 98.1 mg/L D) 37.8 mg/L

Problem 2.29) Activated sludge process is best described as

A) Re-introduce aerated sludge back to the waste stream to increase the level of Bacteria.
B) Re-introduce aerated sludge back to the waste stream to decrease the level of Bacteria.
C) Activate the sludge by adding chemicals
D) None of the above

Problem 2.30): Total density of a soil sample was found to be 110 pcf and moisture content to be 60%. What is the dry density of the soil sample.

A) 125.9 pcf B) 45.6 pcf C) 68.7 pcf D) 98.9 pcf

Problem 2.31) Find the lateral earth pressure at point A.

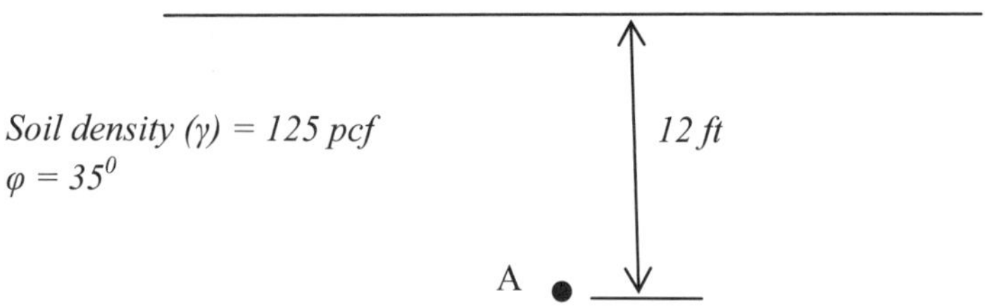

A) 2,131.2 psf B) 245.9 psf C) 639.6 psf D) 412.3 psf

Problem 2.32): Which statement is true regarding SPT (N) value, friction angle, unconfined compressive strength and cohesion?

A) There is a correlation between SPT (N) value and friction angle in sandy soils. There is no correlation between unconfined compressive strength and cohesion in clay soils.

B) There is no correlation between SPT (N) value and friction angle in sandy soil. There is no correlation between unconfined compressive strength and cohesion in clay soils.

C) There is a correlation between SPT (N) value and friction angle in sandy soils. There is a correlation between unconfined compressive strength and cohesion in clay soils.

D) There is no correlation between SPT (N) value and friction angle. There is a correlation between unconfined compressive strength and cohesion.

Problem 2.33): Soil sample has a porosity of 0.46. What is the void ratio of the soil sample?

A) 0.67 B) 0.85 C) 0.32 D) 1.12

Problem 2.34): Figure shows a pile driven in clay soils.

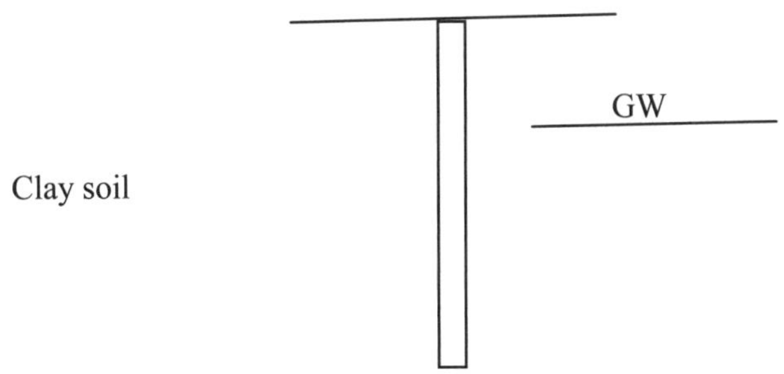

Capacity of the pile is NOT dependant on;

A) Cohesion of soil B) Pile diameter C) Depth to groundwater
D) Soil – Pile adhesion coefficient

Problem 2.35): Consolidation test of a normally consolidated soil is shown below. What is the settlement (ΔH) of the clay layer shown if the initial pressure p_0' = 300 psf and final pressure is p_f' = 600 psf at the center of the clay layer.

Thickness of the clay layer (H) = 12 ft
Gradient = C_c = *0.28*
Initial void ratio (e_0) = 0.65

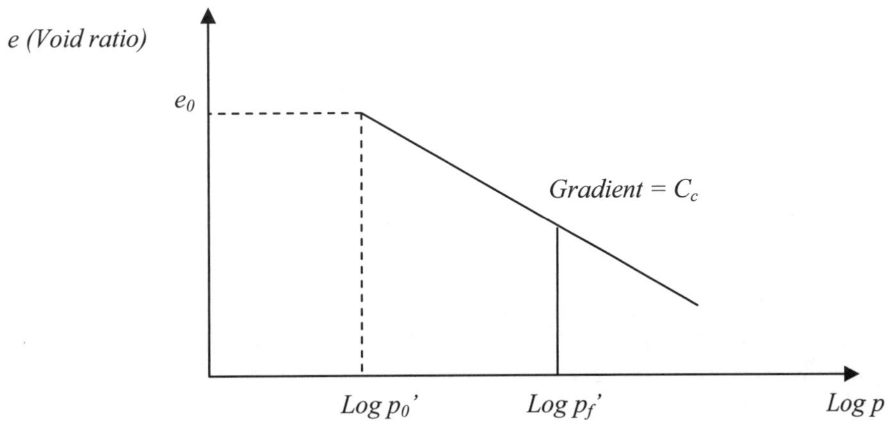

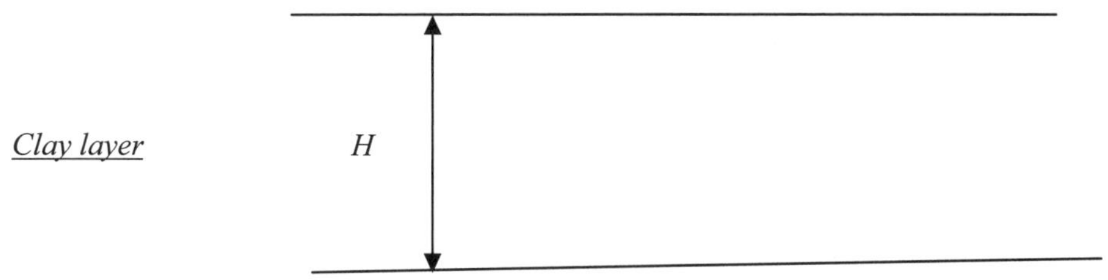

A) 3.21 in. B) 1.89 in. C) 4.98 in D) 7.35 in.

Problem 2.36) Proposed road and existing ground profile is shown below. Volume of soil in each segment is given. Swell factor and shrinkage factor of soil are 18% and 20% respectively. How many trucks of soil need to be brought in from an outside borrow pit if the swell factor and shrinkage factor of borrow pit soil are 15% and 17% respectively. Assume each truck can carry 10 CY of soil.

Bank volumes: V1 = 123 CY; V2 = 330 CY; V3 = 60 CY

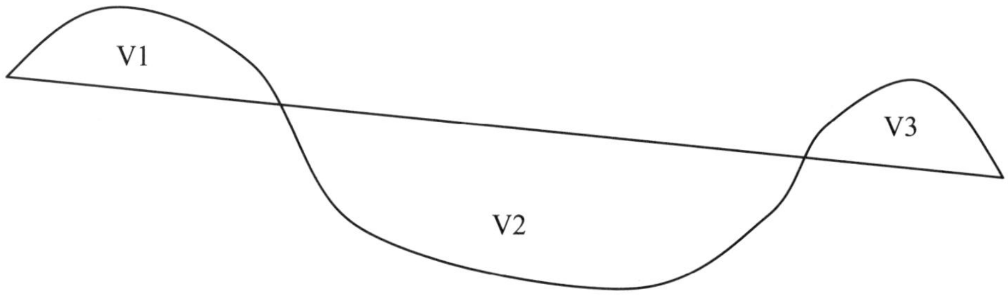

A) 26 B) 23 C) 28 D) 22

Problem 2:37) Contractor needs to erect 200 tons of steel. He has two crews. (crew 1 and crew 2). He decided to use crew 1 to erect 200 tons of steel. What would have happened, if he had used crew 2 instead?

Crew 1:
5 steel workers
3 Laborers
1 foreman
1 welder
Crew 1 productivity = 1.5 tons/LH
Crew 2:
4 steel workers
4 Laborers
1 foreman
2 welders

Crew 2 productivity = 1.7 tons/LH
Wages:

steel workers - \$75/hr
laborers -\$55/hr
foreman - \$90/hr
welders - \$65/hr

A) Contractor would have lost 1,523 dollars
B) Contractor would have lost 912 dollars
C) Contractor would have saved 1,352 dollars
D) Contractor would have saved 2,023 dollars

Problem 2.38) What loads are considered as construction loads?

A) Wind load, dead weight, live load from construction workers, Earthquake load
B) Wind load, dead weight, live load from construction workers, Load due to wet concrete
C) Wind load, dead weight, live load from workers, Earthquake load
D) Wind load, dead weight, live load from workers, Seismic load

Problem 2:39) Find the number of bricks required for the wall shown.
Following information provided;

Length of the wall = 120 ft
Height of the wall = 10 ft
Size of bricks = 8 in x 3 5/8 in x 2 ¼ in
Mortar thickness between bricks = 0.5 inches
The wall is one brick thick.

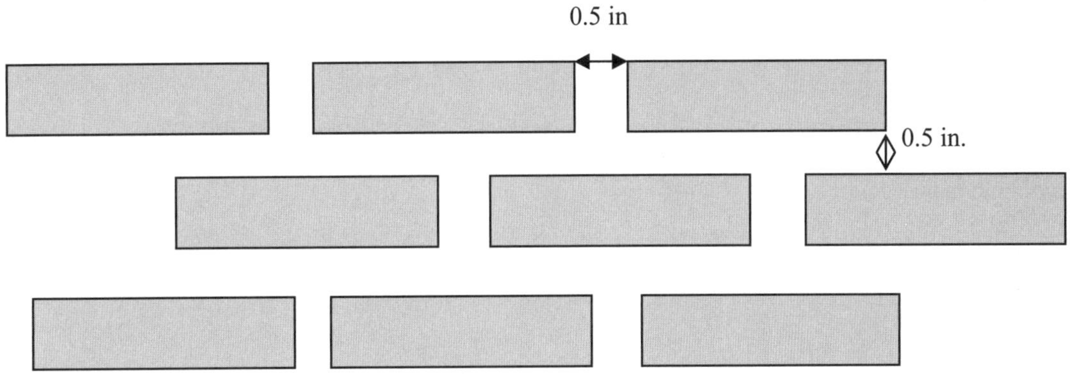

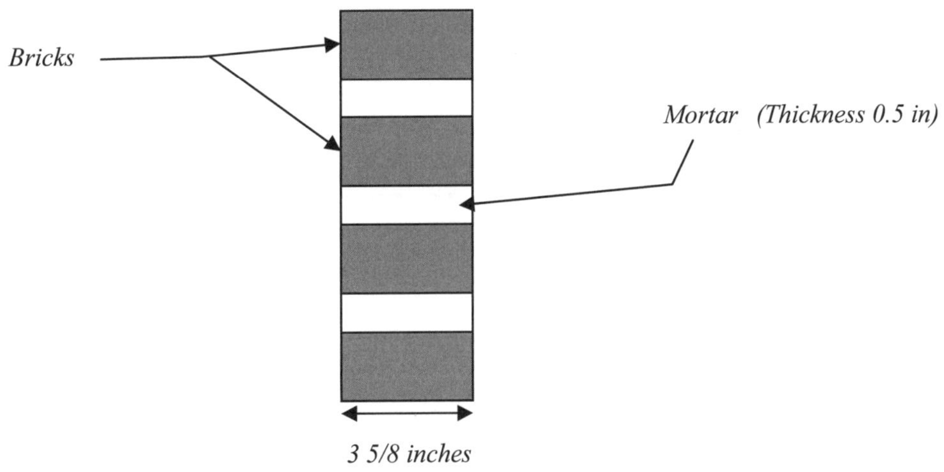

Wall - Side View

A) 7,392 bricks B) 6,823 bricks C) 4,987 bricks D) 3,651 bricks

Problem 2:40) Find the volume of mortar required for the wall given in the previous problem.

A) 1.25 CY B) 3.98 CY C) 2.89 CY D) 3.09 CY

SAMPLE EXAM 2 (SOLUTIONS)

Water Resources:

Problem 2.1) Find the hydraulic radius of the channel shown.

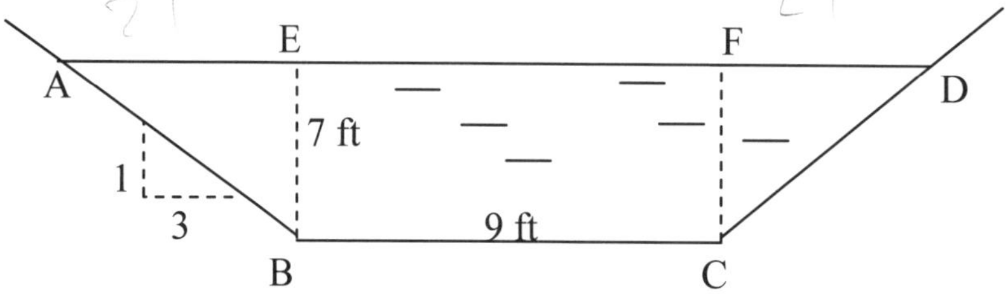

A) 3.95 ft B) 6.54 ft C) 5.31 ft D) 2.10 ft

Solution 2.1):

Hydraulic radius (R) = A/P

A = Area;
P = Wetted Perimeter

STEP 1: Find the area (A);

BE = 7ft; Hence AE = 3 x 7 = 21 ft (Due to 3:1 slope)

FD = 21
AD = 21 + 9 + 21 = 51

Area of a trapezoid is given by the following formula.

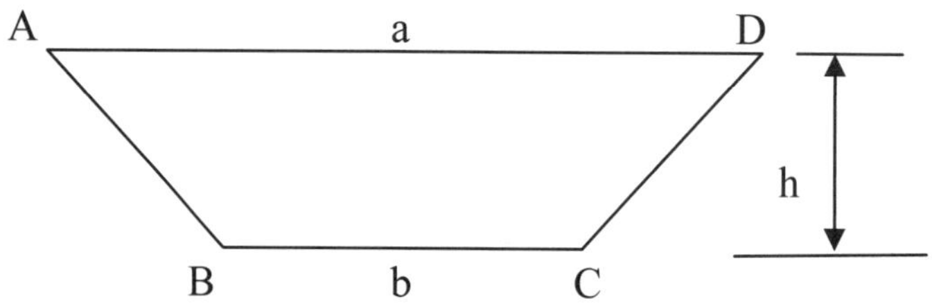

Area = (Base + Top)/2 x h

Area of the channel shown = (AD + BC)/2 x 7
= (51 + 9)/2 x 7 = 210 sq. ft

STEP 2: Find the wetted perimeter (P);

P = AB + BC + CD

From Pythagoras theorem;

$AB^2 = BE^2 + AE^2$
$AB^2 = 7^2 + 21^2$
$AB = 22.1$
$P = AB + BC + CD = 22.1 + 9 + 22.1 = 53.2$

$R = A/P = 210/53.2 = 3.95$ ft (Ans A)

Problem 2.2) Water is flowing thru a trapezoidal channel at a depth of 6 ft. The slope of the channel is given to be 0.4% and Manning roughness coefficient is given to be 0.002. What is the flow rate (cu. ft/sec) of the channel?

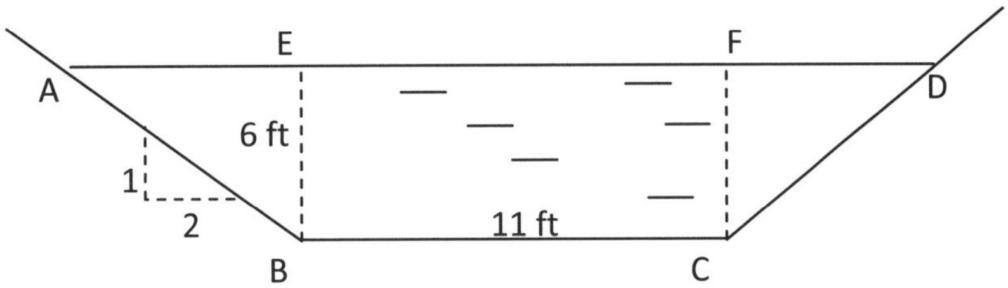

A) 12,340 B) 15,414 C) 19,089 D) 16,679

Solution 2.2):

Manning Equation:

$$v = \frac{1.49 \times R^{2/3} S^{1/2}}{n}$$

v = Velocity,
R = Hydraulic Radius = A/P
A = Area;
P = Wetted perimeter
S = slope
n = Manning coefficient

STEP 1: Gather all the given data.

Slope = S = 0.4% = 0.004

Area = (Base + Top)/2 x h

$$\text{Area (A)} = \frac{(\text{Base} + \text{Top})}{2} \times \text{Height} = \frac{(AD + BC)}{2} \times 6$$

EB = 6 ft.
Hence AE = 12 ft due to 2:1 slope.

Hence AD = 12 + 11 + 12 = 35 ft

$$\text{Area (A)} = \frac{(AD + BC)}{2} \times 6 = \frac{(35 + 11)}{2} \times 6 = 138 \text{ sq. ft}$$

P = Wetted perimeter = AB + BC + CD

$AB^2 = AE^2 + BE^2$
$AB^2 = 12^2 + 6^2 = 180$
$AB = 13.4$ ft

Wetted perimeter = AB + BC + CD = 13.4 + 11 + 13.4 = 37.8 ft

R (Hydraulic radius) = A/P = 138/37.8 = 3.651

STEP 2: Apply the Manning equation:

$$v = \frac{1.49 \times R^{2/3} S^{1/2}}{n} = \frac{1.49 \times 3.651^{2/3}\ 0.004^{1/2}}{0.002} = 111.7 \text{ ft/sec}$$

Flow Rate = Area x Velocity = 138 x 111.7 = 15,414 cu. ft/sec (Ans B)

Problem 2.3) During a hydraulic jump, upstream depth found to be 2.5 ft and downstream depth found to be 8.4 ft. What is the energy dissipation due to the jump?

A) 2.445 ft B) 1.887 ft C) 5.617 ft D) 9.238 ft

Solution 2.3):

Energy dissipation during a hydraulic jump is given by the following equation.

Energy dissipation during a hydraulic jump = E1 – E2 = $(y_2 - y_1)^3/(4.y_1y_2)$

Source: Open channel hydraulics, Ven Te Chow;

Energy dissipation during the hydraulic jump = $(8.4 - 2.5)^3/(4 \times 8.4 \times 2.5)$

Energy dissipation during the hydraulic jump = 2.445 ft (Ans A)
Energy loss is given in head of water.

Problem 2.4: Find the pressure at point B in the pipe shown below. Assume there is no energy loss due to friction.

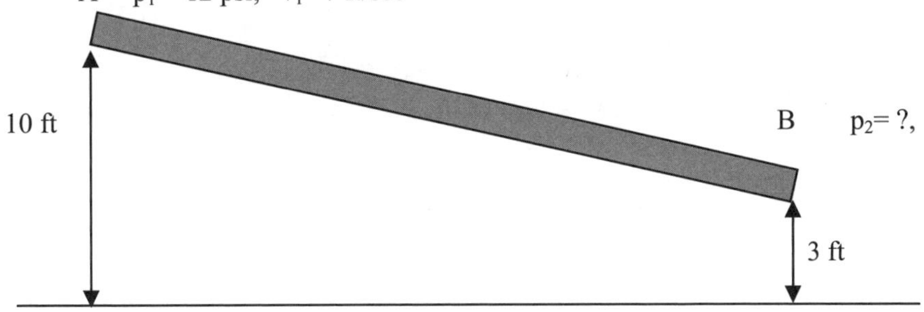

A) 15.03 psi B) 13.03 psi C) 2.89 psi D) 17.98 psi

Solution 2.4) Use the energy equation;

$$h_1 + v_1^2/2g + p_1/\gamma = h_2 + v_2^2/2g + p_2/\gamma$$

Water coming from point A has to come out from point B. Hence $v_1 = v_2$.

$h_1 = 10$ ft; $v_1 = 7$ fps; $p_1 = 12$ psi $= 12 \times 144$ psf
$h_2 = 3$ ft; $v_2 = 7$ fps; $p_2 = ?$
$v_1 = v_2$.
(v_1 is equal to v_2 due to continuity of flow. When the diameter does not change, velocity would be the same along a pipe line).

Pressure is given in psi. Multiply the pressure value by 144 to convert to psf.

$10 + 7^2/(2 \times 32.2) + (12 \times 144)/(62.4) = 3 + 7^2/(2 \times 32.2) + p_2/62.4$

$p_2 = 2{,}164.8$ psf
$= 15.03$ psi (Ans A)

Problem 2.5: A pipe is shown below. Find the pressure at point B. (Use Hazen Williams formula to calculate the head loss due to friction).

Hazen Williams coefficient = 120
Diameter of the pipe = 6 in (0.5 ft)
Length of the pipe = 35 ft

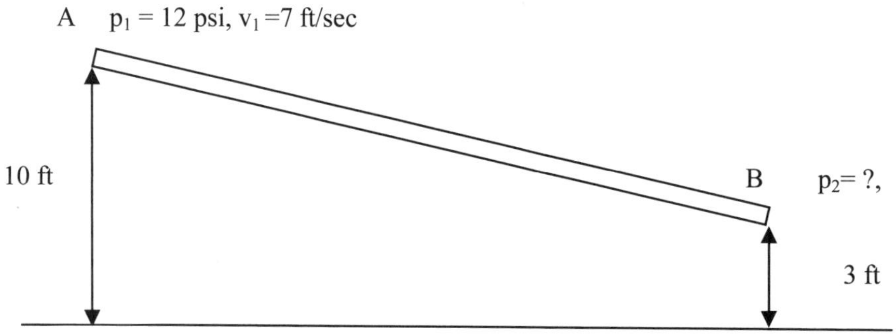

A) 18.2 psi B) 14.5 psi C) 1.2 psi D) 5.1 psi

Solution 2.5:

Closed Pipe Flow Equations Including Hazen-Williams and Darcy-Weisbach Equation:

As we all know energy is lost due to friction. Head loss due to friction is calculated using Hazen Williams equation and Darcy – Weisbach equation. When we have head loss due to friction, Bernoulli's equation has to be modified.

Energy at point A = Energy at point B + Head Loss due to Friction

$$h_1 + v_1^2/2g + p_1/\gamma = h_2 + v_2^2/2g + p_2/\gamma + h_L$$

h_L = Head loss due to friction

Hazen William formula can be used to compute h_L (head loss due to friction).

Hazen Williams Equation:

$$h_L = [4.73\ C^{-1.852}\ L\ .\ D^{-4.87}]\ .\ Q^{1.852}$$

h_L= Head loss due to friction in ft
C = Hazen Williams Coefficient
L = Length of the pipe (ft);
D = Diameter of the pipe (ft)
Q = Flow (cu. ft/sec)

All the parameters in this equation are self-explanatory.

Use the Bernoulli's equation with a term for head loss.

Energy at point A = Energy at point B + Head loss

When water flows from A to B, there is a loss of head. The loss of head is given by h_L.

$$h_1 + v_1^2/2g + p_1/\gamma = h_2 + v_2^2/2g + p_2/\gamma + h_L \text{.------------------------(1)}$$

<u>STEP 1:</u> <u>Find Head Loss</u>:

$$h_L = [4.73\ C^{-1.852}\ L.\ D^{-4.87}].\ Q^{1.852}$$

C = 120;
D = 0.5 ft;
L = 35 ft

$$h_L = [4.73 \text{ x } 120^{-1.852} \text{ x } 35 \text{ x } 0.5^{-4.87}]\ .\ Q^{1.852}$$

Since velocity is given, Q can be computed.

$$Q = \pi.\ D^2/4 \text{ x v } = \pi.\ 0.5^2/4 \text{ x } 7 = 1.374 \text{ cu.ft/sec}$$

Insert the value of Q into above equation;

$$h_L = [4.73 \text{ x } 120^{-1.852} \text{ x } 35 \text{ x } 0.5^{-4.87}]\ Q^{1.852}$$
$$h_L = [4.73 \text{ x } 120^{-1.852} \text{ x } 35 \text{ x } 0.5^{-4.87}] \text{ x } 1.374^{1.852} = 1.229 \text{ ft}$$

STEP 2: Find the pressure at point B:

$$h_1 + v_1^2/2g + p_1/\gamma = h_2 + v_2^2/2g + p_2/\gamma + h_L$$

h_1 = 10 ft;
h_2 = 3 ft
v_1 = 7 ft/sec
$v_1 = v_2$ (Since the diameter is the same at two points)
p_1 = 12 psi
p_1 needs to be converted to psf.
p_1 = 12 x 144 psf
γ = 62.4 pcf (Density of water)
h_L = 1.229 ft (Found in step 1)

$$10 + 7^2/2g + (12 \text{ x } 144)/62.4 = 3 + 7^2/2g + p_2/\gamma + h_L$$

p_2 = 2,088 psf = 2,088 /144 psi
= 14.5 psi (Ans B)

<u>Note:</u>

Pressure at point B was 15.03 psi when the head loss of the pipe was ignored. When the head loss was considered, pressure dropped to 14.5 psi. Some of the energy has dissipated due to friction.

Problem 2.6: Pitot tube is shown in the figure. Find the flow velocity.

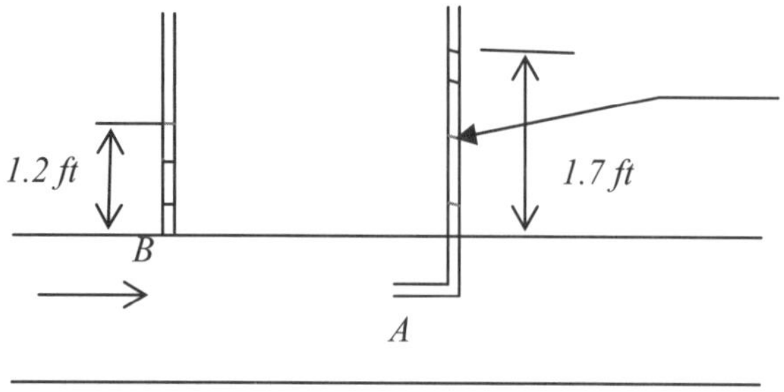

A) 12.3 ft/sec B) 5.67 ft/sec C) 3.45 ft/sec D) 9.4 ft/sec

Solution 2.6:

It is easy to measure the pressure head or the velocity head in a pipe than the flow or velocity. Imagine a fire hydrant. As soon as you open the valve, water would start to gush out of the pipe.

Fire hydrant – Pressure energy converted to kinetic energy

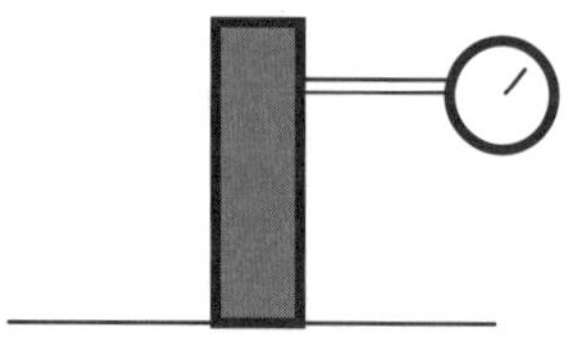

Closed fire hydrant - preessure gauge

If you open a valve of a pressurized pipe and attach a pressure gauge you would be able to measure the pressure easily. Pitot tubes and Venturi meters are developed to measure the pressure at two points. Then it is possible to compute the flow or the velocity using the measured pressure values.

When water is flowing thru a pipe, tube at point B measures only the pressure head. On the other hand, water is jamming into the tube at point A.

Head at point A is due to both pressure and velocity.
Head at point B is only due to pressure.
Since point B is closer to the wall, velocity of water particles are negligible.
Hence velocity head at point B is ignored.

Head at point A = 1.7 ft = $v^2/2g + P/\gamma$ -------------------------(1)
Head at point B = 1.2 ft = P/γ ------------------------------(2)

Equation (1) – (2)

$1.7 - 1.2 = v^2/2g$

$0.5 \text{ ft} = v^2/2g$

$v^2 = 2 \times 32.2 \times 0.5$

$v = 5.67$ ft/sec (Ans B)

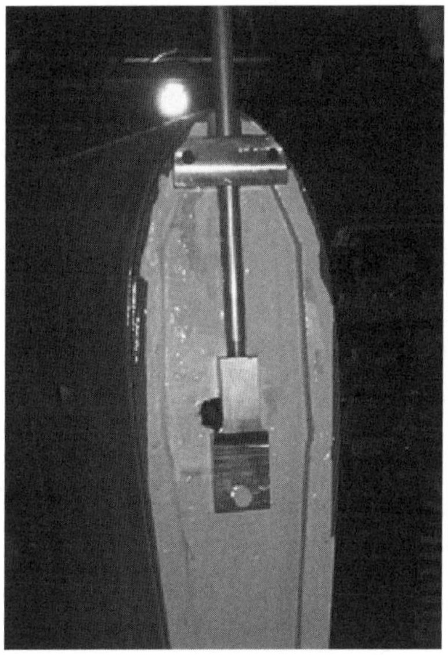

Pitot tube (Tube at the center of the pipe)

Water Resources: Hydrology:

Problem 2.7): Hydrologists have developed the following equation for a given region for 25 yr rainfall intensity.

I = 105/Duration;

I = Rainfall intensity (in/hr);
Duration (min)

Time of concentration of the drainage basin was found to be 35 minutes. Hydrologists believe that water will not start to runoff during the first 10 minutes of the rainfall due to percolation into the ground. Runoff coefficient of the drainage basin is 0.67. The area of the drainage basin is 7.5 acres. Find the design flow of the drainage basin.

A) 12.34 Acre.in/hr B) 32.67 Acre in/hr C) 23.12 Acre. in/hr D) 11.71 Acre. in/hr

Solution 2.7):

The problem needs to be solved using the rational formula.

$$Q = C.I.\ A$$

Q = Maximum flow
C = Runoff coefficient (C = 0.67)
A = Area of the drainage basin (A = 7.5 acres)

STEP 1: Find the design rainfall intensity;

During first 10 minutes of the rainfall, water is percolated into the ground and has no runoff. Runoff starts after first 10 minutes of the rainfall.

Time of concentration (t_c) = 35 minutes
Rainfall duration that would produce maximum flow = 35 + 10 = 45 minutes

Why 45 minutes? Imagine the furthest point from the culvert. Water from the furthest point from the culvert would take 35 minutes to reach the culvert. But water will not start to flow till 10 minutes after rainfall had started. Hence to get the full flow, one needs to wait for a rainfall that last for 45 minutes or more.

Design intensity (I) $= \frac{105}{45} = 2.33$ in/hr

Apply the rational formula; Q = C. I. A
Q = 0.67 x 2.33 x 7.5 = 11.71 Acre in/hr (Ans D)

Problem 2.8) Two areas are draining into a drain pipe as shown. Information of the two areas are given below.

Area 1 = Area is 12 acres, time of concentration 22 minutes, runoff coefficient 0.76
Area 2 = Area is 17 acres, time of concentration 27 minutes, runoff coefficient 0.68

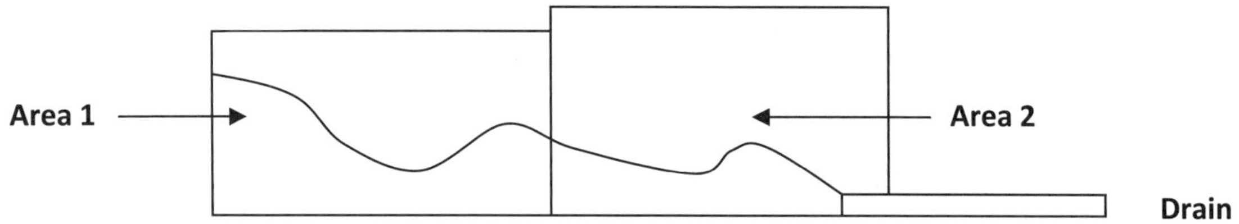

Maximum Intensity of 30 year, 49 minute rainfall is 2.3 in/hr.
Find the maximum flow for a 30 year rainfall.
A) 47.8 cu.ft/sec B) 65.0 cu.ft/sec C) 21.4 cu.ft/sec D) 34.5 cu.ft/sec

Solution 2.8):

This problem has to be solved using the rational formula. Q = C. I. A.
Q = Flow;
C = Runoff coefficient;
I = Intensity of rainfall (in/hr);

A = Area

You need to find C, I and A to find Q.
Since there are two areas, composite C needs to be found.

STEP 1: Find the composite C (also known as average C)

$$C_{composite} = \frac{C_1A_1 + C_2A_2}{A_1 + A_2}$$

$$C_{composite} = \frac{0.76 \times 12 + 0.68 \times 17}{12 + 17} = 0.71$$

STEP 2: Find the relevant rainfall intensity (I) :
Time of concentration is the time that water from most distant point to reach the culvert or the drain pipe. Time of concentration for the area 1 is 22 minutes. Time of concentration for area 2 is 27 minutes.
Total time of concentration = 22 + 27 = 49 minutes.
The problem states that rainfall intensity for a 30 year 49 minute storm is 2.3 in/hr.
Hence I = 2.3 in/hr.

STEP 3: Apply the rational formula to obtain the maximum flow (Q):
Q = C.I.A
A = 12 + 17 = 29 acres
I = 2.3 in/hr;
C = 0.71;

Q = 0.71 x 2.3 x 29 = 47.4 acre. in/hr
C is dimensionless and I is in in/hr. A (Area) is in acres.

The answer needs to be converted to cu.ft/sec.
47.4 acre.in/hr = 47.4/12 acre.ft/hr = 3.95 acre.ft/hr
1 Acre = 43,560 sq.ft
3.95 acre.ft/hr = 3.95 x 43,560 cu.ft/hr = 3.95 x 43,560/(60 x 60) cu.ft/sec = 47.8 cu.ft/sec (Ans A)

Problem 2.9) Mountain region drains into a road as shown. Engineers have designed a trench 2,000 ft long and move the water thru a pipe across the road. Time of concentration for each point of the trench is as shown in the figure. Area of the drainage basin is 5 acres and the runoff coefficient is 0.55.
Velocity of flow in the trench is 8 ft/sec.

Rainfall intensity of the region is given by the following equation.

$$I = 100/(t_c + 15)$$

I = Rainfall intensity (in/hr);
t_c = Time of concentration

Time of concentration for three points in the trench is shown.
Find the maximum flow thru the culvert.

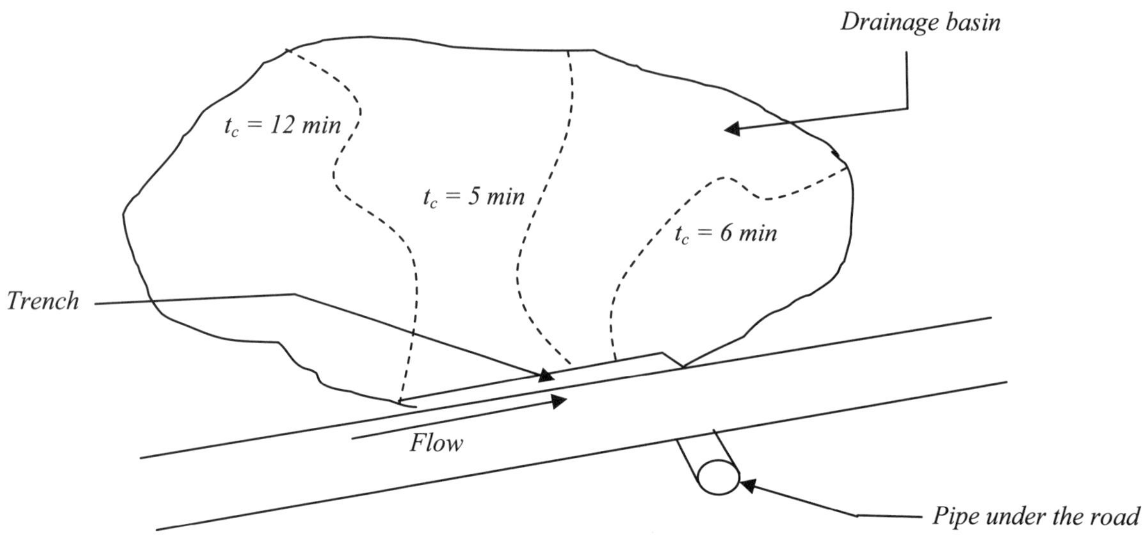

A) 12.35 cu.ft/sec B) 8.87 cu.ft/sec C) 3.45 cu.ft/sec D) 4.98 cu.ft/sec

Solution 2.9):

STEP 1: Write the rational formula;

Q = C. I . A

Q = Flow (cu.ft/sec)
C = Runoff coefficient (Given to be 0.55)
I = Rainfall intensity (in/hr)
A = Area in acres (Given to be 5 acres)

STEP 2: Find the rainfall intensity;

Maximum time of concentration is 12 min + time taken for the water to reach the pipe.
(If you look at the figure given, time of concentration for the longest path is 12 minutes).

Length of the trench = 2,000 ft
Flow velocity in the trench = 8 ft/sec.

Time to travel 2,000 ft = 2,000/8 = 250 seconds
= 250/60 min = 4.2 min

Maximum time of concentration = 12 + 4.2 = 16.2 min

$I = 100/(t_c + 15)$
$I = 100/(16.2 + 15) = 3.2$ in/hr

Find the maximum flow using Q = C. I. A

Q = 0.55 x 3.2 x 5 Acre. in/hr
Q = 8.8 acre. in/hr

Q = 8.8 x 43,560/12 cu.ft/hr
Q = 319,440/3,600 cu.ft/sec
Q = 8.87 cu.ft/sec (Ans B)

Problem 2.10) Find the maximum flow in the drain pipe (cu.ft/sec) for a unit hydrograph. Area of the drainage basin is 4 acres.

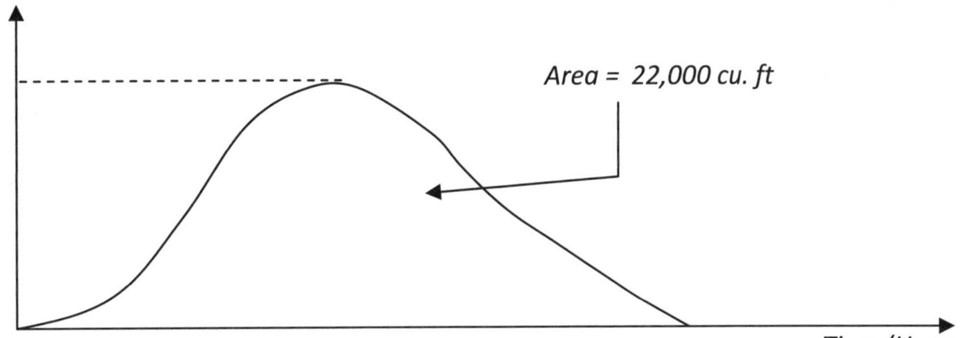

A) 7.86 cu.ft/sec B) 8.27 cu.ft/sec; C) 2.78 cu.ft/sec; D) 7.97 cu.ft/sec

Solution 2.10)

STEP 1: Find the net precipitation;

Runoff = Net precipitation x Area of the drainage basin

22,000 cu. ft = Net precipitation x (4 x 43,560)

Area of the drainage basin is 4 acres. Acres are converted to square feet.
1 Acre =43,560 sq. ft

Net precipitation = 0.1263 ft
= 1.51 inches

STEP 2: Find the maximum flow in a unit hydrograph;

The hydrograph shown has a maximum flow of 12.5 cu.ft/sec.
Unit hydrograph is developed for a net precipitation of 1.0 inches.

Hence by proportioning;

$$\frac{\text{Maximum flow of a given hydrograph}}{\text{Net precipitation of that hydrograph}} = \frac{\text{Maximum flow of the unit hydrograph}}{\text{1 inch}}$$

Unit hydrographs by definition has a net precipitation of 1.0 inches.

12.5/1.51 = Maximum flow of the unit hydrograph/1.0

Maximum flow of the unit hydrograph = 12.5/1.51 = 8.27 cu.ft/sec (Ans B)

TRANSPORTATION:

Problem 2.11) A car was travelling on an asphalt road and ran out of the road and collided with a car parked away from the road in a sandy area. Impact speed was found to be 30 mph. Friction coefficient (μ) of asphalt is 0.6 and friction coefficient (μ) of sand is 0.5. Length of travel in asphalt was 200 ft and length of travel in sand was 120 ft. What was the speed of the vehicle just before applying brakes?

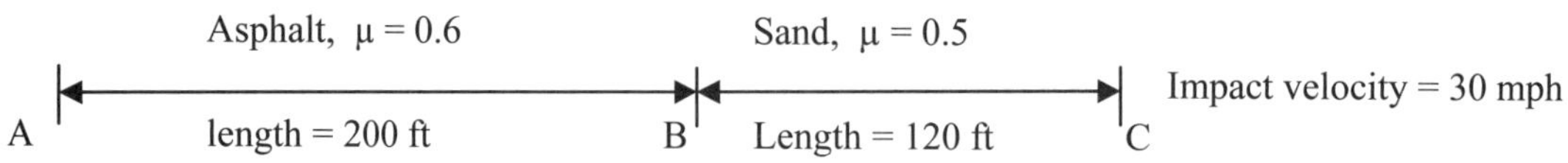

A) 116.3 ft/sec B) 231.2 ft/sec C) 78.9 ft/sec D) 43.9 ft/sec

Solution 2.11) In this problem, a car was travelling on asphalt roadway. At point A, the driver applied brakes to stop the car. Then the car skidded for 200 ft in asphalt and went out of the road to a sandy soil area. Then the car skidded another 120 ft.
Impact velocity was found to be 30 mph.
Impact occurred at point C. (See the figure above). You are required to find the velocity at point A.

First find the velocity at point B. To do this, you need the deceleration from point B to point C.
Once you know the velocity at point B, you can find the velocity at point A. For that you need the deceleration between points A and B.

STEP 1: Find the deceleration from point B to point C.

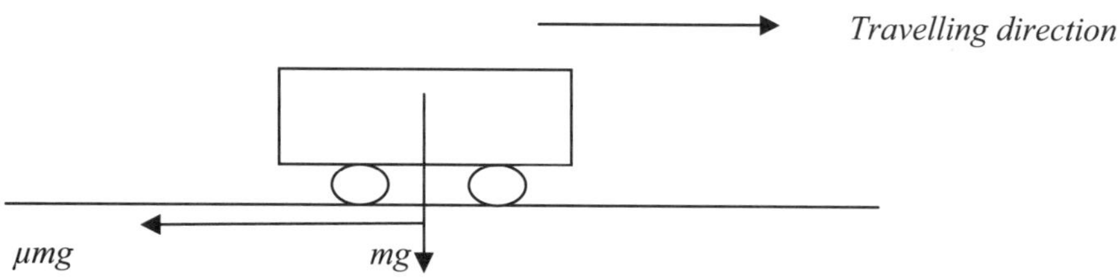

Weight of the car is mg and the frictional force is μmg. μ for asphalt road is given to be 0.6.

Apply Newton's second law (P = mf)

P = m x f
P = Acceleration force (lbf); f = acceleration (ft/sec^2)

Acceleration force = -μ mg
Acceleration force is negative since the force is acting against the direction of movement.

-μ mg = m. f
- μ g = f (Divide by m)
g = 32.2 and μ = 0.5 for sandy soil. (Note that point B to point C is sandy soil).
Hence f = -16.1 ft/sec^2.

"f" is negative since the velocity is reducing. It is a deceleration.

STEP 2: Find the velocity at point B;
Consider movement from point B to point C.

Three equations of dynamics;

$v = u + ft$ ----------------------------(1)

$s = ut + ½ ft^2$ ------------------------ ---(2)

$v^2 = u^2 + 2fs$ ------------------------- --(3)

u = Initial velocity; v = Final velocity; s = Distance; t = time

In this problem, final velocity or the impact velocity is given. (v = 30 mph).
Travel distance is given to be 120 ft from B to C. Hence s = 120 ft.
velocity "u" can be found using the above third equation.

$v^2 = u^2 + 2fs$

v = 30 mph = 44 ft/sec; f = -16.1 ft/sec^2. s = 120 ft

$44^2 = u^2 + 2 \times (-16.1) \times 120$

$u^2 = 5{,}800$;

u = 76.2 ft/sec

Velocity at point B is 76.2 ft/sec.

Next you need to find the velocity at point A. To do that you need to find the deceleration from point A to point B.

STEP 3: Find the deceleration from point B to point C.

Apply Newton's second law (P = mf)

$P = m \times f$

P = Acceleration force (lbf); f = acceleration (ft/sec^2)

$-\mu\, mg = m.\, f$

Note that the force is negative since the force is acting against the direction of movement.

$- \mu\, g = f$ (Divide by m)

g = 32.2 and μ = 0.6

Hence f = -19.3 ft/sec^2.

"f" is negative since the velocity is reducing. It is a deceleration.

STEP 4: Find the velocity at point A;
Consider movement from point A to point B.

$v^2 = u^2 + 2fs$

v = Velocity at point B = 76.2 ft/sec; f = -19.3 ft/sec^2. s = 200 ft

u = Velocity at point A (Need to be found)

$76.2^2 = u^2 + 2 \times (-19.3) \times 200$

$u^2 = 13{,}526$

u = 116.3 ft/sec

Velocity at point A is 116.3 ft/sec. Ans (A)

Problem 2.12) Stopping sight distance of a road is 460 ft. Radius of the horizontal curve is 600 ft. Find the horizontal sight offset required for the horizontal curve.

A) 123.56 ft B) 25.89 ft C) 91.00 ft D) 43.55 ft

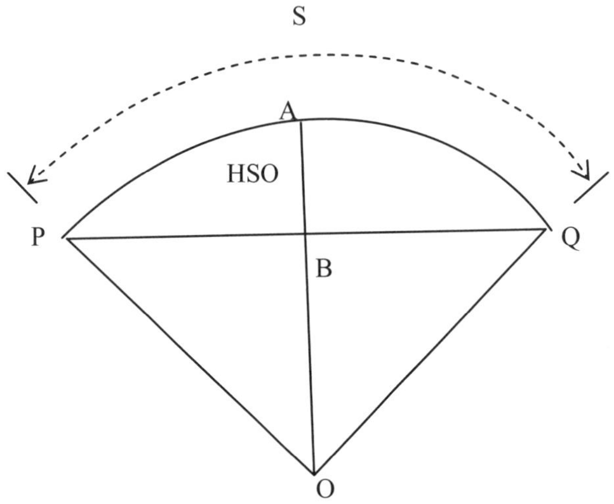

AB = HSO (Horizontal sight offset)

Solution 2.12): HSO is given by the following equation.

$$HSO = R\,[1 - \cos(28.65.\ S/R)]$$

HSO = Horizontal sight offset (ft)
R = Radius of the horizontal curve (ft)
S = Stopping sight distance (ft)

STEP 1: Write down the parameters given;
R = Radius = 600 ft
S = Stopping sight distance = 460 ft

STEP 2: Apply the above equation;
HSO = R [1 – cos (28.65. S/R)]
HSO = 600 [1 – cos (28.65 x 460 /600)]
HSO = 43.55 ft
Ans D

Theory:
Let us look at a horizontal curve.

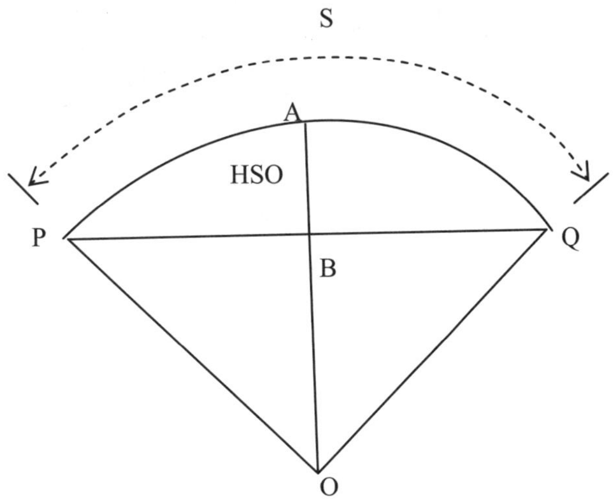

Assume there is a driver at point P. If he happens to see a stopped disabled car at point Q, then he should be able to stop the car prior to reaching point Q. Stopping sight distance is curve length from P to Q.

What is HSO? The length AB is considered to be HSO.
HSO is the horizontal sight offset. In other words, no object should be covering the AB line. If there is a house or tree in HSO, the driver may not be able to see the disabled vehicle at point Q.

Let's assume there is a house as shown below.

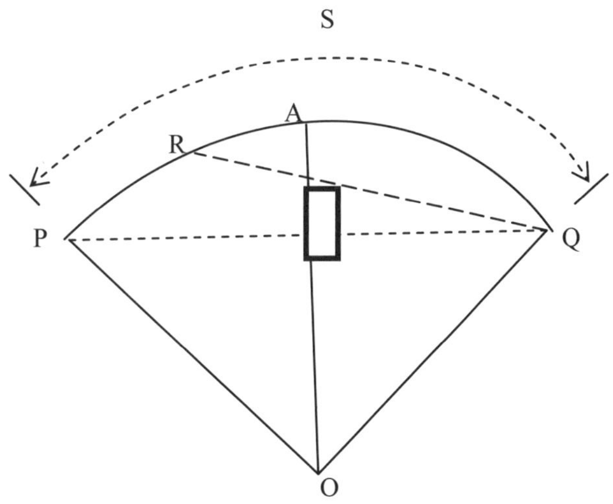

If there is a house as shown, the driver will not see the disabled vehicle until he reaches the point "R". Then he may not be able to stop the car prior to reaching point Q since stopping sight distance is "S" which is longer than curve length from point R to Q. You would have an accident.

Hence no trees, houses or any other object should be along HSO. That line should be kept clear.
Now let us develop a relationship between HSO and radius.

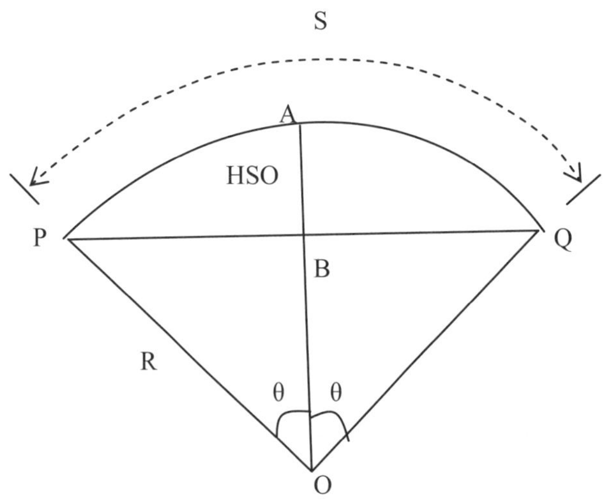

Let us call angle POA as θ.
Then QOA = θ as well.

AO = R = Radius
BO = R cos θ
AB = AO – BO = R – R cos θ
AB = R (1 – cos θ)
AB = HSO
HSO = R (1 – cos θ) ---------------------------(1)

There is a relationship between the curve length of a circle and angle at the center of the circle. That relationship is;

S = Radius x Angle at the center of the circle measured in radians

Hence;

$S = R. (2\ \theta_{radians})$ -------(2) (θ should be measured in radians).

Convert θ to degrees.

180 degrees = π radians
1degree = π/180 radians

$\theta_{radians} = \theta_{degrees}\ x\ (\pi/180)$

From above (2)
$S = R. (2\ \theta_{radians})$

Substitute $\theta_{radians}$.

$S = R. (2.\theta_{degrees}\ x\ \pi/180)$
$\theta_{degrees} = (S/R)\ x\ (90/\pi)$

Insert this in above equation (1).

HSO = R (1 – cos θ)
HSO = R [1 – cos {(S/R) x (90/π)}]

HSO = R [1 – cos (28.65 S/R)

Problem 2.13): A traverse is done as shown. Internal angles of the traverse is as given. What is the total error of internal angles?

A) 4.662^0 B) 3.632^0 C) 4.169^0 D) 6.167^0

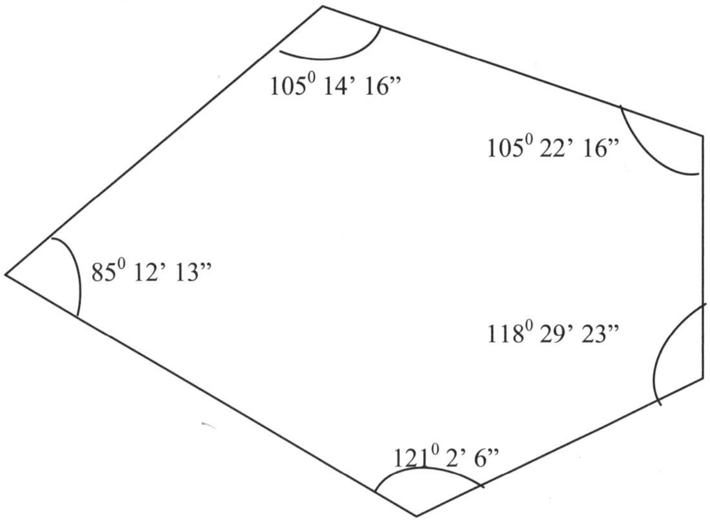

Solution 2.13):

STEP 1: Find the summation of internal angles;

Summation of internal angles of any polygon is given by the following equation;

> Summation of internal angles of a polygon = 180 x (N – 2)
>
> N = Number of legs

This is a five sided polygon.
Summation of internal angles = 180 x (5 – 2) = 540^0

STEP 2: Add all the measured angles;
Total of measured angles = 85^0 12' 13" + 105^0 14' 16" + 105^0 22' 16" + 118^0 29' 23" + 121^0 2' 6"

Total of measured angles = 85.2036 + 105.2377 + 105.3711 + 118.4897 + 121.035 = 535.3371
Error = 540 – 535.3371 = 4.6629^0
Ans A

Problem 2.14) Two vehicles are travelling in the same direction in a highway next to each other. First vehicle is travelling at a speed of 50 mph and the second vehicle is travelling at a speed of 40 mph. Both drivers apply brakes at the same time. First vehicle decelerates at 12 ft/sec^2 and the second vehicle decelerates at 8 ft/sec^2. When both vehicles come to a complete stop, what's the distance between the two vehicles?

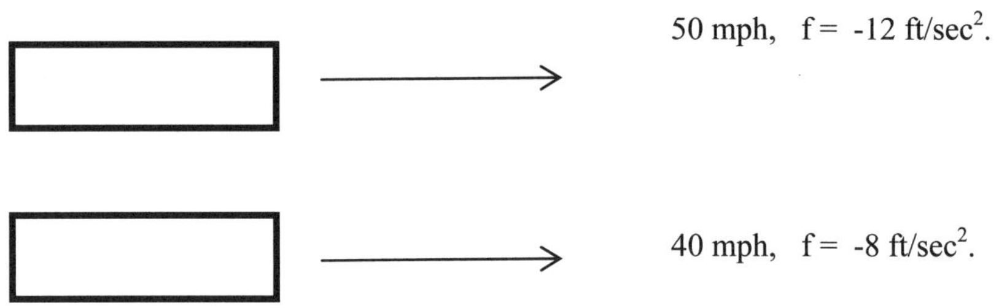

A) 12.34 ft B) 8.97 ft C) 32.45 ft D) 4.56 ft

<u>Solution 2.14):</u>

STEP 1: There are three dynamic equations of motion.

$v = u + f.t$ ----------(1)

$s = u.t + \frac{1}{2}.f.t^2$ ----------(2)

$v^2 = u^2 + 2.f.s$ ----------(3)

u = Initial velocity
v = Final velocity
f = Acceleration (Negative value for deceleration)
s = Distance travelled.
t = Time elapsed

Use equation 3.
Final velocity (v) = 0;
Initial velocity = u

<u>First vehicle;</u>

v = Final velocity = 0;
u = Initial velocity = 50 mph = 73.33 ft/sec;
f = -12 ft/sec^2.

$v^2 = u^2 + 2.f.s$
$0 = u^2 + 2.f.s$
$0 = 73.33^2 + 2.(-12).s$
$s = 224.07$ ft

<u>Second vehicle;</u>

v = 0; u = 40 mph = 58.67 ft/sec; f = -8 ft/sec^2.

$v^2 = u^2 + 2.f.s$
$0 = u^2 + 2.f.s$
$0 = 58.67^2 + 2.(-8).s$
$s = 215.1$ ft

Distance between two vehicles (when stopped) = 224.07 – 215.1 = 8.97 ft (Ans B)

<u>Problem 2.15):</u> Station at point of intersection (PI) is given to be 14+70 and the radius of the curve is 1,400 ft. Find the station at PT. (Angle of intersection is 53 degrees).

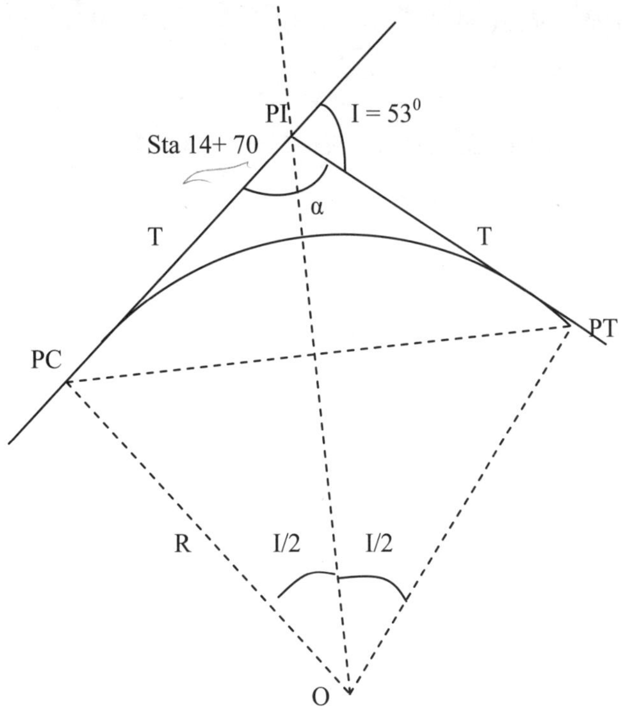

A) 14+66 B) 34+75 C) 20+67 D) 24+81

Solution 2.15):

PC = Start of the curve is known as point of curvature
PT = End of the curve is known as point of tangent
PI = PI is the intersection of two tangents. Intersection angle at PI is generally represented with "I". In this problem it is given to be 53^0.
R = Radius of the curve
Length between PC and PI is normally represented with "T".
In this problem, station at PT is required.
Station to PT is always measured **along** the curve from PC.

Never measure the station of PT from PI.
Always measure the distance to PT along the curve from PC. (Curve length).

Procedure:

- First find the station at PC
- Then find the length of curve from PC to PT.
- Then find the station at PT

In this problem station at PI, radius (R) and intersection angle is given.

STEP 1: Find the distance T

From trigonometry, T = R tan I/2 = 1,400 x tan (53/2) = 698 ft
Station at PC = Station at PI – T
Station at PC is always smaller than station at PI, since surveyors first mark PC and then go to PI.

Station at PI = Sta 14 + 70 = 1,470 ft
Station at PC = 1,470 – 698 = 772 = Sta. 7 + 72

STEP 2: Find the curve length from PC to PT:

Curve length = r. θ
r = Radius
θ = Angle extended at the center. θ should be measured in radians.

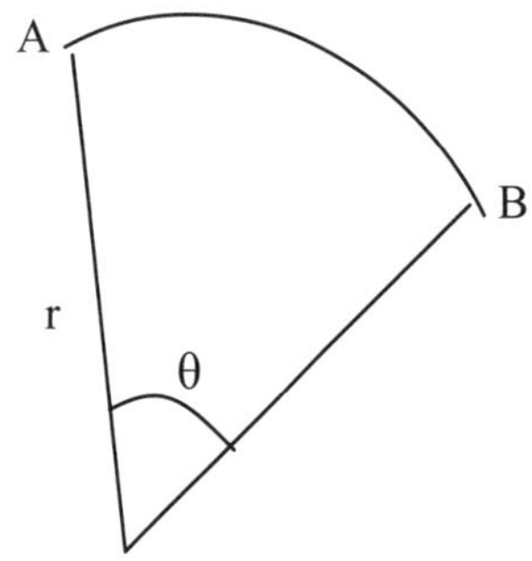

Curve AB = r. θ

Curve length PC. PT = Radius x I
I is the internal angle and should be measured in radians.

I = 53 degrees = 0.9250 Radians
Degrees to radians conversion can be done using any scientific calculator.

Or else you can use the following equation;

$\theta_{radians} = \theta_{degrees} \times \pi/180$

Hence; $I_{radians} = I_{degrees} \times \pi/180$
$I_{radians} = 53 \times \pi/180 = 0.9250$

Curve length PC to PT = Radius x $I_{radians}$ = 1,400 x 0.9250 = 1,295 ft
Hence length along the curve from PC to PT = 1,295 ft

Station at PT = Station at PC + 1,295

Station at PT = 772 + 1,295 = 2,067 = Sta. 20 + 67 (Ans C)

Problem 2.16): Vertical curve is shown in the figure. Upward gradient is 1.8% and downward gradient is 3.2% as shown. BVC elevation is 132.71 and BVC station is 11+53. Station of point P is given to be 20 +00. EVC station is 22+48. Find the elevation at point P.

A) 105.94 B) 113.87 C) 106.79 D) 121.97

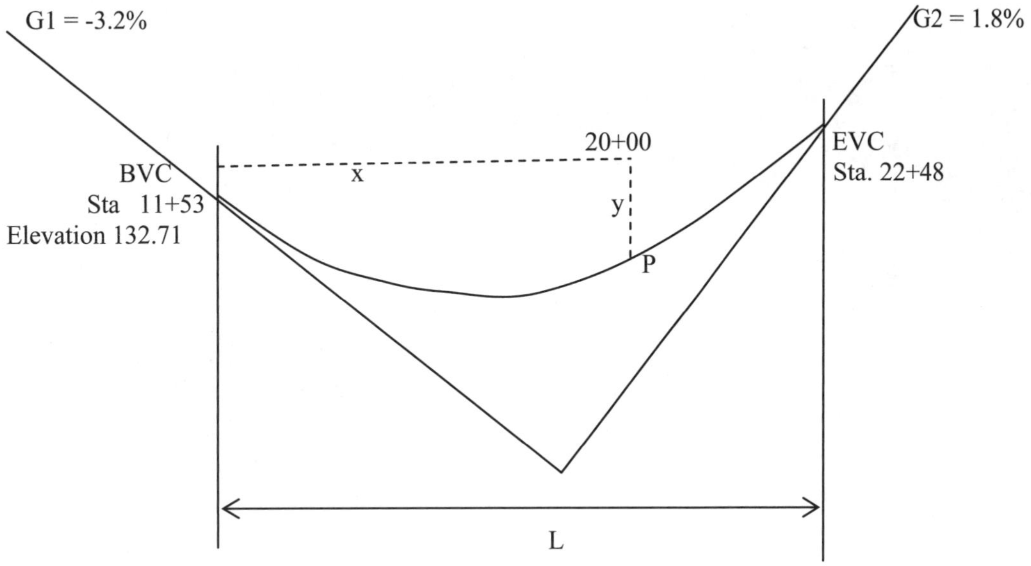

BVC = Beginning of vertical curve
EVC = End of vertical curve

Solution 2.16):

Vertical Curve Equation;

$$y = \frac{(G2 - G1)}{2L} \cdot x^2 + G1 \cdot x$$

y = Vertical height to point in the curve from BVC in ft
x = Horizontal distance between two points
G1 and G2 are slopes as shown in the figure.
G1 and G2 should be in decimals. Percentages need to be converted to decimals. (example: 3% = 0.03)
Downward gradient is taken to be negative (-).
Upward gradient is taken to be positive (+)
L = Horizontal distance between BVC and EVC in ft.
G1 = -0.032;
G2 = 0.018

L = 2248 – 1153 = 1095 ft
y = Vertical height to any point in the curve from BVC in ft
x = 2000 – 1153 = 847 ft

Apply the equation;

$$y = \frac{(G2 - G1)}{2L} \cdot x^2 + G1\,x$$

$$y = \frac{(0.018 - -0.032)}{2 \times 1095}\, 847^2 + (-0.032) \cdot 847$$

y = 16.37 + (-27.104) = -10.734

Elevation at P = Elevation at BVC + (-10.734) = 132.71 – 10.734 = 121.97 ft (Ans D)

Vertical sag curve to permit rail road to go over the road

Problem 2.17) Engineers are designing a roadway horizontal curve. Side friction factor of the roadway is 0.2 and the speed of vehicles is 60 mph. Radius of the horizontal curve is 700 ft. What is the superelevation required at the curve to maintain the speed.

A) 12.3% B) 14.3% C) 3.2% D) 7.9%

Solution 2.17)

Superelevation is provided at horizontal curves to maintain the speed of vehicles.

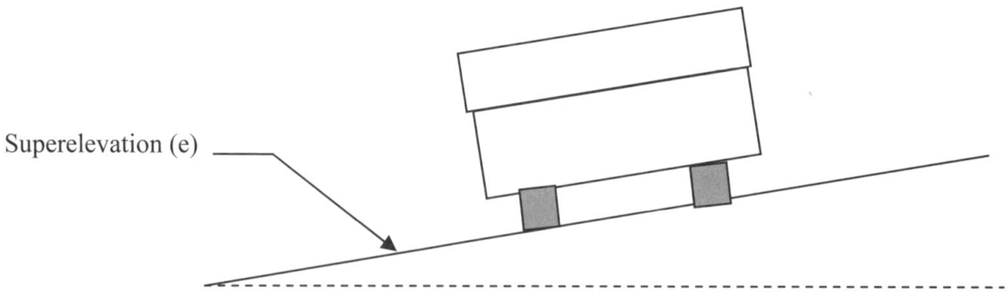

Following equation provided by AASHTO Geometric Design of Highways known as green book.

$$f = V^2/15.R - e/100$$

f = Side friction coefficient (decimals)
V = Velocity in mph
R = Radius of the horizontal curve (ft)
e = Superelevation in percent

In this example;

f = 0.2; V = 60 mph; R = 700 ft

$$0.2 = 60^2/(15 \times 700) - e/100$$

e = 14.3%

Superelevation is provided at horizontal curves to maintain the speed of vehicles.

Superelevation at a highway horizontal curve

Problem 2.18) Horizontal curve is designed with a superlevation for a highway. The road has a normal crown and after the superelevated horizontal curve it switches back to the normal crown. What is the correct process of changing a normal crown to superelevation and then back to a normal crown?

A) Normal crown, Tangent runout, superelevation runoff, full superelevation, superelevation runoff, Tangent runout, Normal crown

B) Normal crown, superelevation runoff , Tangent runout, full superelevation, superelevation runoff, Tangent runout, Normal crown

C) Normal crown, superelevation runoff , full superelevation, superelevation runoff, Tangent runout, Normal crown

D) superelevation runoff, superelevation runout, Tangent runout, Normal crown, superelevation, superelevation runoff, Tangent runout, Normal crown

Solution 2.18):

First let's look at the nomenclature.

Normal Crown:

Highways are built with a crown in the middle to facilitate drainage. Highway cross section is shown below with a crown at the center.

Typical Normal crown (Raised at the center)

Typical Normal Crown

Tangent Runout: Outside lane is made horizontal. The distance required to remove the normal crown is known as tangent runout.

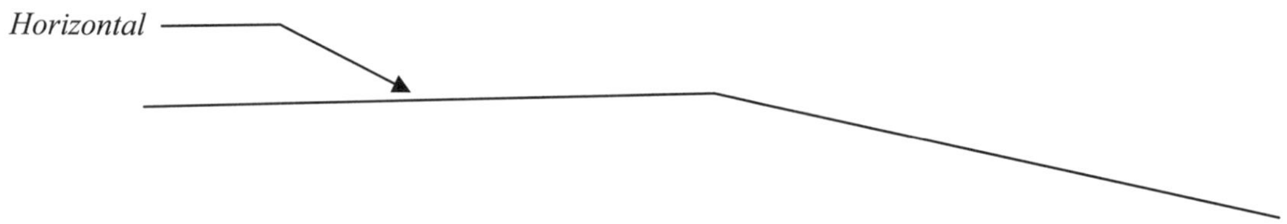

Superlevation Runoff: The distance required to achieve full superelevation from a horizontal section known as superelevation runoff.

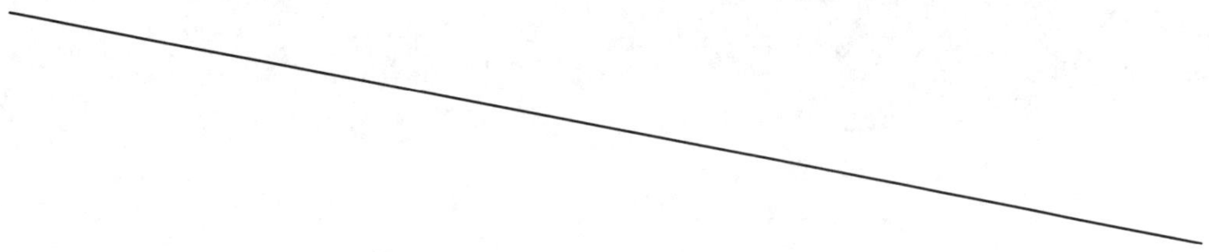

Full Superelevation

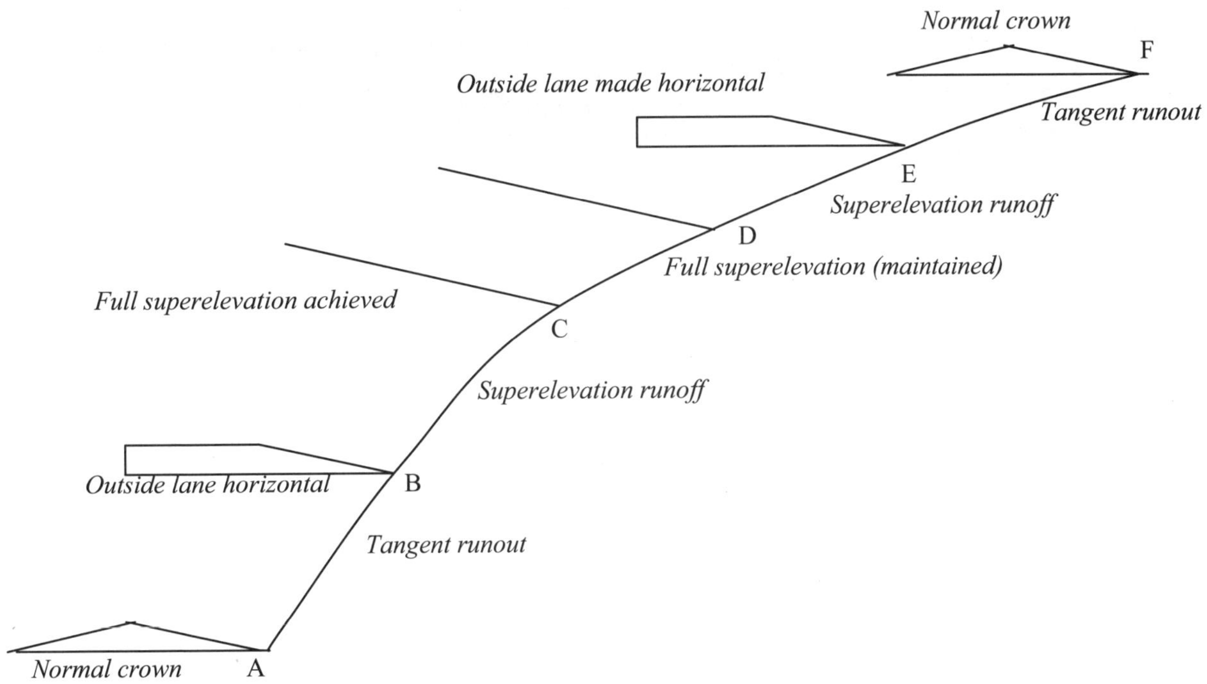

A - In straight sections, roads have crown sections as shown.
A to B – Crown is taken away from the outside lane. Outside lane becomes horizontal. This is done gradually.
A to B is known as "*Tangent runout*"
B to C - Full superelevation is achieved. B to C is known as "*Superelevation runoff*".
C to D – Full superelevation is maintained.
D to E - Superelevation is removed from the outside lane. Outside lane is horizontal again. D to E is known as "*Superelevation runoff*"
E to F - Crown section is achieved again. E to F is known as "*Tangent runout*".
The road is straight after F and has a normal crown.

(Ans A)

Normal crown section transforms to a full superelevation and then gets back to a normal crown. Full superelevation is seen in this image

STRUCTURAL:

Problem 2.19) Find the deflection at the end of the hollow rectangular section shown. Outside dimensions of the section is 12 x 12 inches and the thickness is 0.25 inches. The Young's modulus of steel is 29 x 10^6 psi.

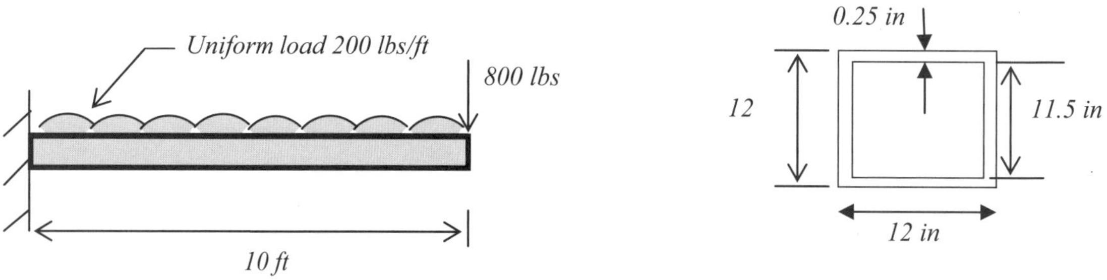

A) 0.149 in B) 1.23 in C) 0.453 in D) 1.67 in

Solution 2.19):

Deflection of the beam is caused by the uniform load and the concentrated load at the end.

Deflection (y_1) at the end of a cantilever section with a uniform load is given by

$$y_1 = \frac{w L^4}{8 E I}$$

w = uniform load = 200 lbs/ft = 200/12 lbs/in = 16.67 lbs/in
I = Moment of inertia
L = Length of the beam = 10 ft = 120 inches.
E = Young's modulus

Deflection (y_2) at the end of a cantilever section with a concentrated load at the end is given by

$$y_2 = \frac{P L^3}{3 E I}$$

P = Concentrated load at the end = 800 lbs
I = Moment of inertia

$$\text{Total deflection} = y_1 + y_2 = \frac{w L^4}{8 E.I} + \frac{P L^3}{3 E.I}$$

y_1 = Deflection due to uniform load
y_2 = Deflection due to concentrated load at the end
Moment of inertia of a hollow rectangular tube is given by (I) = $\frac{b . (h_2^3 - h_1^3)}{12}$

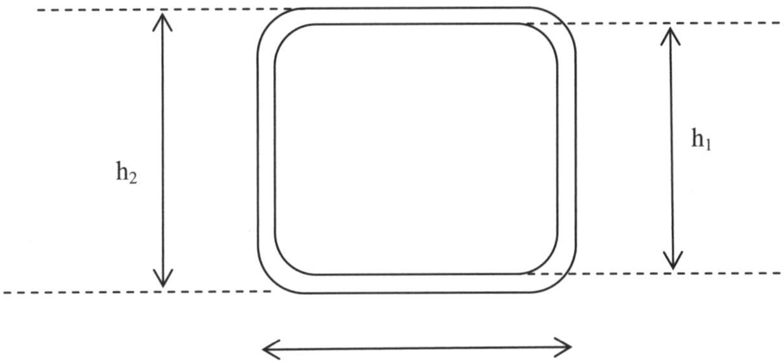

$$I = \frac{b . (h_2^3 - h_1^3)}{12} = \frac{12 . (12^3 - 11.5^3)}{12} = 207.125$$

E = 29 x 10^6 psi

w = Uniform load = 200 lbs/ft = 200/12 lbs/in = 16.67 lbs/in
L = 10 ft = 120 in;
P = 800 lbs

$$\text{Total deflection} = y_1 + y_2 = \frac{w L^4}{8 E.I} + \frac{P L^3}{3 E.I}$$

$$\text{Total deflection} = y_1 + y_2 = \frac{16.67 . 120^4}{8 \times 29 \times 10^6 \times 207.125} + \frac{800 \times 120^3}{3 \times 29 \times 10^6 \times 207.125}$$

$$\text{Total deflection} = y_1 + y_2 = 0.072 + 0.077 = 0.149 \text{ in}$$ (Ans A)

Problem 2.20): Seismic design categories are developed
A) Based on soil type and building type
B) Based on risk and importance of the structure with regard to loss of life
C) Based on soil type only
D) Based on the probability of an earthquake in the region

Solution 2.20): Ans B.

Seismic design categories are

Category A - Buildings located in regions having a very small probability of earthquake

Category B - Buildings for ordinary occupancy in areas with moderate risk of earthquakes.

Category C - Buildings for ordinary occupancy in areas with high risk of earthquakes or important buildings in regions with moderate earthquake risk.

Category D - Buildings for ordinary occupancy in areas with very high risk of earthquakes or important buildings in regions with high earthquake risk.

Category E - Buildings for ordinary occupancy located few kilometers of active fault lines.

Category F - Critically important buildings such as earthquake shelters, hospitals located few kilometers of active faults.

Problem 2.21): Generalized spectral response curves in seismic engineering prepared by USGS (adopted by building codes in USA) S_1 and S_s represents;
A) S_1 represents the ground motion for one second earthquake and S_s represents the ground motion for short period earthquakes.
B) S_1 represents the spectral response acceleration for one second earthquake and S_s represents the spectral response acceleration for short period earthquakes.
C) S_1 represents the spectral response acceleration for buildings with one second natural period and S_s represents the spectral response acceleration for short buildings with natural period less than 1 second.
D) None of the above

Solution 2.21): Ans C

Discussion:

During an earthquake event, short buildings would vibrate with smaller natural period. Also short buildings would undergo high spectral accelerations compared to tall buildings. In the case of tall buildings, natural period is longer. Tall buildings typically undergo much smaller accelerations compared to short buildings.

Short buildings - Short natural period and high spectral acceleration
Tall buildings - Long natural period and low spectral acceleration

Spectral Acceleration

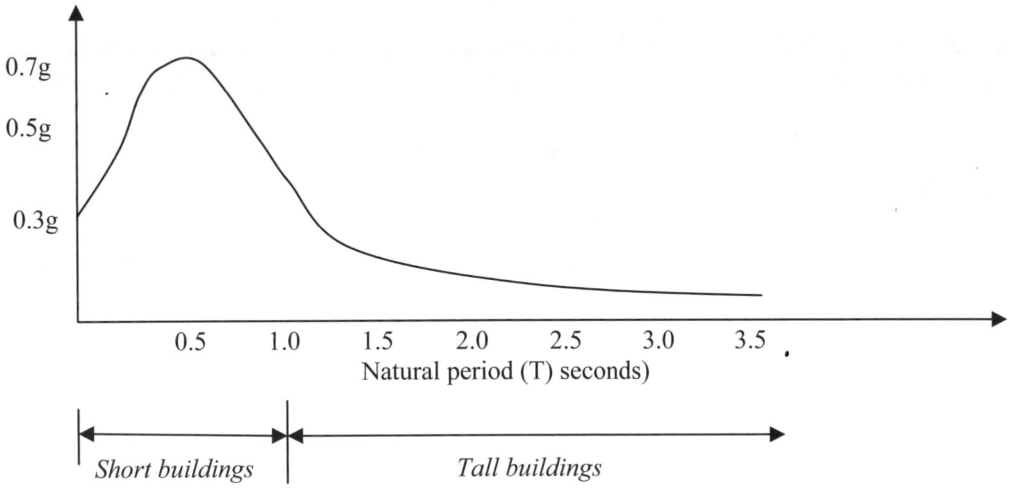

Above figure shows spectral accelerations of short and tall buildings for the same earthquake. Natural period for tall buildings are much higher.
Above figure was simplified by ASCE. (See below).

Spectral Acceleration (S_a)

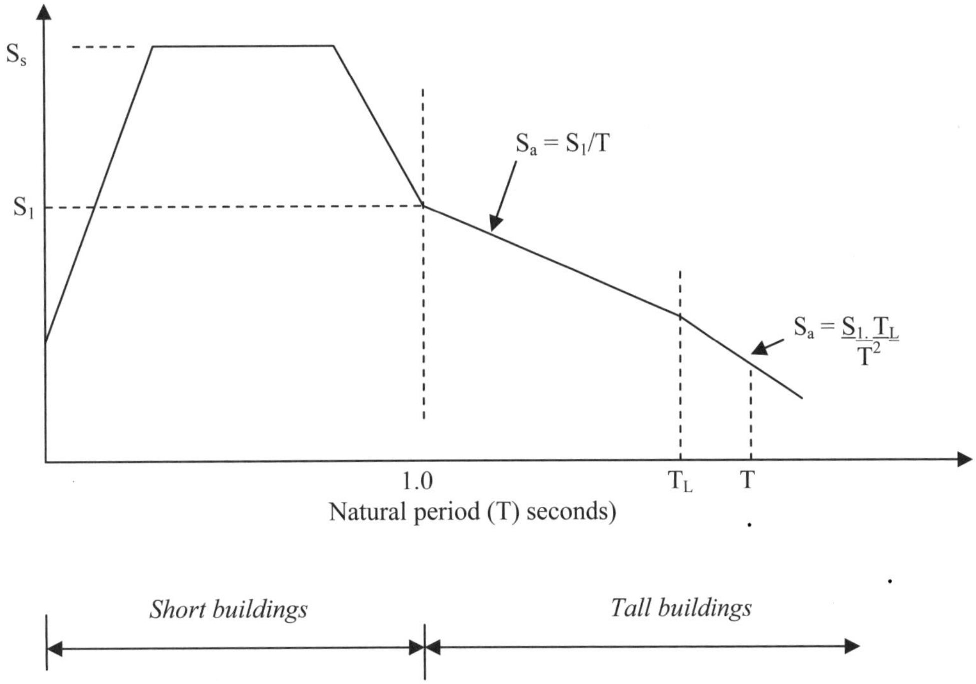

Generalized spectral acceleration of buildings (ASCE 7.1)

We can derive the following from the above figure.

- Spectral acceleration for buildings with natural period 1 second = S_1
- Spectral acceleration for buildings with natural period less than 1 second = S_s

- Spectral acceleration for buildings with natural period greater than 1 second but less than $T_L = S_1/T$
- Spectral acceleration for buildings with natural period greater than $T_L = S_1 . T_L/T^2$

USGS have developed spectral response acceleration spectrums (S_1, S_s) and long period transition period (T_L) for the continental United States. These maps are adopted by building codes.

Magnitudes of parameters T_L, S_s and S_1 are dependent upon the city.
For an example, following values are obtained from USGS maps.

City	*Short period spectral acceleration (S_s)*	*One second spectral acceleration (S_1)*	*Long period transition period (T_L)*
San Francisco	1.50g	0.75g	16
Denver	0.70g	0.20g	8
Manhattan	0.20g	0.06g	6

If the building is a short building, use S_s. If it is a high rise building, use S_a.

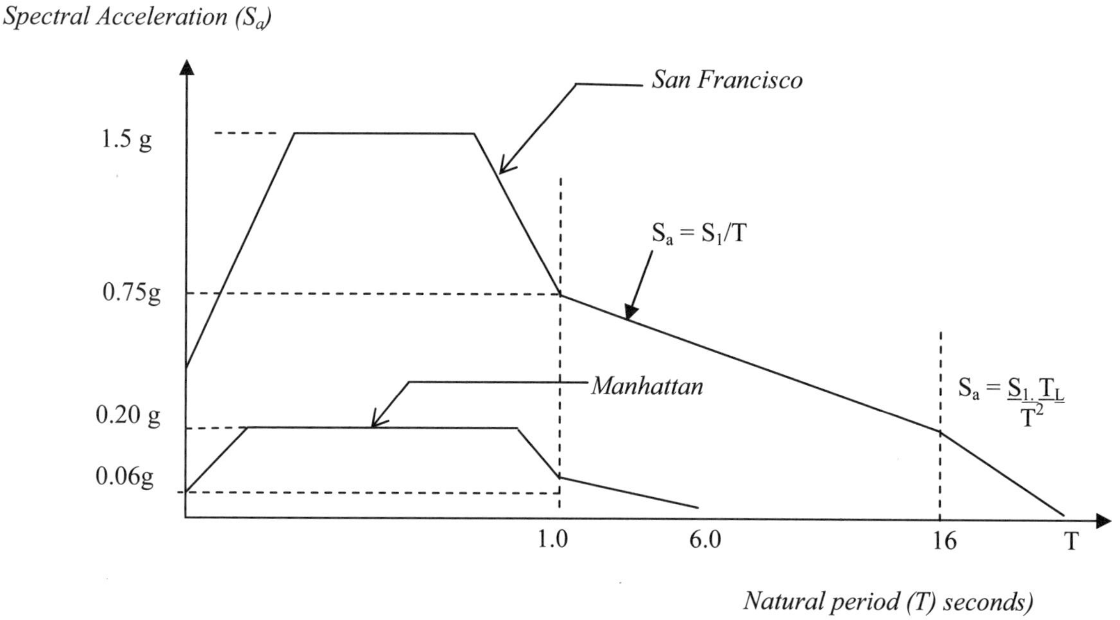

Spectral response curves for San Francisco and Manhattan are shown above. Any structure built as per the curves given above has less than 1% chance of collapsing due to an earthquake within a time period of 50 years. As you could see spectral acceleration expected for a same exact building is much higher in San Francisco than in Manhattan. This is due to the fact that earthquakes expected in San Francisco area within a 50 year period is much stronger.

Problem 2.22) What is not true about moment frame buildings?

A) Need less steel compared to braced frame structures
B) More labor intensive than braced frame structures
C) Larger deflections can be expected compared to braced frame structures
D) Easy to design doorways and windows.

Solution 2.22): Ans (A)

Steel structures could be of many types. Most common types of steel structures are

- Braced frame structures
- Moment frame structures or rigid frame structures

Moment frame structures require more steel than braced frame structures. Hence A is not true. Since there are no braces it is easy to design doorways and windows. Moment connections are much more labor intensive than connecting braces.

Braced frame structure

New world trade center (Freedom tower) – Braced frame structure

Moment frame structure (No braces)

Problem 2:23): Find the nominal moment of the beam shown.

f_y= 60,000 psi
Concrete compression strength (f_c') = 4,000 psi
A_s = 4.3 sq. in

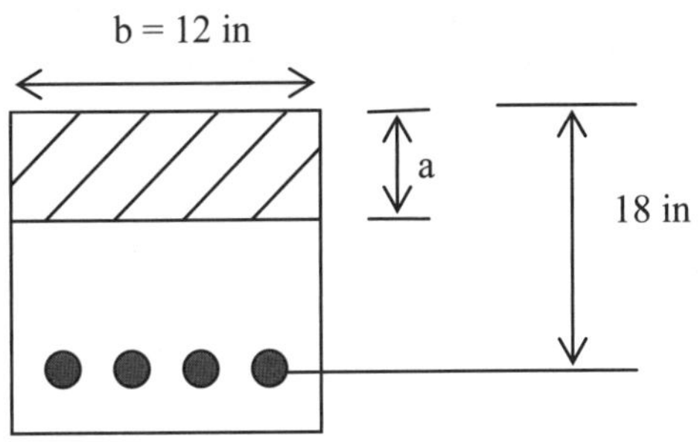

A) 213 kip. ft B) 319 kip. ft C) 154 kip. ft D) 298 kip.ft

Solution 2:23):

STEP 1:

Write the force balance equation;

Force in Concrete = Force in Steel

Force in concrete = $0.85 f_c' A_c$

f_c' = Concrete compressive strength
A_c = Concrete compression area

Force in steel = $f_y . A_s$
f_y = Yield stress of steel
A_s = Steel area

Force in steel = Force in concrete

$0.85 f_c' A_c = f_y . A_s$

$0.85 f_c' A_c = f_y . A_s$

$0.85 f_c' A_c = f_y . A_s$

$0.85 \times 4{,}000 \times A_c = 60{,}000 \times 4.3$

$A_c = 75.88$

$A_c = 75.88 = a. b$
a = depth of the concrete compression area.

(Note: This equation is valid only for rectangular beams. Also the value obtained here is not "c". Typically a = 0.85c)

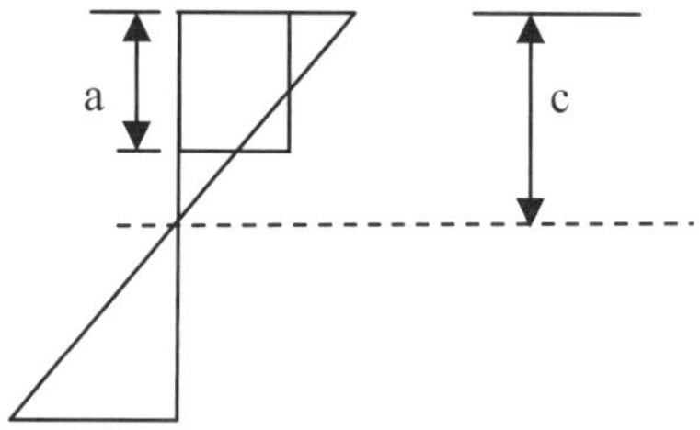

b = Width of the beam = 12 inches;
Hence a = 75.88/12 = 6.32 sq. inches

STEP 2: Find the nominal moment capacity of the beam;

Nominal moment capacity = Force in steel x Moment arm

Moment arm (Z) = The distance from center of gravity of steel to center of gravity of concrete
Force in steel = f_y. A_s = 60,000 x 4.3

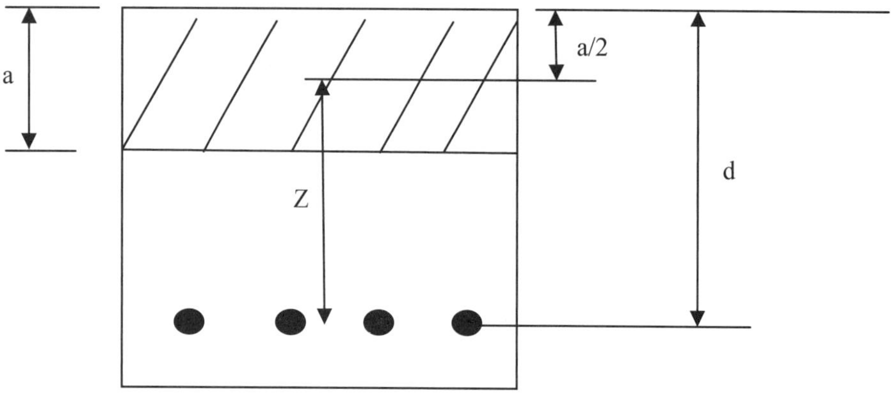

Center of gravity of the concrete compression area is at a/2 distance from the top.

Hence moment arm (Z) = d – a/2
Note that this equation is valid only for rectangular beams.

Moment arm (Z) = d – a/2 = 18 – 6.32/2 = 14.84 sq. in

Moment Capacity of the Beam = Force in steel x Moment arm

Moment Capacity of the Beam = f_y. A_s x Z
Moment Capacity of the Beam = 60,000 x 4.3 x 14.84 lbs. in

Moment Capacity of the Beam = (60,000 x 4.3 x 14.84)/(12 x 1,000) kip. ft
= 319.1 kip. ft (Ans B)

Problem 2:24) What is the maximum steel allowed by ACI 318 for the beam shown.

As per ACI 318, steel strain allowed = 0.004
Concrete crushing strain = 0.003

f_y= 60,000 psi
Concrete compression strength (f_c') = 4,000 psi

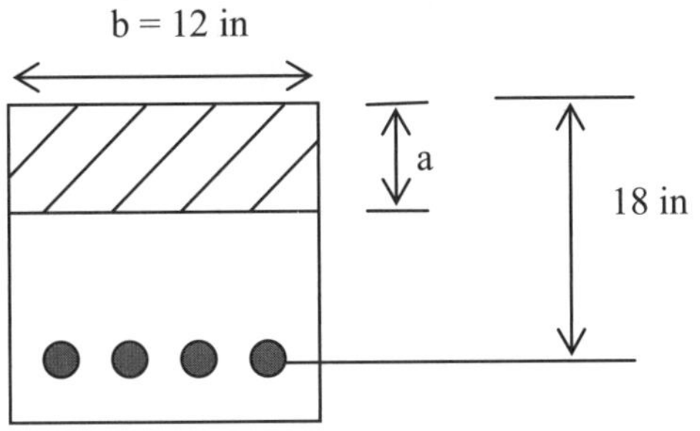

A) 3.45 sq. in B) 4.46 sq. in C) 5.93 sq. in D) 6.12 sq. in

Solution 2:24):

If a beam fails due to concrete crushing, it would be a sudden failure. This is not desirable.

On the other hand, if the beam fails due to steel yielding, the failure would not be sudden. Occupants will be able to see deflection in the beam. Hence ACI 318 provides an upper limit for steel area in a beam.

Minimum steel strain allowed = 0.004

Many designers may use a steel strain of 0.005.

STEP 1: Draw the strain diagram.

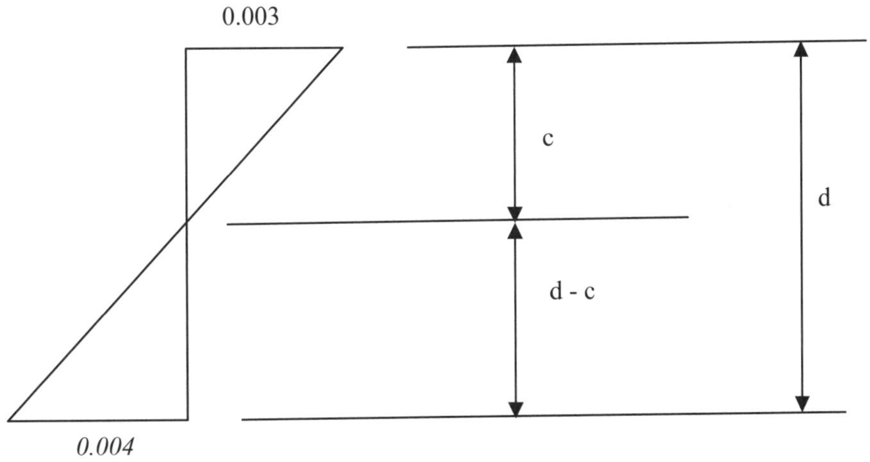

d = Depth (from top of the beam to center of gravity of rebars).
0.003 = Strain in concrete when concrete start to crush
0.004 = Strain in steel

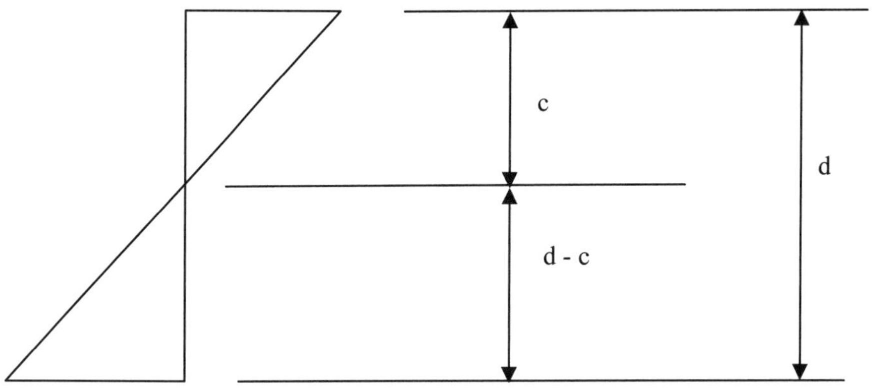

By proportioning;

$$\frac{0.003}{c} = \frac{0.004}{(d-c)}$$

0.003d – 0.003c = 0.004 c
0.003 d = 0.007 c
c = 0.4286d

a = 0.85 c (From ACI 318)
a = 0.85 x 0.4286d

Hence;

a = 0.3643 d

"d" is given to be 18 inches.

a = 0.3643 x 18 = 6.56 inches

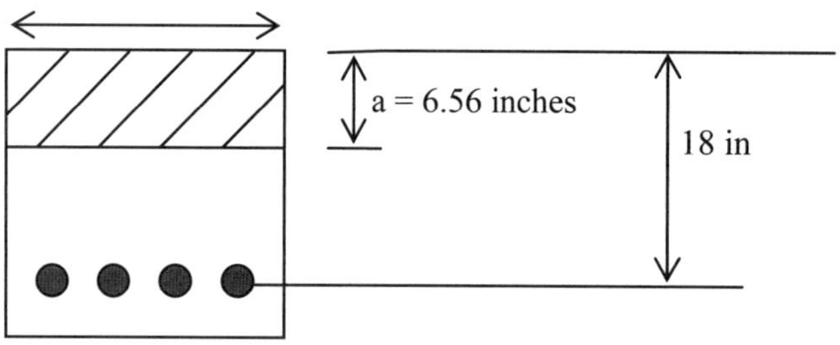

A_c = a x b = 6.56 x 12 = 78.69 sq. in

STEP 2: Find the force in concrete;

Force in concrete = 0.85 f_c' A_c
Force in concrete = 0.85 x 4,000 x 78.69
= 267,546 lbs. in

STEP 3: Equate force in concrete to force in steel;
Force in steel = f_y. A_s
Force in steel = Force in concrete

f_y. A_s = 0.85 f_c' A_c
60,000 x A_s = 0.85 x 4,000 x 78.69

A_s = 4.46 sq. in (Ans B)

Problem 2:25) A beam is loaded with a uniform load of 5 kips/ft and a concentrated load of 30 kips. Find the shear force at point P.

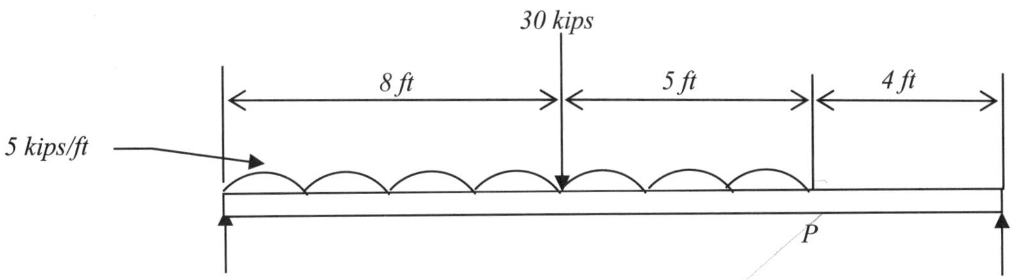

A) 34 kips B) 46 kips C) 39 kips D) 62 kips

Solution 2:25):

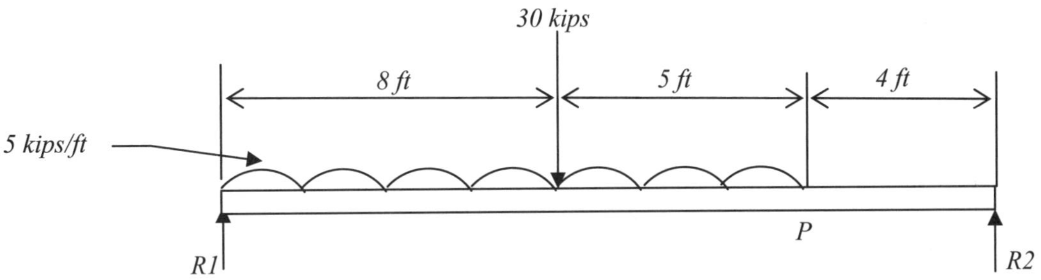

STEP 1: Find R1 and R2:

Resolve forces;

R1 + R2 = 30 + 5 x (8 + 5)

R1 + R2 = 95 kips ---------------------------(1)

Obtain moments from R2.

R1 x (8 + 5 + 4) = 30 x (5 + 4) + 5 x (13) x (4 + 13/2) -----------------------------(2)

Above last term needs to be explained further;

5 is the uniform load.
5 x (13) = This is the total load due to uniform load.
(4 + 13/2) = Distance to the center of gravity of the uniform load from R2.

We can simplify above equation (2) further;

17 R1 = 270 + 682.5

R1 = 56

Hence from (1)

R2 = 95 – R1 = 95 -56

R2 = 39 kips

STEP 2: Draw the shear force diagram;

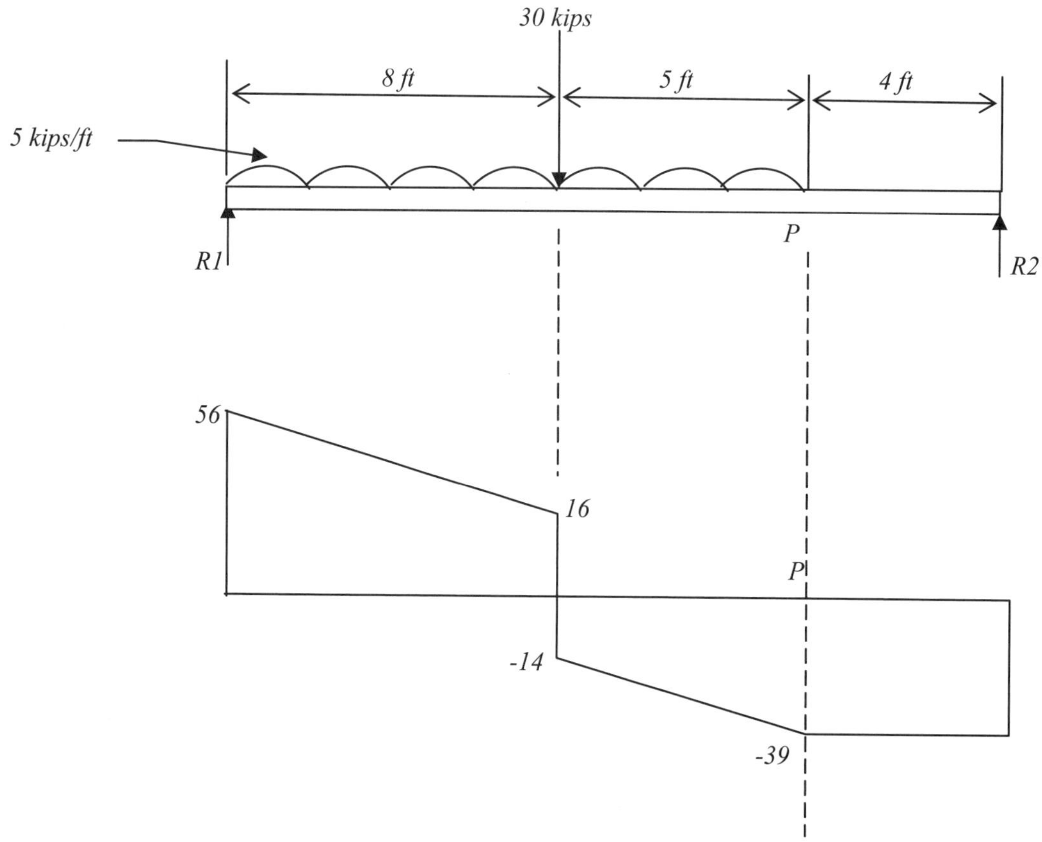

Shear force at point P = 39 kips. (sign does not matter for the shear force). (Ans C)

<u>How to draw the shear force diagram</u>:

STEP 1: Start from a corner. We found R1 to be 56 kips. R1 is upwards. Hence go up 56 kips.

STEP 2: Uniform load is downward. Come down 5 kips/ft along the beam.

After 8 ft, the shear force should come down 40 kips. (5 x 8).

At 30 kip load we end up at 16 kips.
Then go down 30 kips to represent the 30 kip load that is acting downward.
We end at -14 kips.

STEP 3: The go down 5 kips per foot. At point P, we would go down 25 kips. (5 x 5).

Shear force at point P = -14 – 25 = -39 kips

After point P, there is no load. Hence go straight to the end.

R2 was found to be 39 kips. Go up 39 kips. This is a good check to see whether your shear force diagram is correct.

Problem 2:26) Find the maximum bending moment of the beam given in the previous problem.

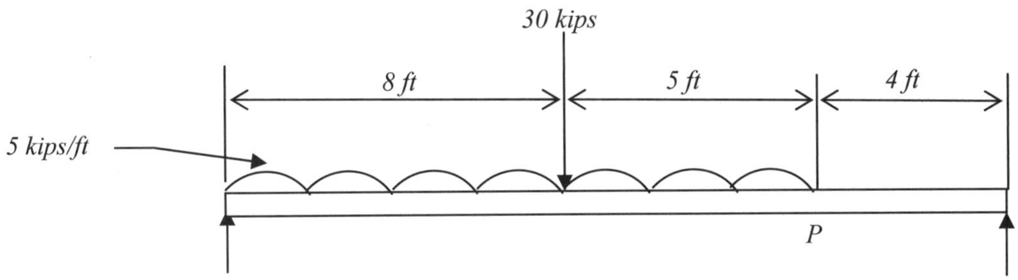

A) 288 kip.ft B) 146 kip.ft C) 239 kip.ft D) 621 kip.ft

Solution 2:26):

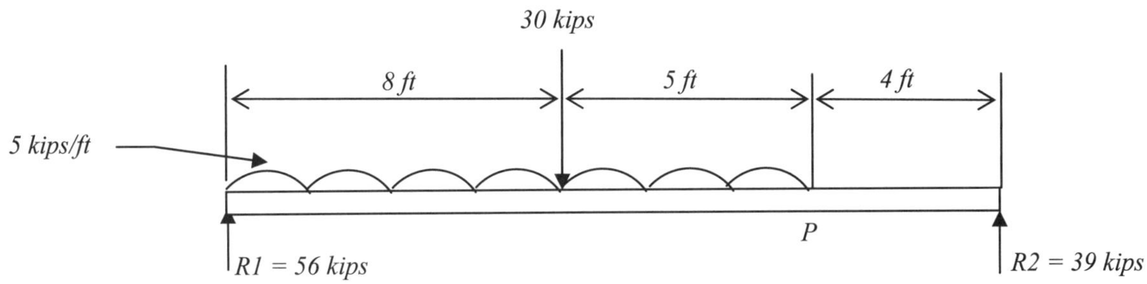

STEP 1: Maximum bending moment occurs at zero shear location.

Zero shear occurs under the load 30 kips. (See the previous problem).

STEP 2: Take moments at 30 kip load for the left hand section of the beam.

M = R1 x 8 – 5 x 8 x (8/2)

M = 56 x 8 – 5 x 8 x (8/2) = 288 kip. ft

56 x 8 = Moment due to the support reaction.
5 x 8 x (8/2) = Moment due to uniform load.
5 kips/ft is the uniform load.
8 is the length of the uniform load.
8/2 = Distance to the center of gravity of the uniform load. (Ans A)

You can double check your answer by taking moments for the right hand section as well.
M = R2 x (5 + 4) - 5 x 5 x (5/2)

M = 39 x 9 – 25 x 2.5
M = 288.5 kip. ft
There is a slight difference since R1 is really 56.02 kips.

WASTEWATER:

Problem 2.27): A solution has 12 grams of dissolved Oxygen. How many moles of Oxygen are dissolved in the solution?

A) 0.375 moles B) 0.618 moles C) 0.750 moles D) 1.204 moles

Solution 2.27):

Atomic weight of Oxygen = 16
One mole of Oxygen molecule (O_2) = 2 x 16 grams = 32 grams

$$\text{Number of moles} = \frac{\text{Weight in grams}}{\text{Molecular weight}} = 12/32 = 0.375$$ (Ans A)

Problem 2.28): 100 ml wastewater sample was diluted with 800 ml of water. Initial dissolved Oxygen level (DO_i) was found to be 14.8 mg/L. After 5 days of incubation, final dissolved Oxygen level (DO_f) was found to be 3.9 mg/L. What is the 5 day BOD of the sample?

A) 211.3 mg/L B) 93.4 mg/L C) 98.1 mg/L D) 37.8 mg/L

Solution 2.28): BOD_5 is defined as follows.

$$BOD_5 = \frac{DO_i - DO_f}{\dfrac{V_{sample}}{V_{sample} + V_{Dilution}}}$$

DO_i = Initial dissolved Oxygen level = 14.8 mg/L.
DO_f = Final dissolved Oxygen level = 3.9 mg/L
V_{sample} = Volume of wastewater sample = 100 ml
$V_{Dialution}$ = Volume of dilution = 800 ml

$$BOD_5 = \frac{DO_i - DO_f}{\dfrac{V_{sample}}{V_{sample} + V_{Dilution}}} = \frac{14.8 - 3.9}{\dfrac{100}{100 + 800}} = 98.1 \text{ mg/L}$$

(Ans C)

Discussion:

BOD$_5$ Test Procedure:

STEP 1: Wastewater sample is taken and diluted with distilled water. Dilution is necessary for various practical reasons.
STEP 2: Measure the dissolved oxygen level of the sample. This is the DO_i or dissolved oxygen initial value.

STEP 3: Seal the sample and wait for 5 days. One can wait two days or six days. But typically 5 days are selected so that values can be compared.
What happens during this stage? Wastewater has bacteria. This bacteria would start to use up the oxygen in the wastewater sample. If there is more bacteria, more oxygen would be used up. If the sample has less bacteria, less oxygen would be used up.
STEP 4: Measure the dissolved oxygen level after 5 days, (DO_f), or dissolved oxygen final.

- If there is more bacteria, then DO_f would be much lower than DO_i.
  Why? Bacteria will be using dissolved oxygen to survive.
- If there is less bacteria, then DO_f value would be closer to DO_i.
- If there is no bacteria, then DO_f value would be same as DO_i.
  Why? Since there is no bacteria, there is nobody there to use the dissolved oxygen.

Problem 2.29) Activated sludge process is best described as
A) Re-introduce aerated sludge back to the waste stream to increase the level of Bacteria.
B) Re-introduce aerated sludge back to the waste stream to decrease the level of Bacteria.
C) Activate the sludge by adding chemicals
D) None of the above

Solution 2.29): (Ans A)

All life forms need food. Food of humans and animals consist of plant matter and animal flesh. Both plants and animal flesh are organic compounds. Organic compounds are basically built with a Carbon backbone. Other atoms and molecules such as Nitrogen, Hydrogen and Oxygen are attached to the Carbon backbone.

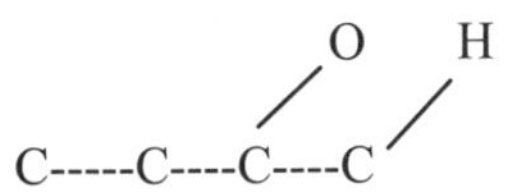

Typical organic molecule contains a Carbon backbone

Wastewater contains organic molecules. Organic molecules are used by Bacteria and other microorganism as food. Where ever there is organic matter, one can expect to have microrganisms. Some of these microorganisms promote diseases. Hence it is important to get rid of organic matter in wastewater.

One can introduce Bacteria to a wastewater stream. Bacteria would use organic matter as food and give out water and CO_2.

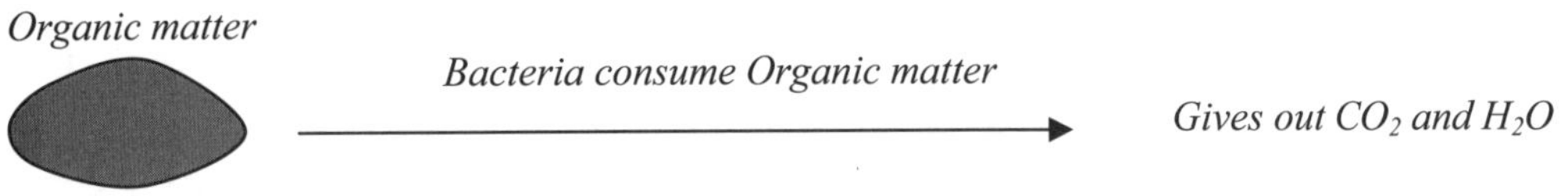

- Bacteria use organic matter as food and leave CO_2 and H_2O as end products.
- Whole idea in wastewater engineering is to get rid of organic matter.
- If one can increase the number of Bacteria, organic matter will be broken down sooner.
- Introduction of O_2 into wastewater increase the Bacteria population. Hence Bacteria would break down the organic matter faster.
- Once all the organic matter is used up by Bacteria they will not have anything to consume. Then Bacteria die as well. Other methods such as UV radiation are also used to kill Bacteria after they have converted organic matter in to CO_2 and H_2O.

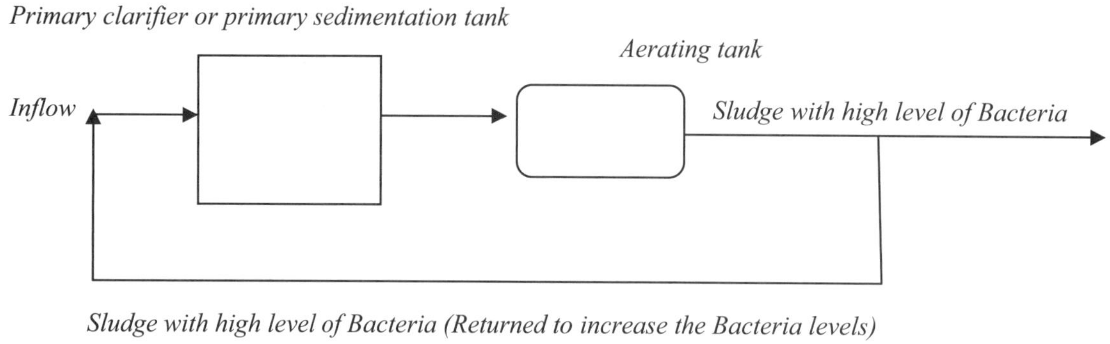

Activated Sludge Process

STEP 1: Wastewater is introduced to the primary sedimentation tank.
STEP 2: Then wastewater is sent to an aerator where air is introduced. Introduction of air would increase the Bacteria levels. Hence this sludge is known as activated sludge.
STEP 3: Some of the activated sludge is returned to the primary sedimentation tank to increase the bacteria in wastewater.
STEP4: High levels of bacteria would consume more organic matter as food. Hence more organic matter would be broken down into H_2O and CO_2.

All what you need to remember is that Bacteria in wastewater is increased by providing air. Providing air (O_2) tend to increase the Bacteria population. Increase of Bacteria would speed up use of organic matter. Once all the organic matter is used up, Bacteria would die due to lack of food.

Wastewater aeration tank (Cups are for the purpose of sending air)

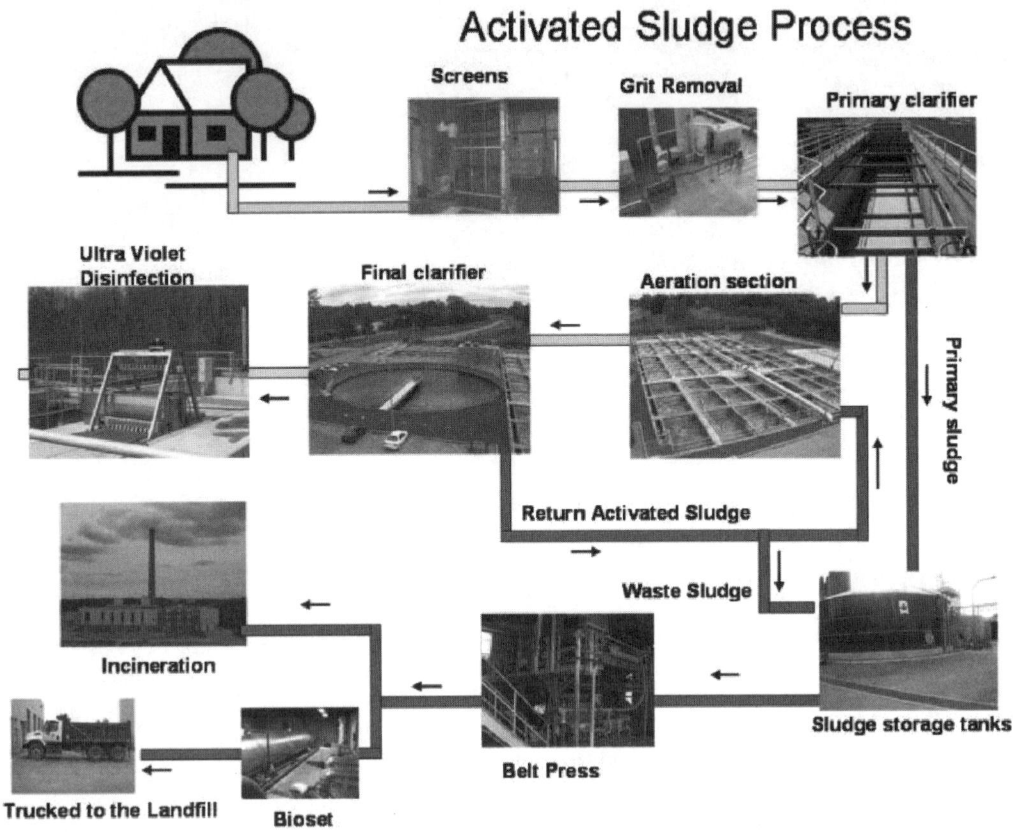

GEOTECHNICAL:

Problem 2.30): Total density of a soil sample was found to be 110 pcf and moisture content to be 60%. What is the dry density of the soil sample.

A) 125.9 pcf B) 45.6 pcf C) 68.7 pcf D) 98.9 pcf

Solution 2.30):

See the appendix for complete description of soil phase relationships.

In this problem total density and moisture content is given. You are required to find the dry density of the soil sample.

$$\gamma_d = \gamma_{wet}/(1 + w)$$

Total density = γ_{total} (Also known as wet density = γ_{wet}) = 110 pcf

Dry density = γ_d

Moisture content = w = 0.60

$\gamma_d = 110/(1 + 0.6) = 68.75$ pcf (Ans C)

Problem 2.31) Find the lateral earth pressure at point A.

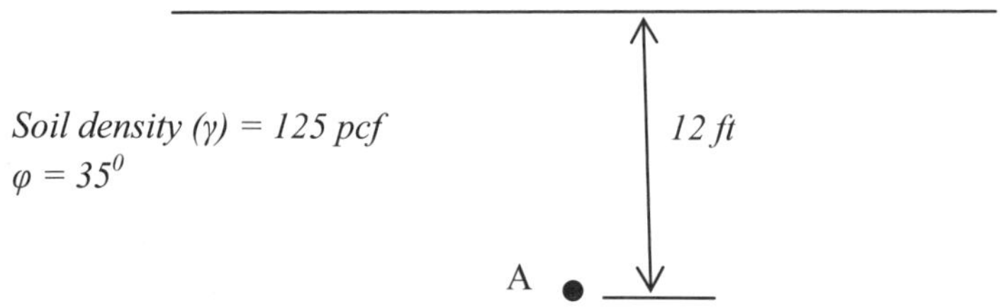

A) 2,131.2 psf B) 245.9 psf C) 639.6 psf D) 412.3 psf

Solution 2.31)

At rest condition applies for this problem.

$K_0 = 1 - \sin \varphi$
$= 1 - \sin 35 = 0.4264$
$P_H = K_0 . \gamma . h = 0.4264 \times 125 \times 12 = 639.6$ psf

Problem 2.32): Which statement is true regarding SPT (N) value, friction angle, unconfined compressive strength and cohesion?

A) There is a correlation between SPT (N) value and friction angle in sandy soils. There is no correlation between unconfined compressive strength and cohesion in clay soils.

B) There is no correlation between SPT (N) value and friction angle in sandy soil. There is no correlation between unconfined compressive strength and cohesion in clay soils.

C) There is a correlation between SPT (N) value and friction angle in sandy soils. There is a correlation between unconfined compressive strength and cohesion in clay soils.

D) There is no correlation between SPT (N) value and friction angle. There is a correlation between unconfined compressive strength and cohesion.

Solution 2.32):

Friction between soil particles is a physical process. On the other hand, cohesion between soil particles is an electro chemical process.

Following can be said of sandy soils and clay soils.

- Sandy soils considered to have friction. Cohesion in sandy soils generally ignored.
- Clay soils considered to have cohesion. Friction in clay soils generally ignored.

Also it should be mentioned here that sand particles are visible to the naked eye and clay particles are not visible to the naked eye.

SPT Test: (Standard Penetration Test): SPT (N) value is an important parameter that is been used to find the friction angle of sandy soils. SPT value is obtained by dropping a 140 lb (63.5 kg) hammer from a height of 30 inches (760 mm). Split spoon samples are taken during the process. There is a correlation between SPT (N) value and friction angle in sandy soils.

Unconfined Compressive Strength Test: Unconfined compressive strength test is done on clay soils. Soil sample is simply placed between two plates and compressed. Unconfined compressive strength test is designed to measure the shear strength or cohesion of clay soils. This is the easiest and most common test done to measure the shear strength. Shear strength of soil is equal to the cohesion of soil.

Ans (C)

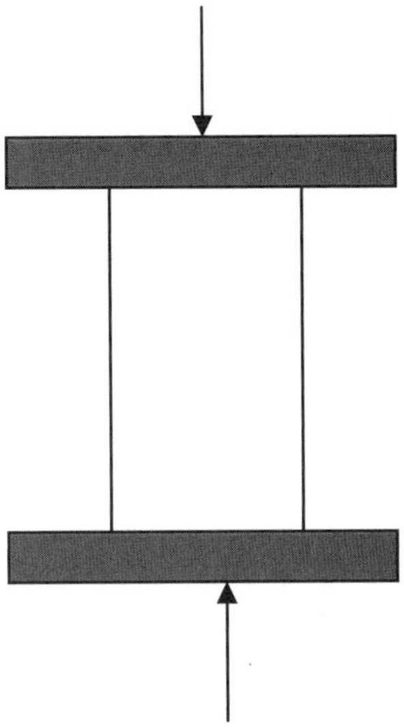

Unconfined Compressive Strength Test.

Problem 2.33): Soil sample has a porosity of 0.46. What is the void ratio of the soil sample.

A) 0.67 B) 0.85 C) 0.32 D) 1.12

Solution 2.33): Relationship between porosity and void ratio is given by the following equation.

$$n = e/(1 + e)$$

n = Porosity = V_v/V
e = Void ratio = V_v/V_s

V_v = Volume of voids
V_s = Volume of soilds

Total volume (V) = Volume of solids + Volume of water + Volume of air

Apply the equation;

$n = e/(1 + e)$
$0.46 = e/(1 + e)$
$0.46 + 0.46e = e$

$0.46 = e - 0.46e$
$e = 0.85$ (Ans B)

Problem 2.34): Figure shows a pile driven in clay soils.

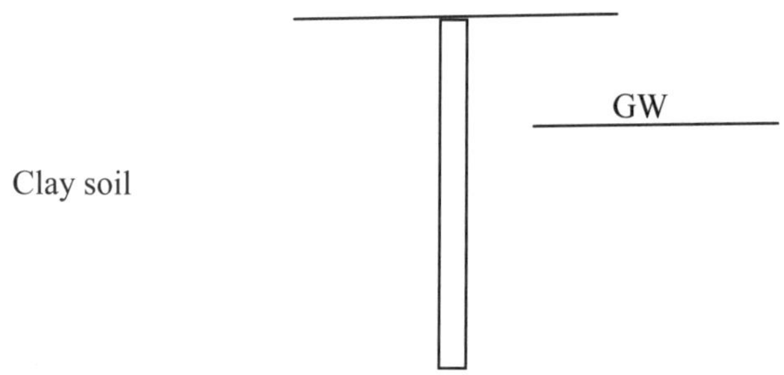

Capacity of the pile is NOT dependant on;

A) Cohesion of soil B) Pile diameter C) Depth to groundwater
D) Soil – Pile adhesion coefficient

Solution 2.34):

Pile capacity is dependant upon the end bearing at the tip of the pile plus the skin friction.

Pile capacity = End bearing + Skin friction

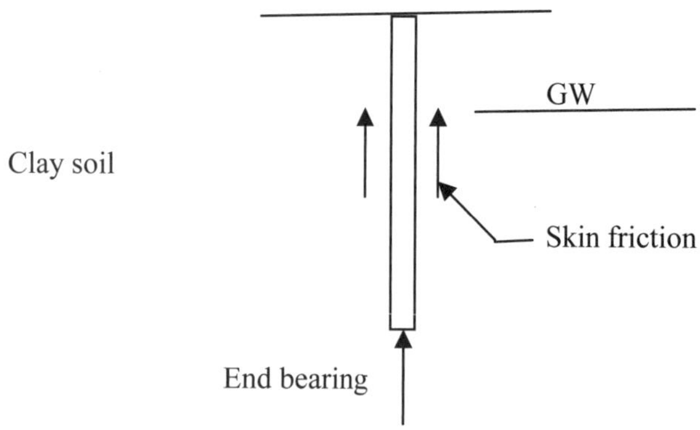

Pile capacity equation for clay soils is given below. (This equation is not valid for sandy soils).

$$P_u = N_c. C. A_c + \alpha. C. A_p$$

$N_c. C. A_c$ = End bearing term
$\alpha. C. A_p$ = Skin friction term

$N_c = N_c$ is taken to be equal to 9.0
P_u= Ultimate pile capacity
C = Cohesion of the soil
A_c = Cross sectional area of the pile
A_p = Perimeter area of the pile
α = Soil pile adhesion coefficient

Cohesion, pile diameter, soil pile adhesion coefficient contribute to pile capacity. Depth to groundwater has no impact on capacity of piles in clay soils. On the other hand groundwater has a huge impact on piles installed in sandy soils.

Steel piles

Steel piles installed

Timber piles

Timber piles installed

Problem 2.35): Consolidation test of a normally consolidated soil is shown below. What is the settlement (ΔH) of the clay layer shown if the initial pressure $p_0' = 300$ psf and final pressure is $p_f' = 600$ psf at the center of the clay layer.

Thickness of the clay layer (H) = 12 ft
Gradient = C_c = *0.28*
Initial void ratio (e_0) = 0.65

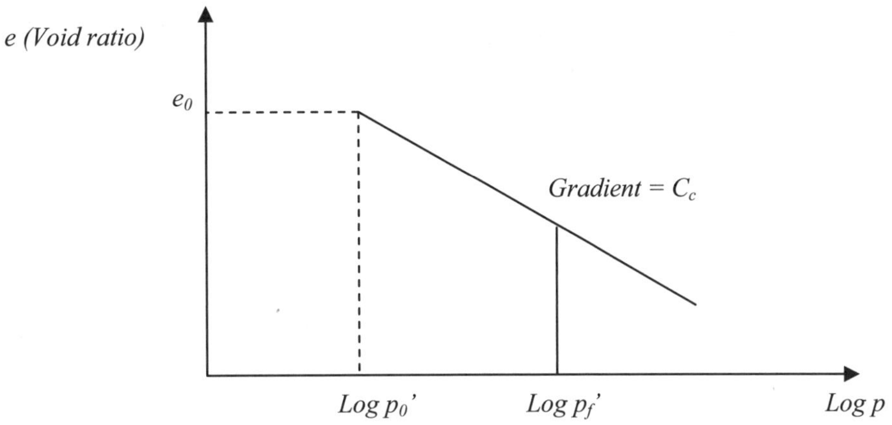

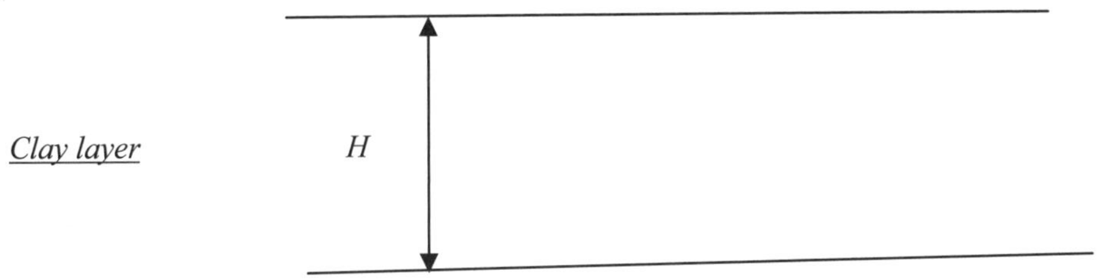

A) 3.21 in. B) 1.89 in. C) 4.98 in D) 7.35 in.

Solution 2:35)

Settlement due to Consolidation of a clay layer is given by the following equation.

$$\Delta H = \frac{C_c H}{(1 + e_0)} \text{Log}\,(p_f'/p_0')$$

ΔH = Settlement due to consolidation
C_c = Compression index
H = Thickness of the clay layer
p_f'= Final pressure at the mid point of the clay layer
p_0' = Initial pressure at the mid point of the clay layer
e_0 = Initial void ratio

$C_c = 0.28$
$H = 12$ ft
$p_f' = 600$ psf
$p_0' = 300$ psf
$e_0 = 0.65$

Apply the consolidation equation;

$$\Delta H = \frac{C_c \ H}{(1 + e_0)} \ \text{Log} \ (p_f'/p_0')$$

$$\Delta H = \frac{0.28 \times 12}{(1 + 0.65)} \ \text{Log} \ (600'/300')$$

$\Delta H = 0.613$ ft
$\Delta H = 7.35$ inches (Ans D)

CONSTRUCTION:

Problem 2.36) Proposed road and existing ground profile is shown below. Volume of soil in each segment is given. Swell factor and shrinkage factor of soil are 18% and 20% respectively. How many trucks of soil need to be brought in from an outside borrow pit if the swell factor and shrinkage factor of borrow pit soil are 15% and 17% respectively. Assume each truck can carry 10 CY of soil.

Bank volumes: V1 = 123 CY; V2 = 330 CY; V3 = 60 CY

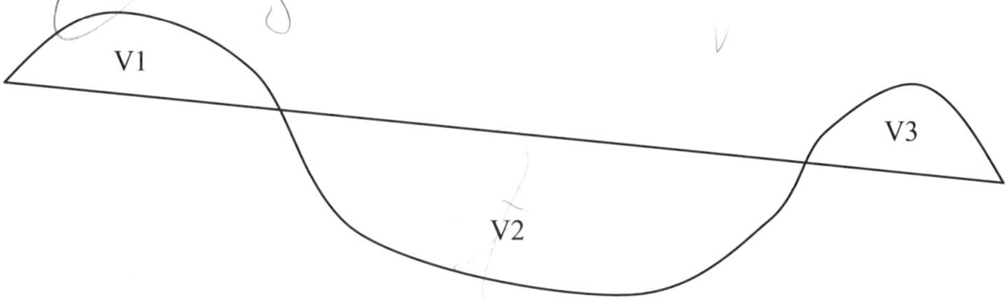

A) 26 B) 23 C) 28 D) 22

Solution 2.36)

Let's look at the process.

- Soil is cut from high locations. The in-situ volume is known as bank volume.
- Then the soil is transported to areas where fill is required. When the soil is in trucks, the soil would swell. This volume is known as loose volume.
- Next the soil is compacted. Once compacted, soil would shrink and new volume is known as compacted volume.

STEP 1: Find the volume of cut (bank volume basis):

Volume of cut = V1 + V3 = 123 + 60 = 183 CY (Bank volume basis)
Volume of fill = V2

STEP 2: Find the volume of cut (compacted volume basis):

183 CY was cut and compacted. When compacted, it would have a different volume.

$$\text{Shrinkage Factor} = \frac{\text{Bank Volume} - \text{Compacted Volume}}{\text{Bank Volume}}$$

Shrinkage factor for soil in the site is 20%.

$$0.20 = \frac{183 - \text{Compacted Volume}}{183}$$

Compacted volume = 146.4 CY

STEP 3: Find the volume of soil needed (Compacted volume basis):

Volume of soil needed (compacted volume basis) = V2 – 146.4
= 330 – 146.4 = 183.6 CY

V2 is given to be 330 CY. (See the figure).

STEP 4: Convert the compacted volume to loose volume:
Contractor needs 183.6 CY of soil compacted. This soil need to be brought in from the borrow pit. Compacted volume needs to be converted to loose volume. That can be done using the following equation.

$$\text{Compacted Volume} = \text{Loose Volume} \times (1 - S_H)/(1 + S_w)$$

Since this soil is brought in from a borrow pit, you need to use S_H and S_w for the borrow pit soil. Swell factor and shrinkage factor of borrow pit soil are 15% and 17% respectively.

Swell factor (S_w) for borrow pit soil = 15%.
Shrinkage factor (S_H) for borrow pit soil = 17%

Compacted Volume = Loose Volume x (1 – 0.17)/(1 + 0.15)

183.6 = Loose Volume x (1 – 0.17)/(1 + 0.15)
Loose volume = 254.39

Number of trucks needed = 254.39/10 = 26 (Ans A)

Large cut section

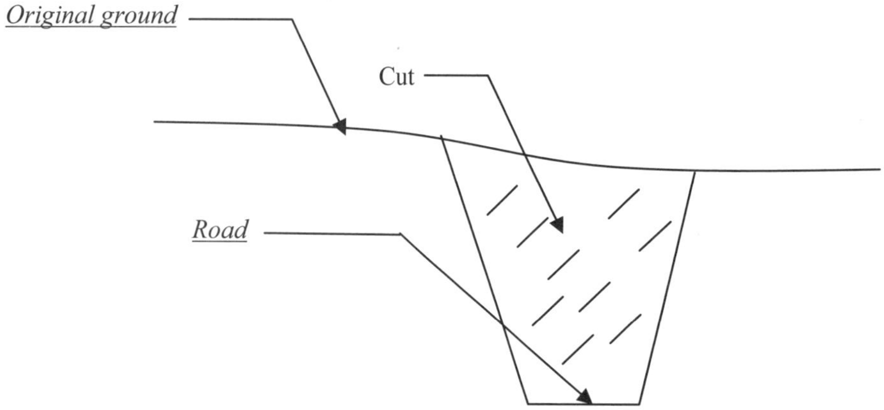

Road fill section

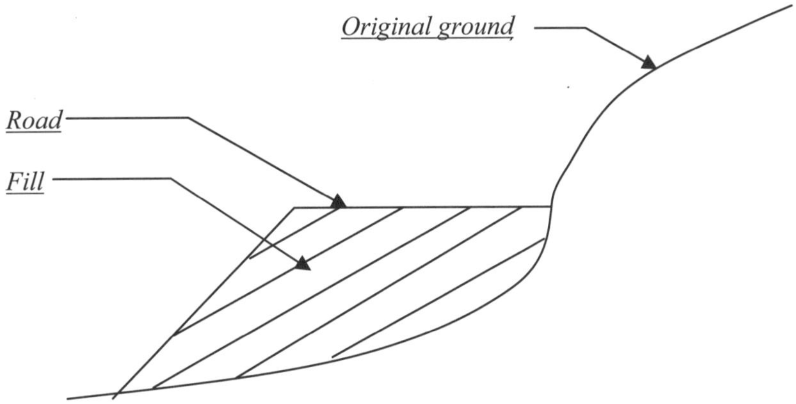

Problem 2:37) Contractor needs to erect 200 tons of steel. He has two crews. (crew 1 and crew 2). He decided to use crew 1 to erect 200 tons of steel. What would have happened, if he had used crew 2 instead?

Crew 1:

5 steel workers
3 Laborers
1 foreman
1 welder

Crew 1 productivity = 1.5 tons/LH

Crew 2:

4 steel workers

4 Laborers
1 foreman
2 welders

Crew 2 productivity = 1.7 tons/LH

Wages:

steel workers	- $75/hr
laborers	-$55/hr
foreman	- $90/hr
welders	- $65/hr

A) Contractor would have lost 1,523 dollars
B) Contractor would have lost 912 dollars
C) Contractor would have saved 1,352 dollars
D) Contractor would have saved 2,023 dollars

Solution 2.37):

STEP 1: Find the cost of crew hour for crew 1;

Crew 1:

Cost of crew hour = (5 x 75) + (3 x 55) + (1 x 90) + (1 x 65)
= $695

STEP 2: Find the cost of labor hour for crew 1;

Cost of labor hour = Cost of crew hour/Number of workers

Cost of labor hour = 695/10 = $69.5

STEP 3: Number of labor hours needed for crew 1 to complete the project;

Number of labor hours needed for crew 1 to complete the project = Total tonnage/Productivity
Number of labor hours needed for crew 1 to complete the project = 200/1.5 = 133.33 labor hrs
Productivity of crew 1 is given to be 1.5 tons per labor hour.

STEP 4: Cost of completion (Crew 1);

Cost to complete the project = Cost of labor hour x Labor hours required to complete the project
Cost to complete the project = $69.5 x 133.33
= $9,266

STEP 5: Find the cost of completion (Crew 2);

Crew 2:

Cost of crew hour = (4 x 75) + (4 x 55) + (1 x 90) + (2 x 65)
= $740

Find the cost of labor hour for crew 2;

Cost of labor hour = Cost of crew hour/Number of workers

Cost of labor hour = 740/11 = \$67.3
Find the number of labor hours needed for crew 2 to complete the project;
Number of labor hours needed for crew 2 to complete the project = Total tonnage/Productivity

Number of labor hours needed for crew 2 to complete the project = 200/1.7 = 117.6 hrs
Productivity of crew 2 is given to be 1.7 tons per labor hour.

Cost of completion (Crew 2);

Cost to complete the project = Cost of labor hour x Labor hours required to complete the project
Cost to complete the project = \$67.3 x 117.6
= \$7,914

Total cost for crew 1 = \$9,266
Total cost for crew 2 = \$7,914

If contractor had used crew 2, he would have saved 1,352 dollars. (Ans C)

Problem 2.38) What loads are considered as construction loads?
A) Wind load, dead weight, live load from construction workers, Earthquake load
B) Wind load, dead weight, live load from construction workers, Load due to wet concrete
C) Wind load, dead weight, live load from workers, Earthquake load
D) Wind load, dead weight, live load from workers, Seismic load

Solution 2.38):

Earthquake occurring during construction is a rare event. Hence earthquake load is not considered when computing construction loads.

Problem 2:39) Find the number of bricks required for the wall shown.
Following information provided;

Length of the wall = 120 ft
Height of the wall = 10 ft
Size of bricks = 8 in x 3 5/8 in x 2 ¼ in
Mortar thickness between bricks = 0.5 inches
The wall is one brick thick.

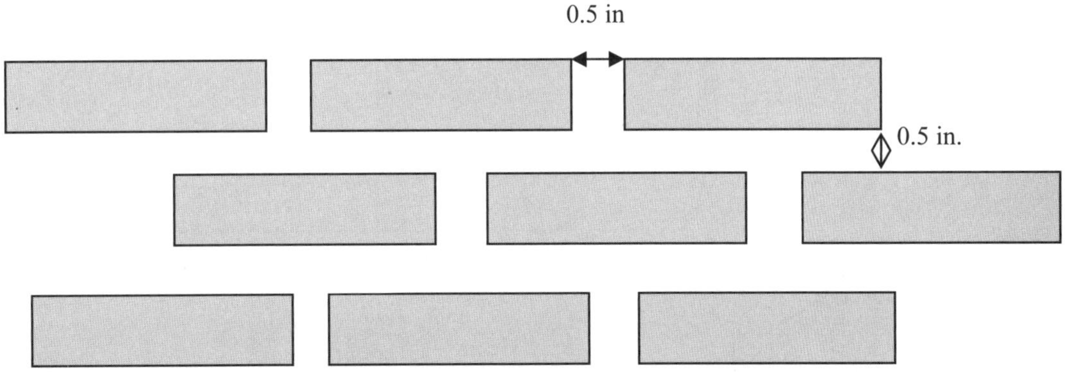

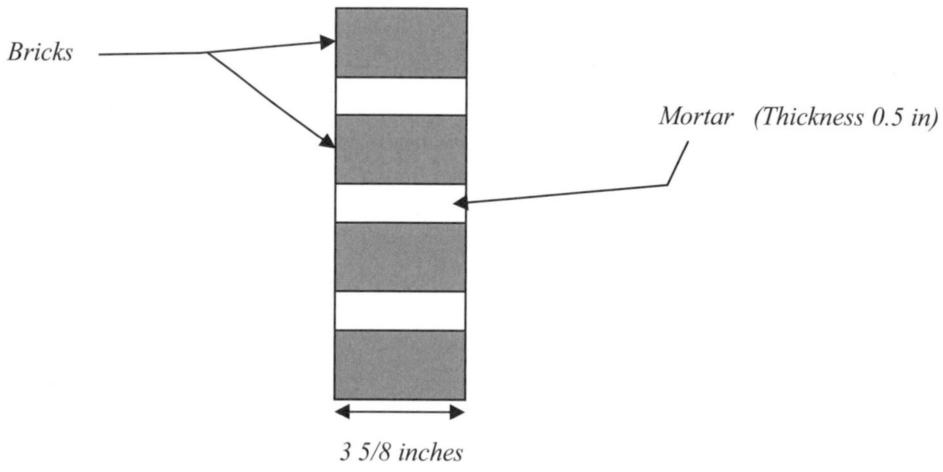

Wall - Side View

A) 7,392 bricks B) 6,823 bricks C) 4,987 bricks D) 3,651 bricks

Solution 2.39): STEP 1: Find the area of one brick with mortar.

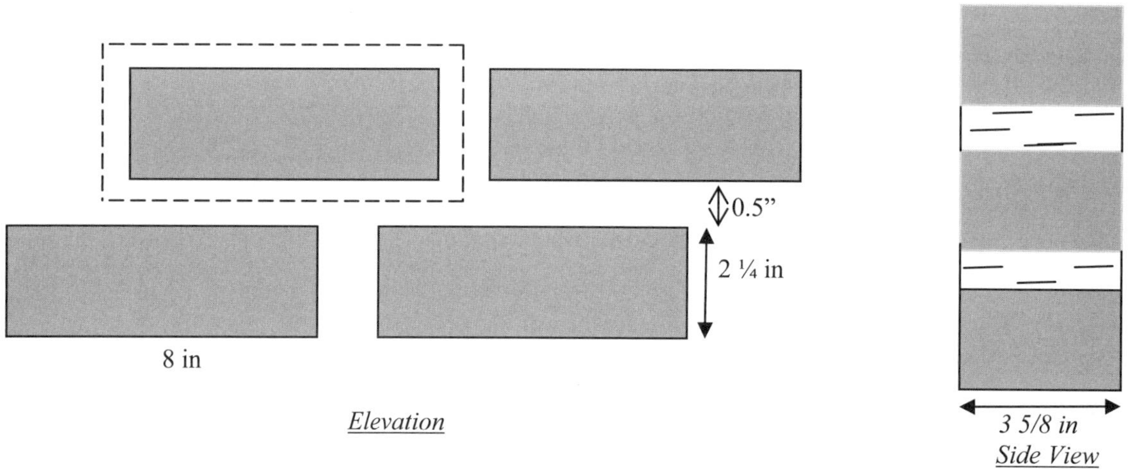

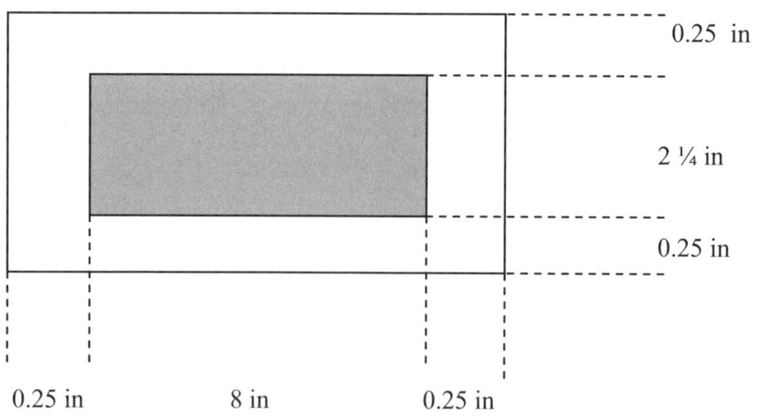

One brick with surrounding mortar

STEP 1: Find the area of one brick with mortar;

Length of one brick with the mortar = 8 + 0.25 + 0.25 = 8.5 in
Height of one brick with the mortar = 2.25 + 0.25 + 0.25 = 2.75 in
Area of one brick with mortar = 8.5 in. x 2.75 in. = 23.375 sq. in

STEP 2: Find the area of the wall;

Area of the wall = 120 x 10 sq. ft
= 1,200 sq. ft = 1,200 x 144 sq. in
= 172,800 sq. in

STEP 3: Find the number of bricks required;

$$\text{Number of bricks required} = \frac{\text{Area of the wall}}{\text{Area of one brick with mortar}}$$

$$\text{Number of bricks required} = \frac{172{,}800}{23.375}$$

Number of bricks required = 7,392.5 bricks (Ans A)

Problem 2:40) Find the volume of mortar required for the wall given in the previous problem.

A) 1.25 CY B) 3.98 CY C) 2.89 CY D) 3.09 CY

Solution 2.40):

STEP 1: Find the total volume of the wall;

Total volume of the wall = Length x Height x Thickness of the wall

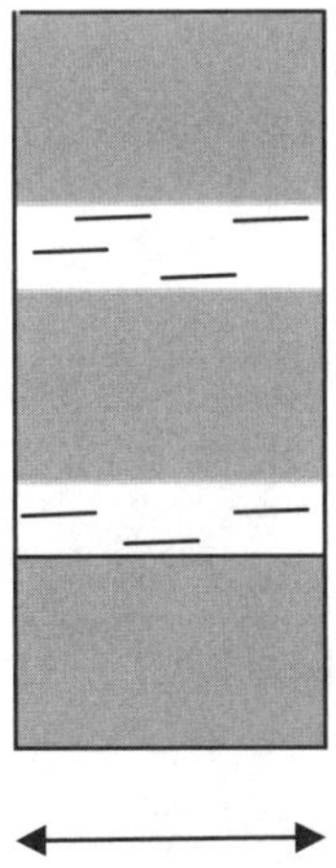

3 5/8 inches (3.625 inches)

The wall is one brick wide.
Width of one brick = 3 5/8 inches

Convert all dimensions to inches.

Total volume of the wall = (120 x 12) x (10 x 12) x (3.625) = 626,400 cubic. inches

STEP 2: Find the volume of one brick

Volume of one brick (without mortar) = 8 x 3.625 x 2.25
= 65.25 cu. inches

STEP 3: Find the number of bricks used;

Number of bricks used = 7,392.5 bricks (Found in the previous problem)

STEP 4: Find the total volume of bricks without mortar;

Total volume of bricks without mortar = 7,392.5 x 65.25 cu. inches
= 482,360.6 cu. inches.

STEP 4: Find the volume of mortar;

Volume of mortar = Total volume of the wall – Volume of bricks
= 626,400 - 482,360.6

Volume of mortar = 144,039 cubic inches

Volume mortar = 144,039/(12 x 12 x 12 x 27) CY = 3.09 CY (Ans D)

Bricks and mortar

SAMPLE EXAM 3

Problem 3.1) Channel cross section is shown in the figure. Manning coefficient (n) of the left section of the channel is 0.13 and Manning coefficient (n) for the right section of the channel is 0.09. If the flow in the channel is 400 cu.ft/sec, what is the slope?

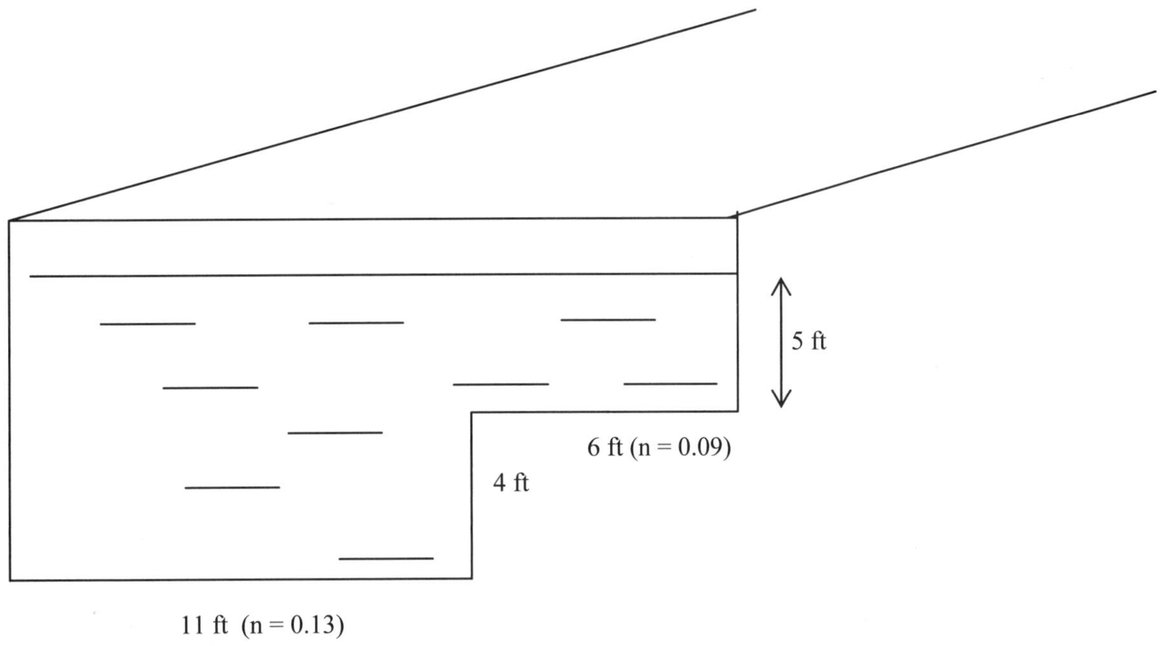

A) 1.0% B) 2.5% C) 3.5% D) 6%

Problem 3.2) Find which section has the highest velocity? All sections have the same Manning roughness coefficient and same slope.

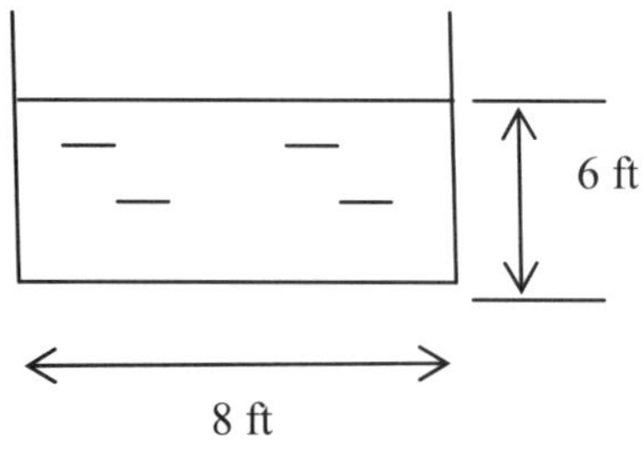

Section A

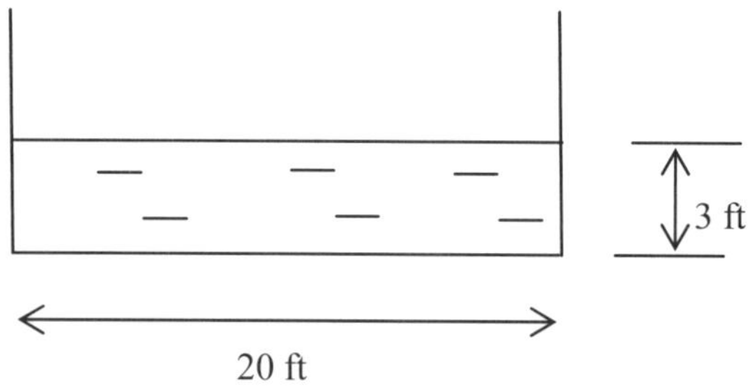

Section B

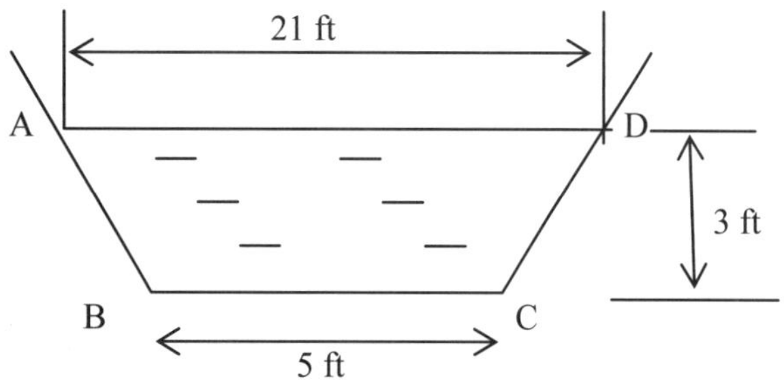

Section C (Symmetrical)

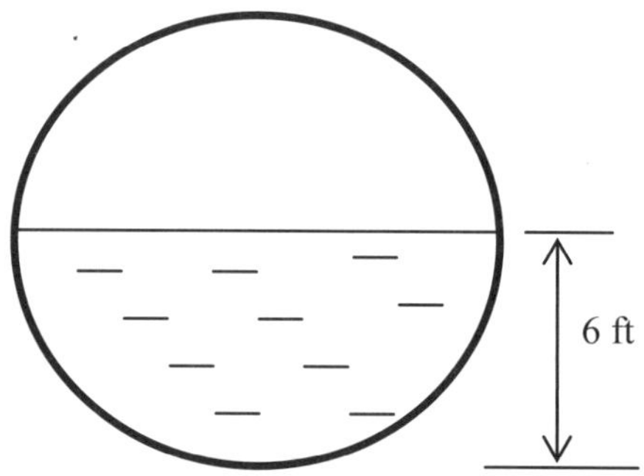

Section D
(Pipe Diameter 12 ft)

A) Section A; B) Section B C) Section C D) Section D

Problem 3.3) Find the Froude number of the trapezoidal channel shown if the flow is 35 cu.ft/sec. Channel width at the bottom is 8 ft and channel depth is 6 ft. Channel slopes are 1 to 2 as shown.

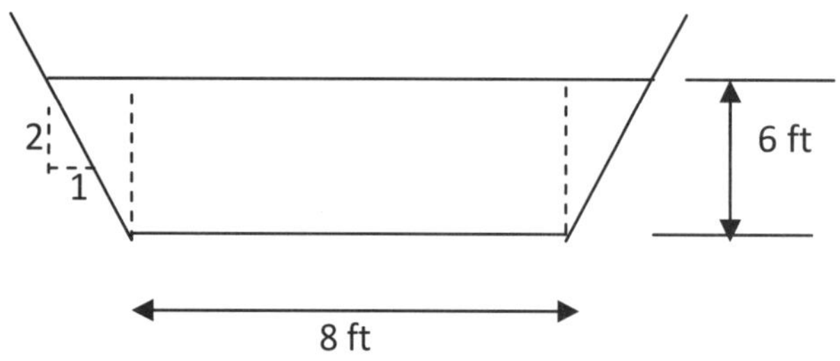

A) 0.034 B) 0.043 C) 0.785 D) 0.676

Problem 3.4) Find the critical depth of channel shown if the flow is 35 cu.ft/sec. Channel width at the bottom is 8 ft. Channel slopes are 1 to 2 as shown.

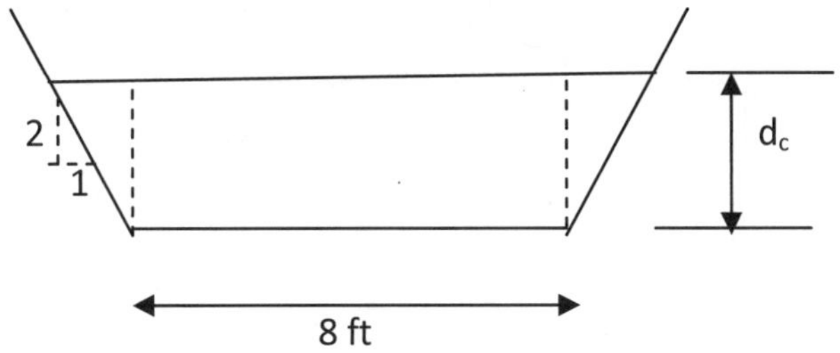

A) 0.034 ft   B) 0.827 ft   C) 0.785 ft   D) 0.676 ft

Problem 3.5) Hydraulic jump in a rectangular channel is shown in the figure. Upstream depth before the jump is 1.1 ft and the velocity is 15 fps. Width of the rectangular channel is 5.5 ft. What can you say of the flow after the jump?

A) Flow is subcritical after the jump
B) Flow is super critical after the jump
C) Flow is critical after the jump
D) Not enough data to answer the question.

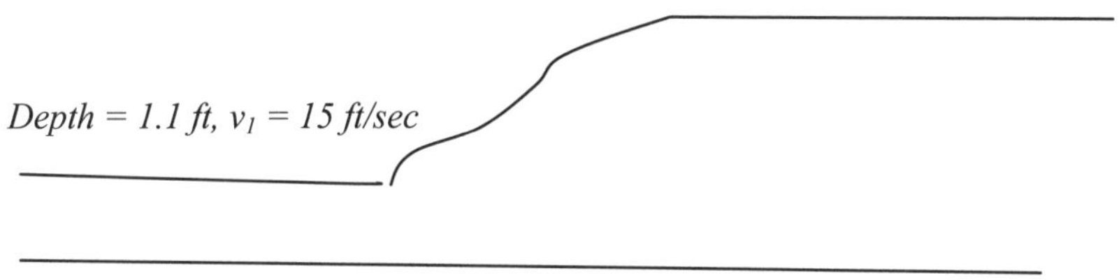

Problem 3.6) Which statement is correct regarding channel flow?
A) Channels with uniform flow have the same depth thru out the channel.
B) Channels with steady flow have the same depth thru out the channel.
C) Channels with uniform flow do not have the same depth thru out the channel.
D) All of the above

Problem 3.7: 1.2 ft diameter pipe is carrying water at a velocity of 2.1 ft/sec and at a pressure of 12 psi. After travelling 900 ft, elevation of the pipe has dropped by 9 ft. Darcy friction coefficient of the pipe is 0.008. What is the pressure in the pipe at the end of the pipe?

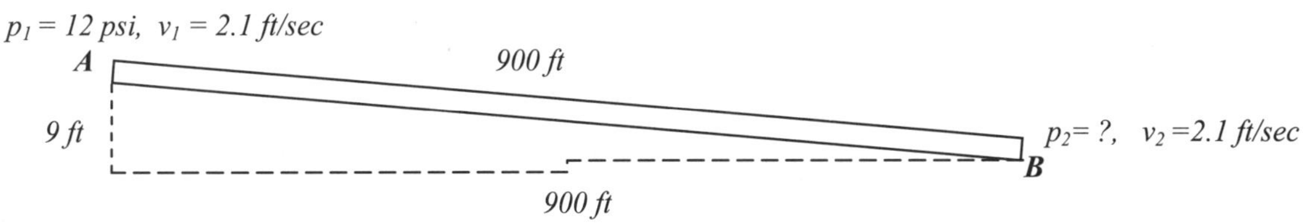

A) 3.4 psi   B) 12.3 psi   C) 15.7 psi   D) 9.8 psi

Problem 3.8: Coefficient of the Pitot tube shown in the figure is 0.9. If the flow velocity is 6 ft/sec, find "y":

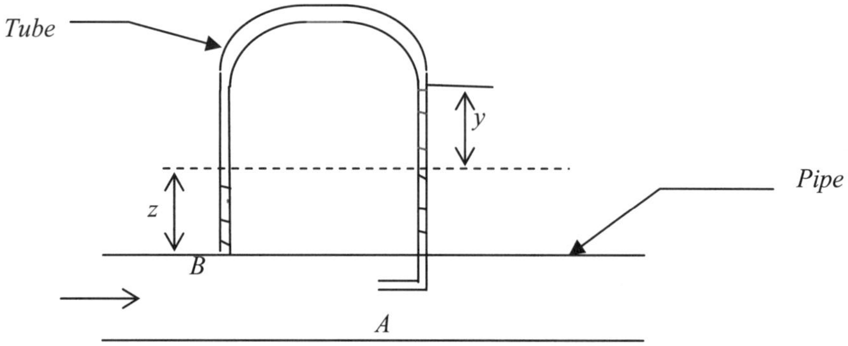

A) 1.3 ft B) 6.7 ft C) 0.69 ft D) 2.1 ft

Problem 3.9: A Venturi meter is shown (with a water column) in the figure. Assuming a discharge coefficient of 1.0 find the flow in the pipe.

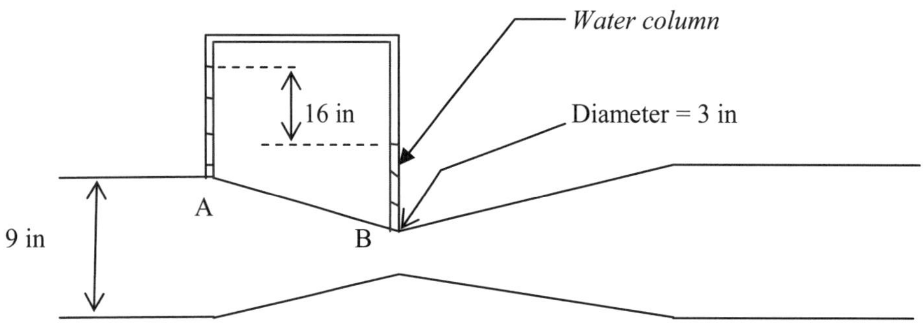

A) 0.01 cfs B) 0.45 cfs C) 0.69 cfs D) 2.1 cfs

Problem 3.10) Two areas are draining into a drain pipe as shown. Information of the two areas is given below.
Area 1 = Area is 9 acres, time of concentration 31 minutes, runoff coefficient 0.51
Area 2 = Area is 3 acres, time of concentration 30 minutes, runoff coefficient 0.40

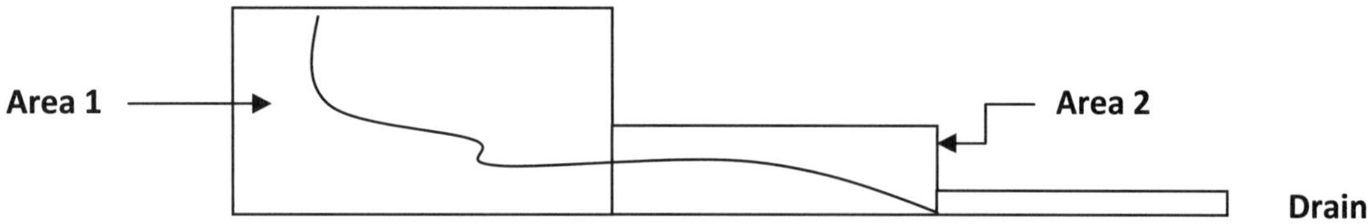

Rainfall intensity for 10 year, 15 year and 30 year rainfalls are given by following graphs.

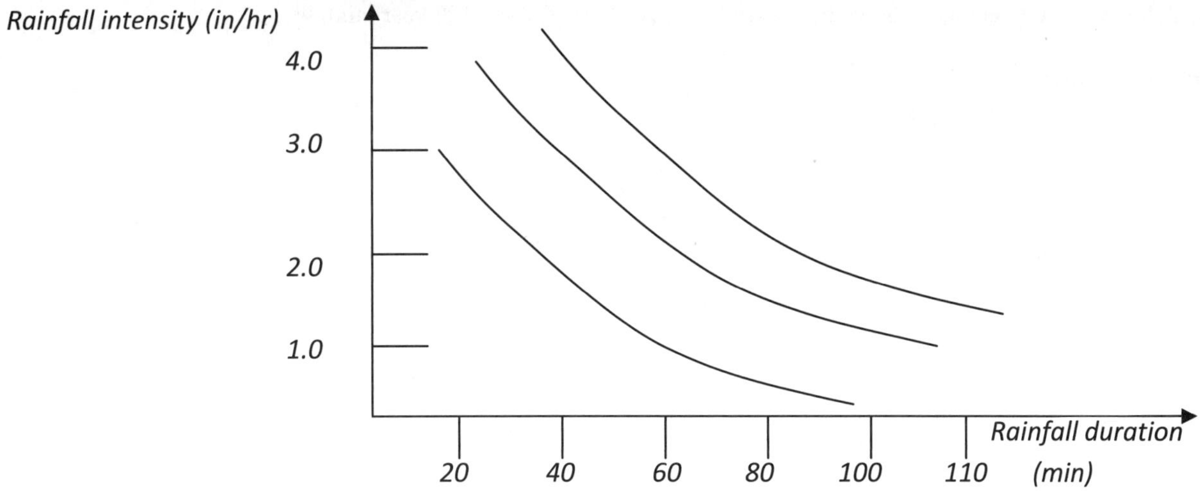

Find the maximum flow in the drain pipe for a 30 year rainfall.

A) 78.9 cu.ft/sec B) 16.84 cu.ft/sec C) 61.3 cu.ft/sec D) 23.5 cu.ft/sec

Problem 3.11) Area 1 and area 2 are discharging into a pipeline as shown. Following information is provided.

	Area	*Runoff coefficient*	*Slope*	*Distance to hydrologically most distant point*
Area 1	23 acres	0.70	8%	D1 = 0.6 miles
Area 2	35 acres	0.55	18%	D2 = 0.4 miles

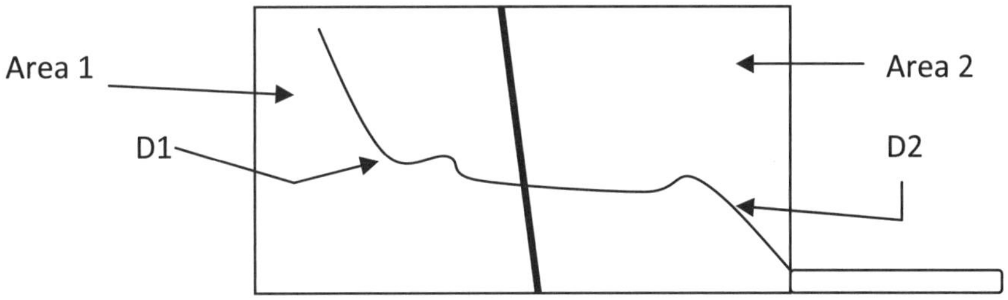

D1 = 0.6 miles
D2 = 0.4 miles

Rainfall data for the region as shown below are for 10 year, 20 year and 100 year rainfalls.

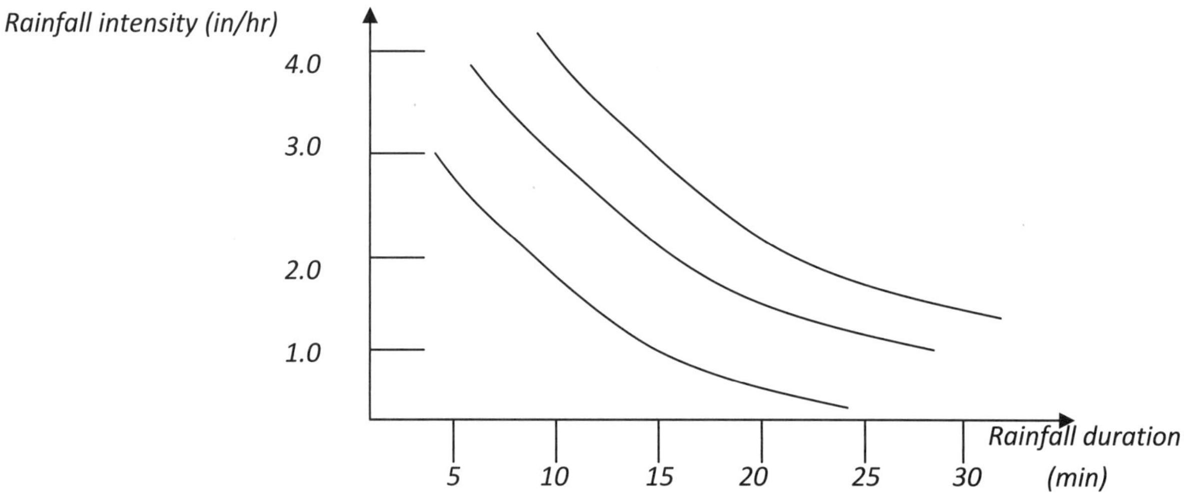

Slope	*Velocity*
5%	*2 ft/sec*
10%	*10 ft/sec*
15%	*13 ft/sec*
20%	*20 ft/sec*

Find the flow in the pipeline for a 20 year rainfall in cu.ft/sec.

A) 78.9 cu.ft/sec B) 71.6 cu.ft/sec C) 141.3 cu.ft/sec D) 110.6 cu.ft/sec

Problem 3.12) Wastewater engineer designed a circular primary sedimentation tank with a detention time of 2.5 hrs. Height of the tank is 5.6 ft. What is the diameter of the tank if the flow is 1.2 MGD?

A) 12.3 ft B) 98.7 ft C) 61.6 ft D) 24.8 ft

Problem 3.13) Certain highway traffic consists of 8% truck traffic and 4% recreational vehicles. The terrain is considered to be rolling terrain. What is the heavy vehicle factor (f_{HV})?

	Type of Terrain		
Factor	**Level**	**Rolling**	**Mountainous**
E_T (Trucks and busses)	1.5	2.5	4.5
E_R (Recreational vehicles)	1.2	2.0	4.0

HCM – Highway Capacity Manual

A) 0.764 B) 0.862 C) 0.145 D) 0.634

Problem 3.14) Traffic analyst have found that trips generated per household per day is to be 0.8 + 1.5P + 0.7A
P = Number of persons per household
A = Number of automobiles per household

If the area under research includes 800 households and 3.5 persons per household and 1.8 autos per household what is the number of trips generated per day.

A) 200 B) 10,065 C) 5,848 D) 4002

Problem 3.15) Four lane highway has an AADT of 120,000 and a k factor of 0.9 and a directional factor of 0.8. If the rate of increase of traffic per year is 6%, what is the DDHV per lane, 5 years from now.

A) 121,345 B) 115,623 C) 45,908 D) 234,900

Problem 3.16) Design traffic volume in a highway is 1,400 vehicles per hour per lane. It is reported that this highway has 3% truck traffic and 2% RVs (Recreational vehicles). The road is in level ground. PHF (Peak hour factor) is 0.85. The traffic is mostly office commuters.
E_T = Truck equivalent factor = 1.5 for level ground
E_R = RV equivalent factor = 1.2 for level ground
(E_T and E_R obtained from table 21-8 (HCM)
f_P = Driver population factor = 1.0
If the free flow speed is 65 mph, what is the level of service of the road?

Density vs LOS table from HCM;

	Level of Service (LOS)				
	A	B	C	D	E
Density (veh/mile)	≤ 11	> 11	> 18	> 26	> 35

HCM – Highway Capacity Manual

A) LOS A B) LOS B C) LOS C D) LOS D

Problem 3.17) Station at point of intersection (PI) is given to be 112 +70 and the degree of curve (arc method) is given to be 2.3^0. Find the station at PT. (Angle of intersection is 43^0 13' 12").

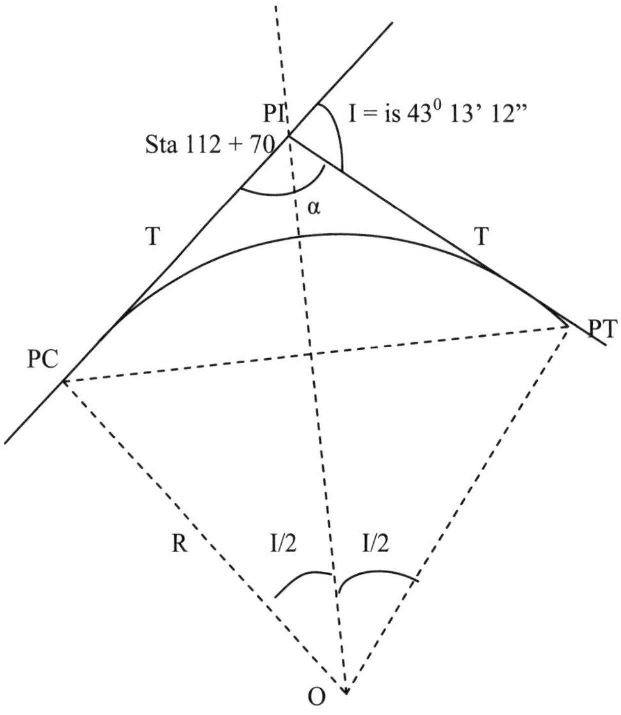

A) 121+ 62

B) 134+75

C) 220+67

D) 234+81

Problem 3.18): Sag vertical curve of a roadway is shown in the figure. Find the elevation at point P.

Station of BVC = 105 + 45,
Elevation of BVC = 124.56 ft
Station of PVI = 109 + 34
Station at point P = 113 + 08

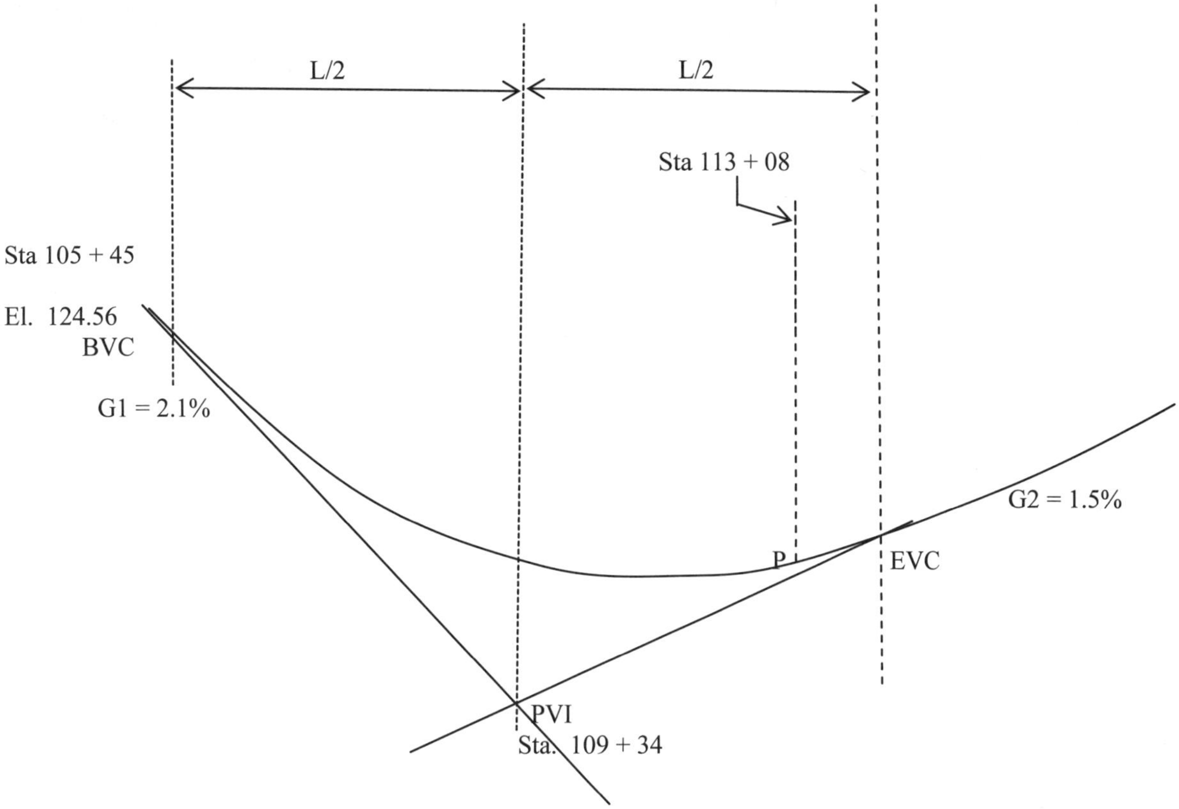

A) 124.5 ft B) 120.7 ft C) 122.01 ft D) 108.9 ft

Use the data given in 3.19 to solve 3.20.

Problem 3.19) Engineers are designing a horizontal curve with a superelevation. Following information is provided regarding the highway.
Number of lanes = 3,
Width of one lane = 12 ft
Lane adjustment factor = 0.67
Superelevation = 4%
Relative gradient = 0.5%
Find the runoff length required prior to achieving the full superelevation.
A) 123 ft B) 231 ft C) 193 ft D) 351 ft

Problem 3.20): Find the tangent runout for the horizontal curve given in the previous problem.
Normal cross slope = 2%
Design superlevation = 4%.

A) 102.3 ft B) 96.5 ft C) 234.8 ft D) 78.9 ft

Problem 3.21) A portion of surveyor's log book is shown below. What is the elevation of point D

Location	**Elevation**	**Backsight Reading**	**Line of sight El.**	**Foresight Reading**
Benchmark X	201.67	4.56		
Point A				3.89
Point B				4.34
Point C		6.89		8.45
Point D				4.21

A) 200.46 B) 106.05 C) 103.56 D) 199.87

Problem 3.22): Bearing of a line is given to be S 25^0 14' 20"W. Find the azimuth of the line.
A) 205^0 14' 20" B) 125^0 14' 13" C) 275^0 14' 33" D) 114^0 24' 33"

Problem 3.23): Following successor relationships are available for a project. Draw an activity on node diagram and find the early start of activity E?

Activity ID	**Duration (days)**	**Successors**
A	6	B
B	12	C, D, E
C	6	D,F
D	12	E
E	11	F
F	4	

A) 22 days B) 26 days C) 36 days D) 40 days

Problem 3.24: Shallow foundation of the building shown is 3 ft wide and 18 in high. Find the square footage of formwork needed.

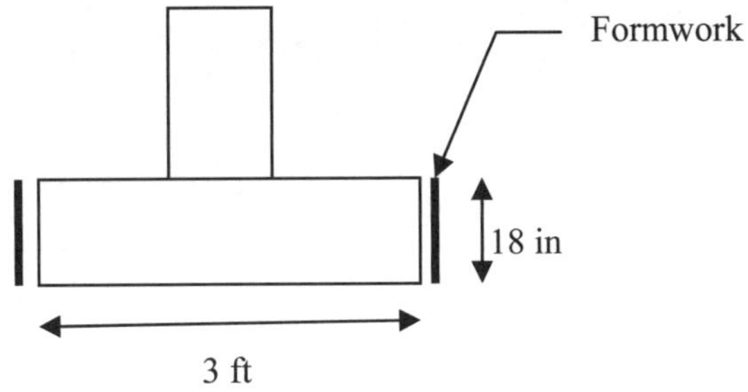

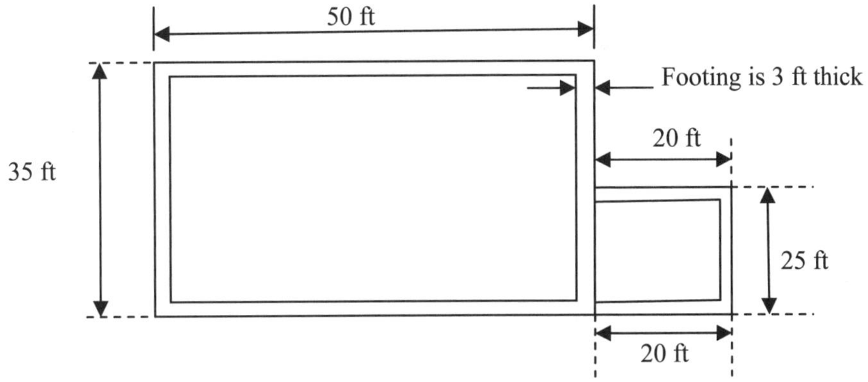

All dimensions are measured from outside.

A) 686 sq. ft B) 651 sq. ft C) 532 sq. ft D) 444 sq. ft

Problem 3.25): A company has 20 painters, 15 metal workers and 40 laborers. Painters and metal workers work 40 hours a week and laborers work 50 hours a week. All workers worked 50 weeks in a given year. There were nine injuries reported by painters, 2 by metal workers and 13 by laborers. What is the incidence rate?

A) 32.4 B) 28.2 C) 86.6 D) 12.3

Problem 3.26) Following information was obtained from a traverse survey. Find the latitude of CA.

Leg	Latitude (ft)	Departure (ft)
AB	20	25
BC	-15	30

Note: Positive latitude is measured from South to North and positive departure is measured from West to East.

A) 5 B) -5 C) -20 D) 20

Problem 3.27): Activity on node network is shown below. Durations are as shown. Find the late start times of activities C and D.

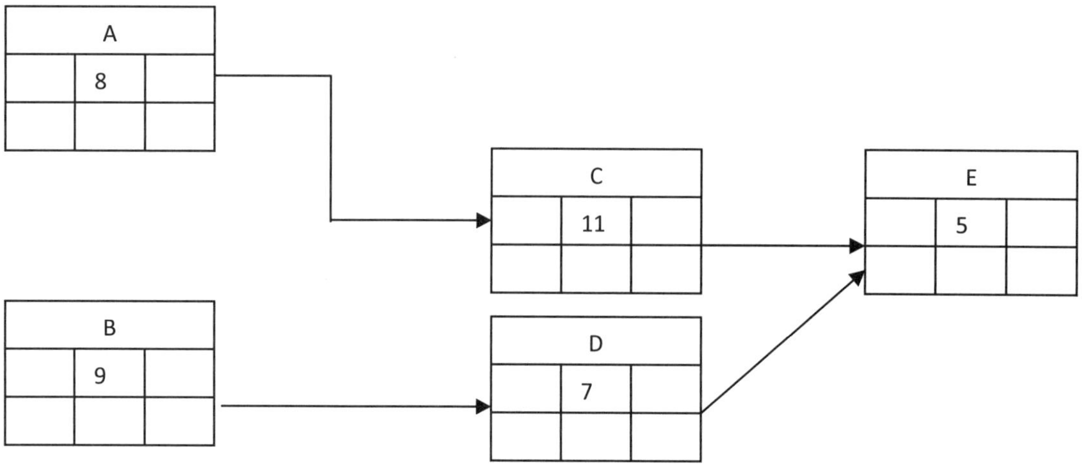

A) LS of activity C = 11, LS of activity D = 15
B) LS of activity C = 9, LS of activity D = 13
C) LS of activity C = 9, LS of activity D = 12
D) LS of activity C = 10, LS of activity D = 12

Key:

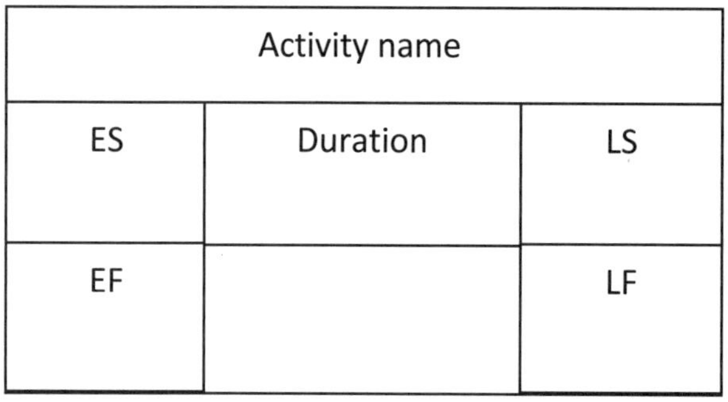

Problem 3.28): Find the total float of activities C and D in the previous problem;

A) Total float of a activity C = 1, Total float of a activity D = 3
B) Total float of a activity C = 0, Total float of a activity D = 3
C) Total float of a activity C = 0, Total float of a activity D = 2
D) Total float of a activity C = 1, Total float of a activity D = 1

Problem 3.29): Find the free float of activities C and D of the previous problem;

A) Free float of a activity C = 1, Free float of a activity D = 3
B) Free float of a activity C = 0, Free float of a activity D = 3
C) Free float of a activity C = 0, Free float of a activity D = 2
D) Free float of a activity C = 1, Free float of a activity D = 1

Problem 3.30) A beam is loaded as shown. A distributed load of 60 lbs/ft is applied and a concentrated load of 570 lbs applied at 5 ft from point A. Find the larger shear force at point M.

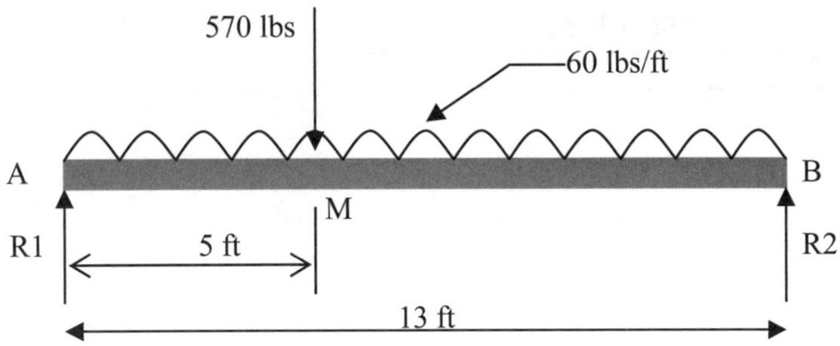

A) 440 lbs B) 590 lbs C) 780 lbs D) 800 lbs

Problem 3.31): Find the maximum bending moment of the beam given in the previous problem.

A) 2440 lbs B) 2954 lbs C) 2780 lbs D) 3800 lbs

Problem 3.32): Find the distance (y) to centroid of the T-beam shown.

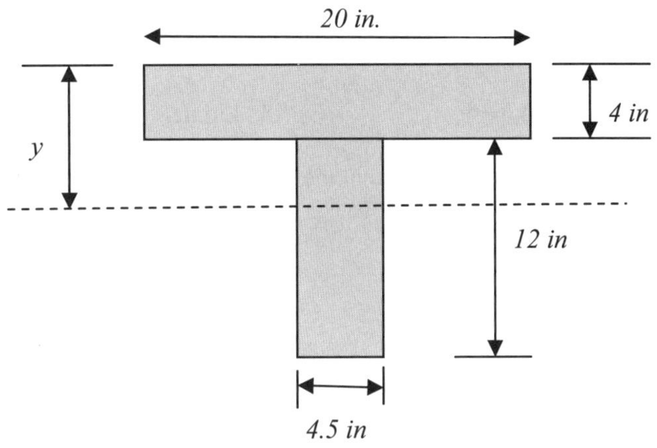

A) 5.22 in B) 3.78 in C) 4.12 in D) 4.81 in

Problem 3.33): Find the nominal moment of the T-beam given.

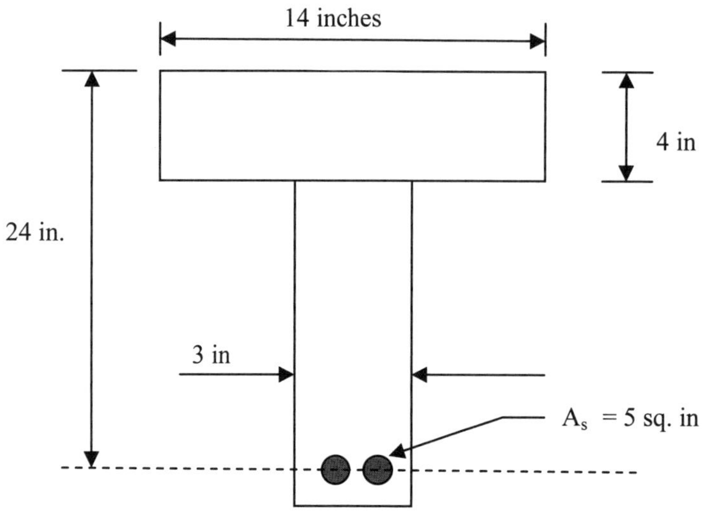

Steel yield stress = 60,000 psi
Concrete compressive strength = 4,000 psi
Steel area = 2 sq. in

A) 483 kip. ft B) 227 kip. ft C) 142 kip. ft D) 331 kip. ft

Problem 3.34): Find the total force in the reinforcements in the beam shown. f_y = 60,000 psi.

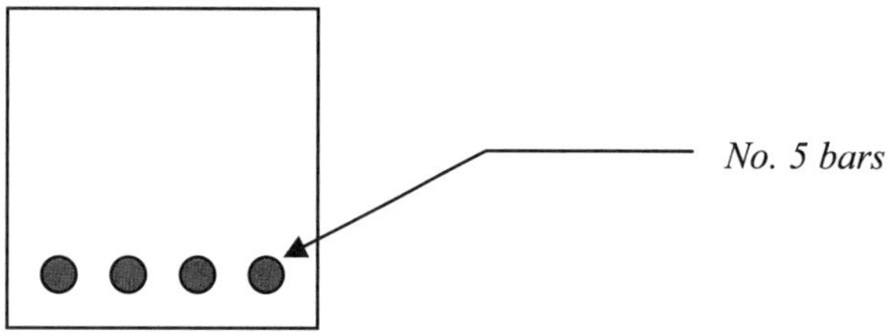

A) 23.8 tons B) 45.7 tons C) 36.8 tons D) 76.9 tons

Problem 3.35): Specific gravity of a soil sample found to be 2.65. Water content of the sample and the degree of saturation was measured to be 0.22 and 0.90 respectively. What is the dry density of the soil sample.?

A) 100.3 lbs/cu.ft B) 121.5 lbs/cu.ft C) 96.9 lbs/cu.ft
D) 116.9 lbs/cu.ft

Problem 3.36): Shelby tubes are used to;
A) Obtain soil samples in clay soils
B) Obtain soil samples in sandy soils
C) Obtain soil samples in silty soils
D) Obtain rock cores

Problem 3.37): Rock core is shown in the figure. What is the RQD of the core if the core length is 5 ft?

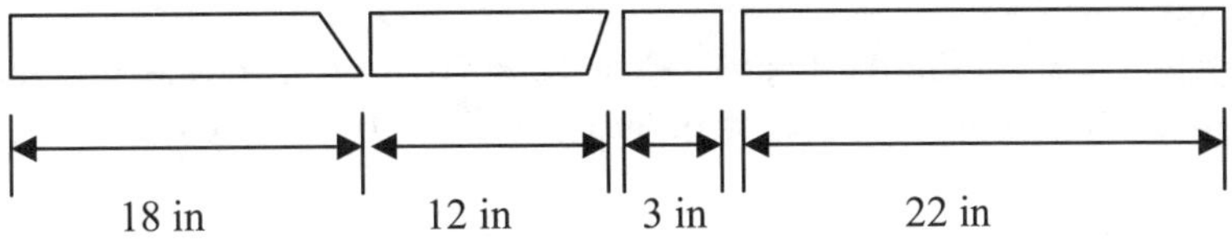

A) 92% B) 87% C) 65% D) 21%

Problem 3.38): Sieve analysis curve is shown below. What is the D_{30} value of the soil.

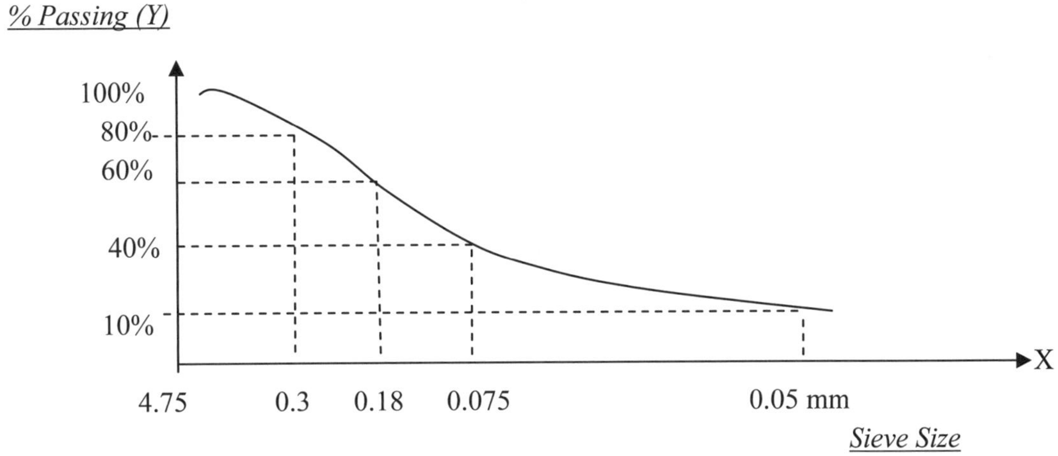

X – axis - Sieve size.
Y - axis - Percent passing.

A) 0.075 mm B) 0.18 mm C) 0.07 mm D) 0.40 mm

Problem 3.39): What is the difference between normally consolidated soils and overconsolidated soils.

A) Normally consolidated soils are saturated and overconsolidated soils are unsaturated.
B) Normally consolidated soils are suitable for geotechnical work while overconsolidated soils are not suitable for geotechnical work.
C) Normally consolidated soils have been subjected to large stresses in the past while overconsolidated soils are not subjected to large stresses in the past.
D) Normally consolidated soils have not been subjected to large stresses in the past while overconsolidated soils have been subjected to large stresses in the past.

Problem 3.40): Road project needs a gravel base. Which gravel do you recommend for the project?

A) Well graded gravel since gravels are of same size and easily compacted.
B) Poorly graded gravel since gravels are of same size and easily compacted.
C) Poorly graded gravel since gravels are of different sizes and easily compacted.
D) Well graded gravel since gravels are of different sizes and easily compacted.

SAMPLE EXAM 3 (SOLUTIONS)

Problem 3.1) Channel cross section is shown in the figure. Manning coefficient (n) of the left section of the channel is 0.13 and Manning coefficient (n) for the right section of the channel is 0.09. If the flow in the channel is 400 cu.ft/sec, what is the slope?

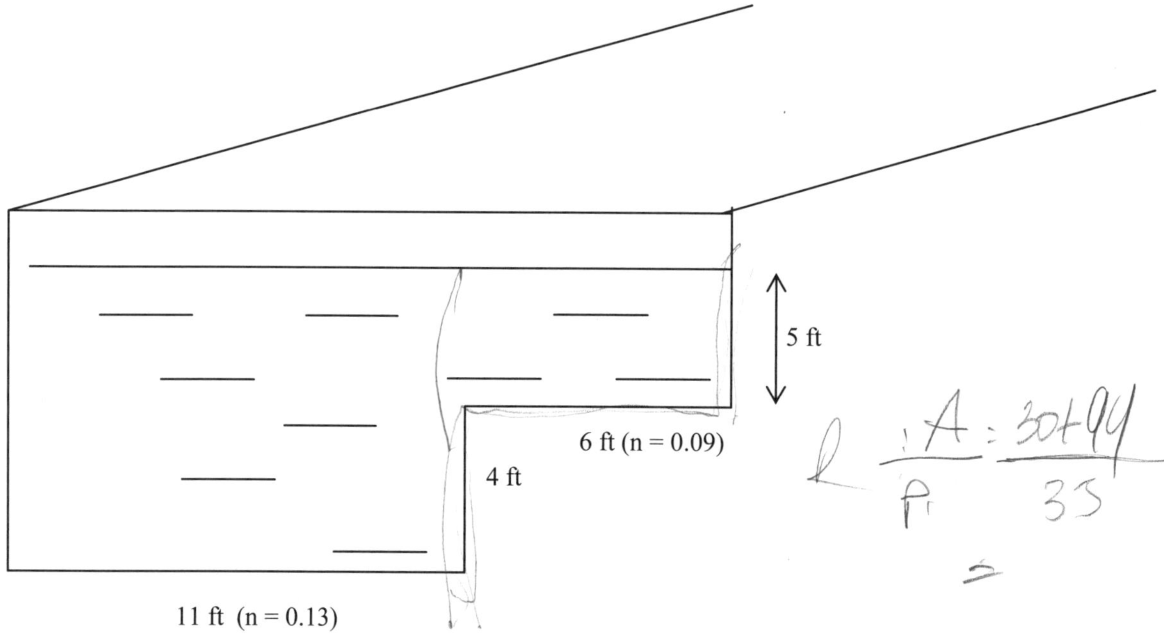

A) 1.0% B) 2.5% C) 3.5% D) 6%

Solution 3.1:

To solve this problem you need to use Manning's formula.

$$Q = 1.49/n.\ A.\ R^{2/3}\ S^{1/2}$$

Q = Discharge or flow;
A = Area of cross section
n = Manning coefficient
R = Hydraulic radius; R = A/P
P = Wetted perimeter;
S = Slope

The problem should be solved in two sections. Divide the channel cross section into two parts. Channel section need to be divided into left hand part and the right hand part.

Left side:

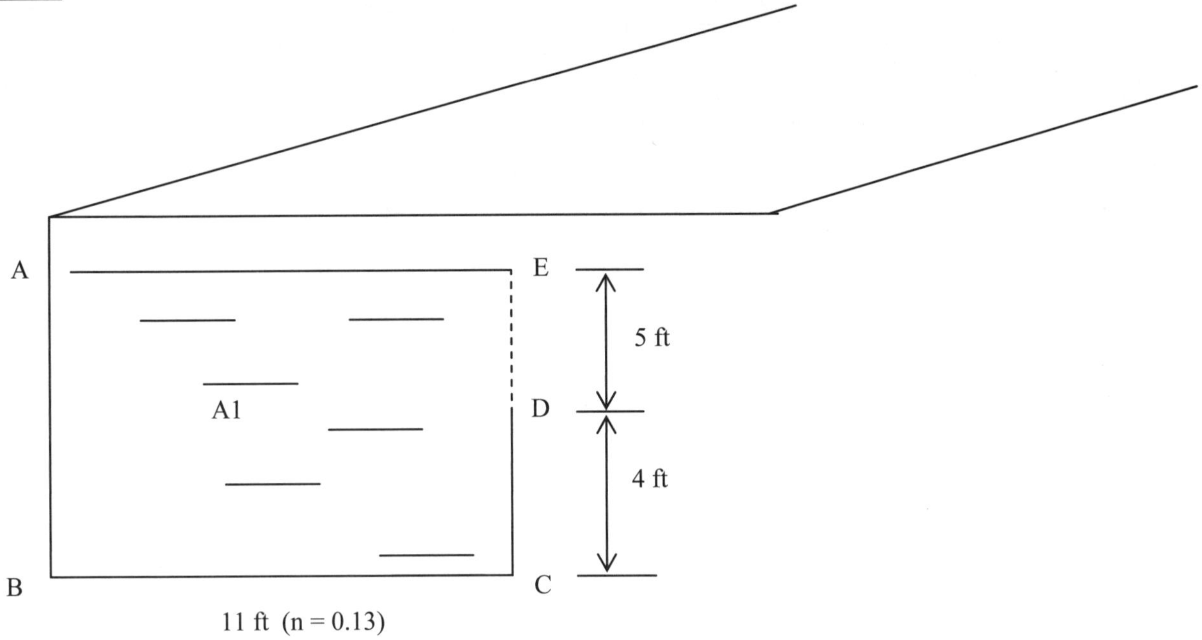

STEP 1: Find area A1

Area (A1) = 9 x 11 = 99 sq. ft

STEP 2: Find the wetted perimeter (P)

Wetted perimeter (P) = AB + BC + CD = 9 + 11 + 4 = 24 ft
Note that DE is not part of the wetted perimeter since it is not a solid boundary.

STEP 3: Find the hydraulic radius (R):

Hydraulic radius (R) = A/P = 99/24 = 4.125

STEP 4: Find the flow (Q1) [Flow on left section of the channel]

Total flow is given to be 400 cu.ft/sec.
Assume flow from left section is Q1 and flow from right section is Q2.

Q1 + Q2 = 400 cu.ft/sec

Flow from left section (Q1) = 1.49/n. A. $R^{2/3}$ $S^{1/2}$
Flow from left section (Q1) = 1.49/0.13 x99 x $4.125^{2/3}$ x $S^{1/2}$ --(1)

Next do the same to the right side of the channel section.

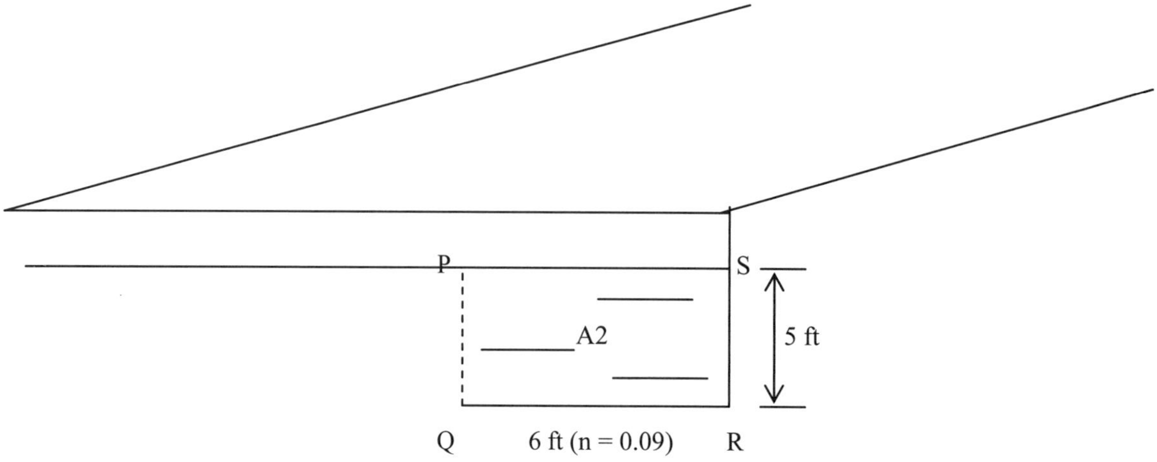

STEP 5: Find the area A2

Area (A2) = 5 x 6 = 30 sq. ft

STEP 6: Find the wetted perimeter (P2)

Wetted perimeter (P2) = QR + RS = 6 + 5 = 11 ft

Note that PQ is not part of the wetted perimeter since that's not a solid boundary.

STEP 7: Find the hydraulic radius (R);

Hydraulic radius (R) = A/P = 30/11 = 2.73

STEP 8: Find the flow (Q2)

Flow from right hand section (Q2) = 1.49/n. A. $R^{2/3}$ $S^{1/2}$

Flow from right hand section (Q2) = 1.49/0.09 x 30 x $2.73^{2/3}$ x $S^{1/2}$ -- (2)

STEP 9: Find the slope;

Total flow = Q1 + Q2 = 400 cu.ft/sec

From eq. 1;

Q1 = 1.49/0.13 x 99 x $4.125^{2/3}$ x $S^{1/2}$

From eq. 2;

Q2 = 1.49/0.09 x 30 x $2.73^{2/3}$ x $S^{1/2}$

Hence;

Q1 + Q2 = 1.49/0.13 x99 x $4.125^{2/3}$ x $S^{1/2}$ + 1.49/0.09 x 30 x $2.73^{2/3}$ x $S^{1/2}$ = 400 cu.ft/sec

$(2,918.5 + 970.1) \times S^{1/2} = 400$ cu.ft/sec

$S = 0.01 = 1.0\%$ (Ans A)

Problem 3.2) Find which section has the highest velocity? All sections have the same Manning roughness coefficient and same slope.

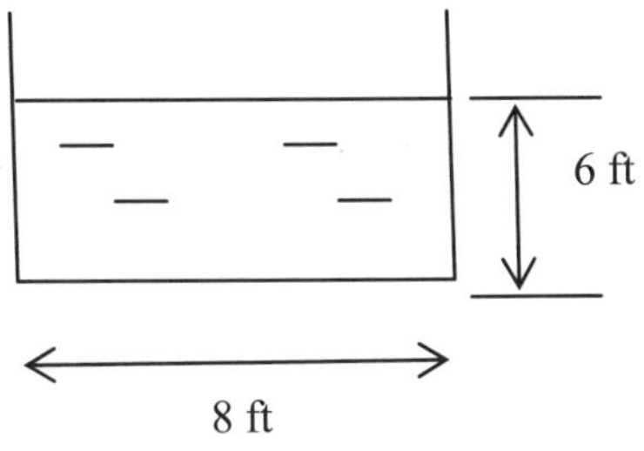

Section A

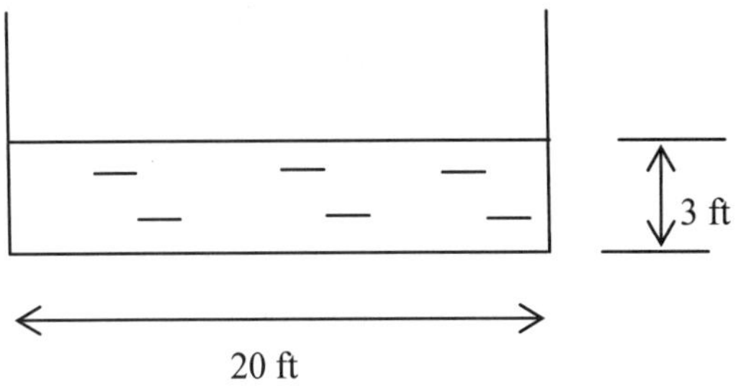

Section B

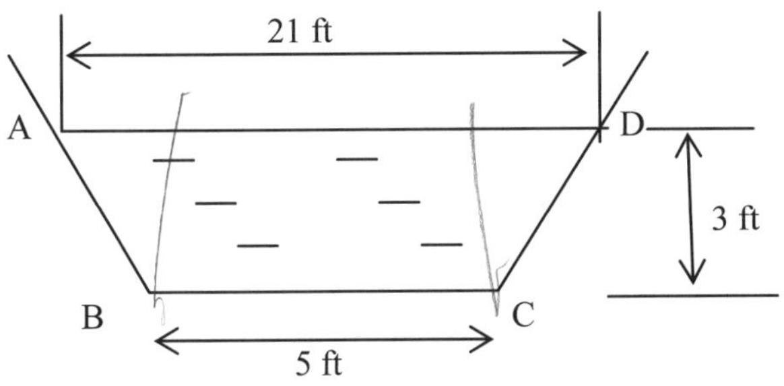

Section C (Symmetrical)

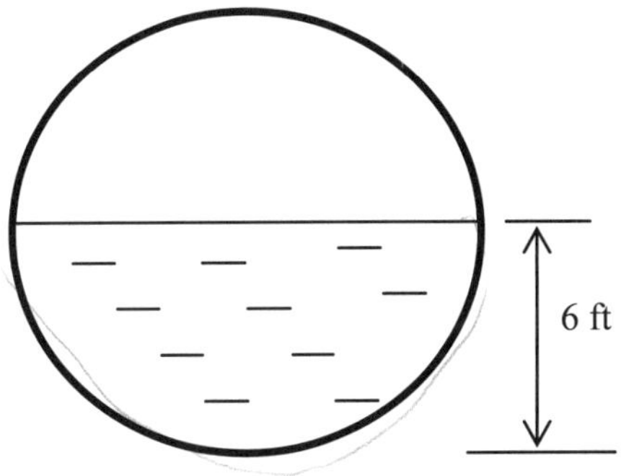

Section D
(Pipe Diameter 12 ft)

A) Section A; B) Section B C) Section C D) Section D

Solution 3.2:

Let's look at the Manning equation;

$$v = 1.49/n.\ R^{2/3}\ S^{1/2}$$

v = Velocity;
n = Manning coefficient
R = Hydraulic radius; R = A/P
A = Area;
P = Wetted perimeter;
S = Slope

Section that has the highest hydraulic radius "R" would have the highest velocity. Hence the section with highest "R" has the highest efficiency.

Hydraulic radius = A/P = Area/Wetted perimeter

Find the hydraulic radius of each section.

Section A: P = 6 + 8 + 6 = 20 ft; A = 6 x 8 = 48
R = A/P = 48/20 = 2.40

Section B: P = 20 + 3 + 3 = 26 ft; A = 3 x 20 = 60
R = A/P = 60/26 = 2.31

Section C: P = AB + BC + CD

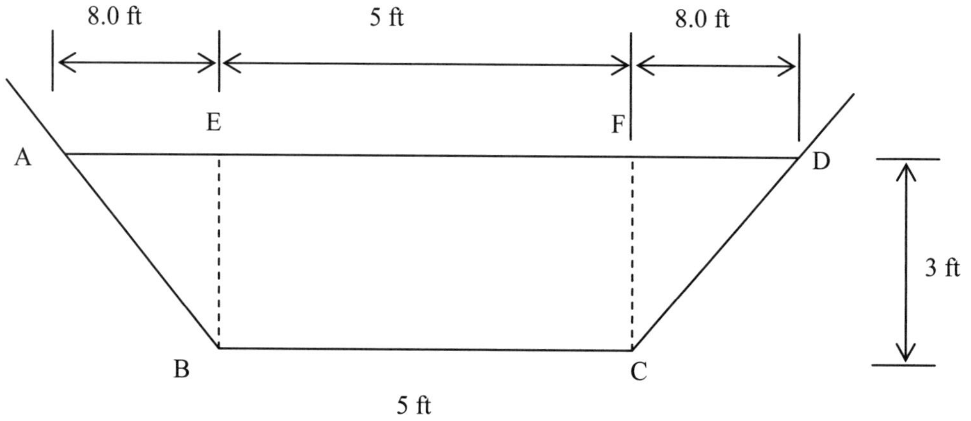

AD is given to be 21 ft.
EF = BC = 5 ft

Hence, AE = (21 – 5)/2 = 8 ft

From Pythagoros theorem;

$AB^2 = BE^2 + AE^2$

$AB^2 = 3^2 + 8^2$
AB = 8.54 ft

P = 8.54 + 5 + 8.54 = 22.1 ft;

$$A = \frac{\text{Base} + \text{Top}}{2} \times \text{Height} = \frac{5 + 21}{2} \times 3 = 39$$

R = A/P = 39/22.1 = 1.76

Section D: Total perimeter area of a circle is given by π x D.

Pipe diameter is 12 ft and the water is filled to a height of 6 ft. Half the pipe is filled with water..

Perimeter of a semi circle = π x D/2 = π x 12/2 = 18.85
Area of the semi circle = ½ x π x $D^2/4$ = 56.55
A/P = 56.55/18.85 = 2.99

Section D has the highest hydraulic radius.
Hence section D has the highest hydraulic efficiency and highest flow. (Ans D)

Finding the Froude Number of a Trapezoidal Channel:

Problem 3.3) Find the Froude number of the trapezoidal channel shown if the flow is 35 cu.ft/sec. Channel width at the bottom is 8 ft and channel depth is 6 ft. Channel slopes are 1 to 2 as shown.

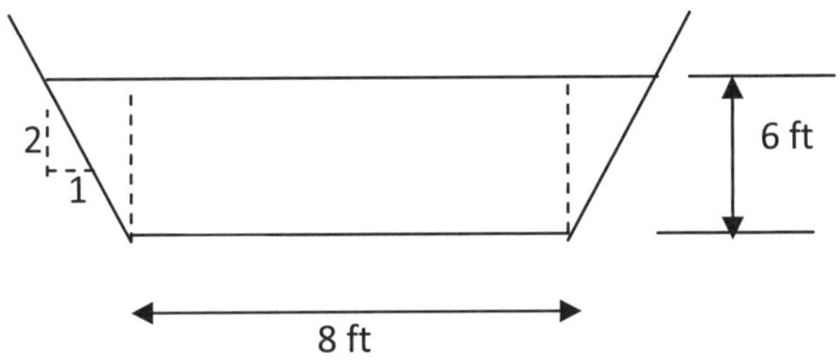

A) 0.034 B) 0.043 C) 0.785 D) 0.676

Solution 3.3):

$$F_R = V/(gD)^{1/2}$$

V = Velocity;
D = Hydraulic depth = A/T,
where A = Area and T = width of the top surface.
For rectangular channels, hydraulic depth becomes equal to regular depth.

g = Gravitational constant (32.2 ft/sec^2 for English units or 9.81 m/sec^2 for metric units).

STEP 1: Find the width of the top surface (T);

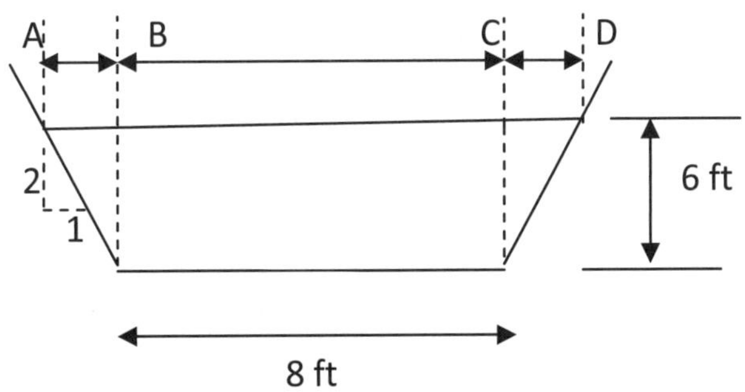

T = Width of the top surface = AD
AD = AB + BC + CD
BC = 8 ft

AB = CD = 6/2 = 3 ft (Since the slope is 1 to 2).
Hence T = 3 + 8 + 3 = 14 ft
D = A/T ;
To find D, you need to find A.

STEP 2: Find the Area (A);
Area of a trapezoid = (Base + Top) x ½ x Height
Area = (8 + 14)/2 x 6 = 66 sq. ft

STEP 3: Find the velocity;

Velocity (V) = Q/Area Q = Flow
Velocity (V) = 35/66 = 0.5303 ft/sec

STEP 2: Find "D"

D = A/T
A = 66 sq. ft
T = Width at water surface = 14 ft
D = 66/14 = 4.714 ft

Hence Froude number (F_R) = $V/(gD)^{1/2}$ = $0.5303/(32.2 \times 4.714)^{1/2}$ = 0.043 (Ans B)

Finding the critical depth of a trapezoidal Channel:

Problem 3.4) Find the critical depth of channel shown if the flow is 35 cu.ft/sec. Channel width at the bottom is 8 ft. Channel slopes are 1 to 2 as shown.

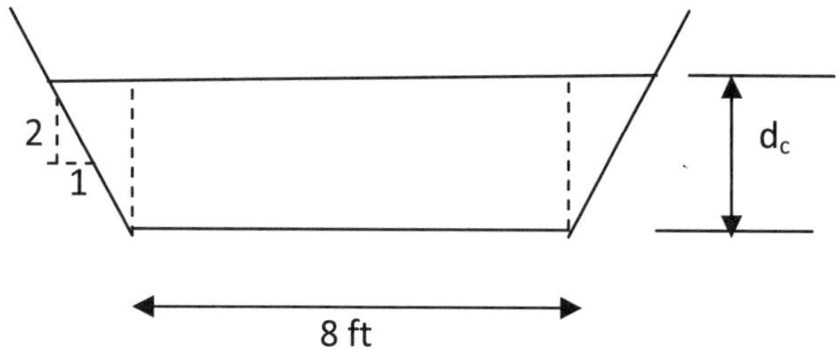

A) 0.034 ft B) 0.827 ft C) 0.785 ft D) 0.676 ft

Solution 3.4): At critical depth, Froude number is equal to 1.0.

$$F_R = V/(gD)^{1/2}$$

V = Velocity;
D = A/T, where A = Area and T = width of the top surface.
g = Gravitational constant (32.2 ft/sec^2 for English units or 9.81 m/sec^2 for metric units).

STEP 1: Find the width of the top surface (T);

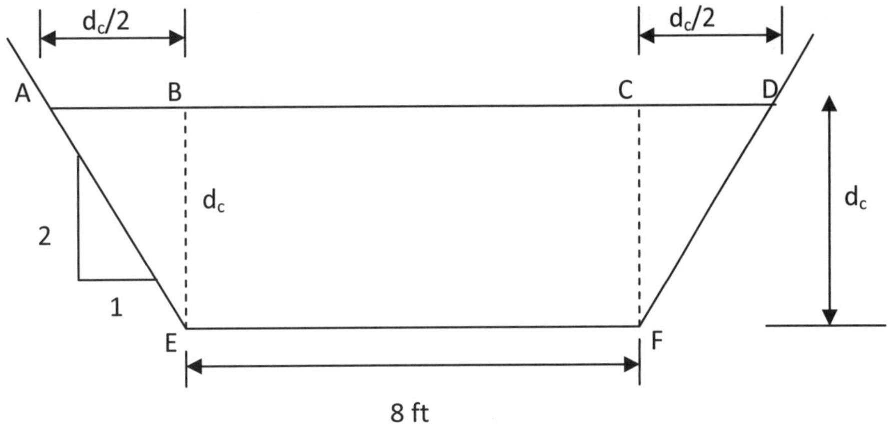

T = Width of the top surface = AD
T = AD = AB + BC + CD
BC = 8 ft
AB = CD = $d_c/2$ (Since the slope is 1 to 2).
T = AB + BC + CD
T = $d_c/2$ + 8 + $d_c/2$ = $(d_c + 8)$ ft

STEP 2: Find the Area (A);

Area of a trapezoid = (Base + Top) x ½ x Height

Area = $(8 + d_c + 8)/2 \times d_c = (16 + d_c)\, d_c/2$ sq. ft

STEP 3: Find "D"

D = A/T
A = $[(16 + d_c)\, d_c/2]$ sq. ft (see step 2)
T = $(d_c + 8)$ ft (see step 1)

$$D = \frac{[(16 + d_c)\, d_c/2]}{(d_c + 8)}$$

STEP 4: Find the velocity;

Velocity (V) = Q/Area Q = Flow
Velocity (V) = $35/[(16 + d_c)\, d_c/2] = 70/[(16 + d_c)\, d_c]$ ft/sec

STEP 5: Find the critical depth

At critical depth, Froude number = 1.0

$$(F_R) = V/(g.D)^{1/2} = 1.0 = \frac{70/[(16 + d_c)\, d_c]}{\{g \times [(16 + d_c)\, d_c/2]/(d_c + 8)\}^{1/2}}$$

Note: There are calculators available to solve such problems. Buying such a calculator and learning the process would save you lots of time in the exam. If not you may have to use trial and error.
$d_c = 0.827$ ft (Ans B)

Problem 3.5) Hydraulic jump in a rectangular channel is shown in the figure. Upstream depth before the jump is 1.1 ft and the velocity is 15 fps. Width of the rectangular channel is 5.5 ft. What can you say of the flow after the jump?

A) Flow is subcritical after the jump
B) Flow is super critical after the jump
C) Flow is critical after the jump
D) Not enough data to answer the question.

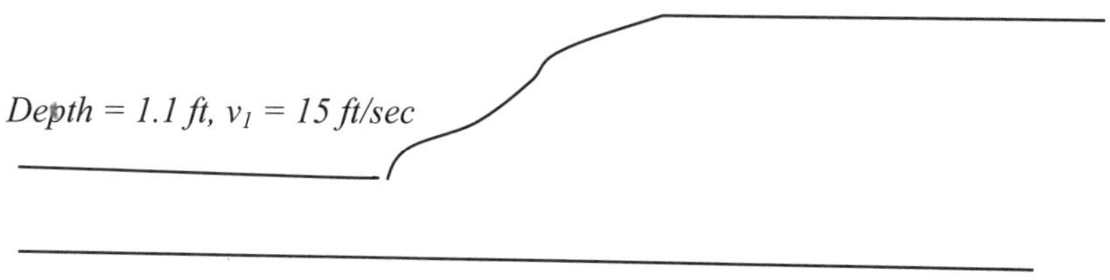

Solution 3.5):

Find the Froude number after the jump.
If (Fr) Froude number is equal to 1.0, the flow is critical.
If Fr < 1.0 then the flow is subcritical.
If Fr > 1.0, then the flow is supercritical.

STEP 1: Find the depth after the jump;

$$y_2/y_1 = ½ [(1 + 8 v_1^2/gy_1)^{1/2} - 1] \text{ --------------------(2)}$$

y_1 = Depth before the jump = 1.1 ft
v_1 = Velocity before the jump = 15 ft/sec
y_2 = Depth after the jump;

$y_2/y_1 = ½ [(1 + 8 v_1^2/gy_1)^{1/2} - 1]$

$y_2/1.1 = ½ [(1 + 8 \text{ x } 15^2/(32.2 \text{ x } 1.1))^{1/2} - 1]$

$y_2 = 3.41$

STEP 2: Find the velocity (v_2) after the jump;

Flow before the jump = v_1 x Area = 15 x Area
= 15 x (5.5 x 1.1)
= 90.75 ft

Flow is same before and after the jump.

Velocity after jump (v_2) = Flow/Area = 90.75/(width x depth)
Depth after the jump = 3.41 ft (see step 1)

(v_2) = 90.75/(5.5 x 3.41) = 4.84 ft/sec

STEP 3: Find the Froude number after the jump;

D_2 = Hydraulic depth after the jump;
$D_2 = A/T = (5.5 \times 3.41)/5.5 = 3.41$

Above 5.5 is the width of the channel and 3.41 is the depth of the channel after the hydraulic jump.

$Fr_2 = v_2/(g.D_2)^{1/2}$

$Fr_2 = 4.84/(32.2 \times 3.41)^{1/2}$
$Fr = 0.46$ (Flow is subcritical). (Ans A)

<u>Discussion</u>:

> If Froude number is equal to 1.0, the flow is critical.
> If $Fr < 1.0$ then the flow is subcritical.
> If $Fr > 1.0$, then the flow is supercritical.

<u>Problem 3.6)</u> Which statement is correct regarding channel flow?
A) Channels with uniform flow have the same depth thru out the channel.
B) Channels with steady flow have the same depth thru out the channel.
C) Channels with uniform flow do not have the same depth thru out the channel.
D) All of the above

<u>Solution 3.6)</u> Answer A:

<u>Uniform Flow</u>: When the flow is uniform, the channel depth is same in the channel at all sections.
<u>Steady Flow</u>: When the channel depth at a given point does not change with time, the flow is known as steady flow. In the case of steady flow, channel depth can change from one location to another.

Above answer "A" is correct. Answer "B" is incorrect since during steady flow, depth can change from place to place.

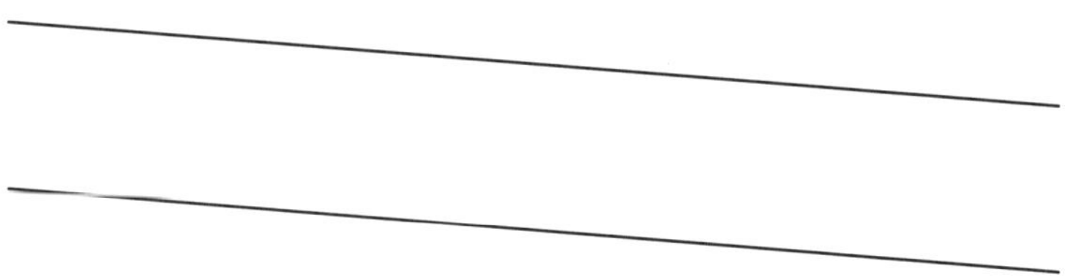

<u>Uniform flow (Depth is same thru out the channel)</u>

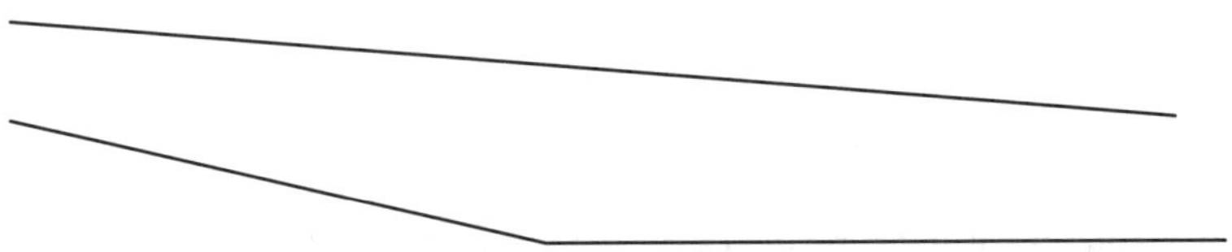

Steady flow (Depth may change from point to point. The depth at a given point will not change with time).

Problem 3.7: 1.2 ft diameter pipe is carrying water at a velocity of 2.1 ft/sec and at a pressure of 12 psi. After travelling 900 ft, elevation of the pipe has dropped by 9 ft. Darcy friction coefficient of the pipe is 0.008. What is the pressure in the pipe at the end of the pipe?

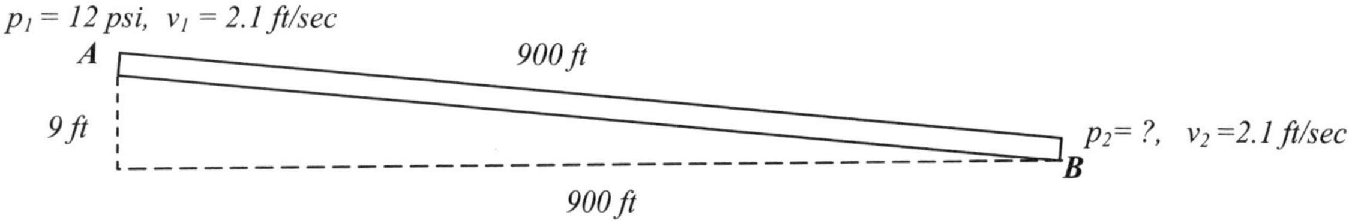

A) 3.4 psi B) 12.3 psi C) 15.7 psi D) 9.8 psi

Solution 3.7):

Darcy - Weisbach Formula: Head loss due to friction in pipes can be computed using either Hazen William formula or Darcy Weishbach formula. We encountered Hazen William formula before.

$$h_L = \frac{f L v^2}{2gD}$$

h_L = Head loss due to friction,
f = Darcy friction coefficient
L = Length of the pipe
D = Diameter of the pipe

In the exam either "C" or "f" would be given. "C" is the Hazen William coefficient. If "C" is given, use the Hazen William formula. If "f" is given use the Darcy formula.

STEP 1: Apply the Bernoulli's equation from A to B.

$h_1 + v_1^2/2g + p_1/\gamma = h_2 + v_2^2/2g + p_2/\gamma + h_L$.------------------------(1)

$h_1 = 9$ ft
$h_2 = 0$ ft
$v_1 = v_2 = 2.1$ ft/sec;
h_L = Head loss due to friction

Velocity at two points has to be equal since flow coming from point A has to come out of point B.

$p_1 = 12$ psi $= 12 \times 144$ psf

From (1) $9 + 2.1^2/(2 \times 32.2) + 12 \times 144/62.4 = 0 + 2.1^2/(2 \times 32.2) + p_2/\gamma + h_L$

$9 + 27.7 = p_2/\gamma + h_L$ --(2)

STEP 2: Find the head loss. Head loss is given by the Darcy equation.

$$h_L = \frac{f L v^2}{2gD}$$

$$h_L = \frac{f L v^2}{2gD} = \frac{0.008 \times 900 \times (2.1)^2}{2 \times 32.2 \times 1.2} = 0.41$$

Hence $9 + 27.7 = p_2/\gamma + 0.41$ $(\gamma = 62.4 \text{ pcf})$

$p_2 = 2264.3$ psf
$p_2 = 2264.3/144 = 15.7$ psi (Ans C)

Discussion:

How to Find the Darcy Friction Coefficient?

In this problem, Darcy friction coefficient is given. Moody diagrams are used to find the Darcy friction coefficient. You need to know the Reynold's number and relative roughness to use the Moody diagrams.

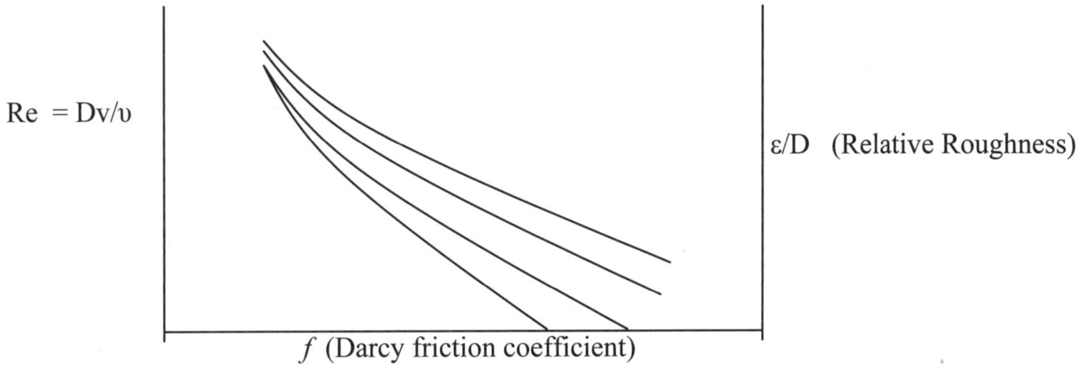

Moody diagrams are not provided in this book but available in many text books.

Using Moody Diagrams:

STEP 1: F Find the Reynold's Number (Re)

Reynold's number is given by $Re = D. v/ \upsilon$

D = Diameter v = velocity of water υ = Kinetic viscosity of the fluid

Usually "υ' is provided in the morning exam.

STEP 2: Find the Relative Roughness: (ε/D)

ε = Roughness of the pipe surface D = Diameter

Roughness (ε) is usually provided in the morning exam.

Once you know the Reynold's number and relative roughness you can find Darcy's friction coefficient using the Moody diagram.

Alaskan Oil Pipe Line;
Alaskan oil pipe line transport 2 million barrels of oil per day. (1 barrel = 42 gallons). Oil is pumped thru 36 pumps located at 12 locations. Nearly 200,000 gallons of oil is needed to operate the pumps. Oil inside the pipeline is at a huge pressure of 1,180 psi. Pipe diameter is 4 ft and length of the pipe is 800 miles.

World famous Alaskan oil pipeline

* As an exercise, find the head loss in the Alaskan pipe line assuming oil to have a friction coefficient of 0.002 and same density as water. Also assume inlet and outflow is at same elevation. (Flow = 2 million barrels per day. Inlet and outlet pressures are the same. Water is flowing due to pumping.)

Problem 3.8: Coefficient of the Pitot tube shown in the figure is 0.9. If the flow velocity is 6 ft/sec, find "y":

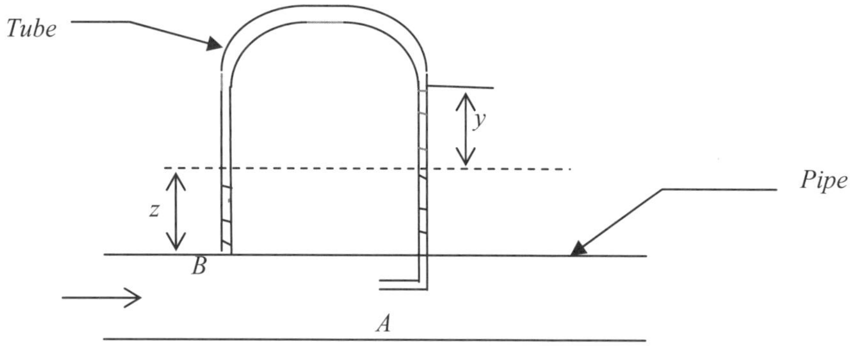

A) 1.3 ft B) 6.7 ft C) 0.69 ft D) 2.1 ft

Solution 3.8:

When water is flowing thru a pipe, tube at point B measures only the pressure head. On the other hand, water is jamming into the tube at point A. Hence tube at point A measures both the pressure head and the velocity head.

Head at point A; $z + y = v^2/2g + p/\gamma$ --------------------(1)

Head at point B; $z = p/\gamma$ ----------------------------------(2)

Equation (1) – (2)

$y = v^2/2g$
$v^2 = 2 \times 32.2\ y$
$v = (64.4\ y)^{1/2}$

Velocity and the flow will be reduced due to energy loss. Pitot tube coefficient (C) is given to be 0.9.

Introduce the Pitot tube coefficient.

$v = 0.9 \times (64.4\ y)^{1/2}$

Velocity of water in the pipe is given to be 6 ft/.sec. (v = 6 ft/sec).

$$6 = 0.9 \times (64.4\ y)^{1/2}$$

Hence y = 0.69 ft (Ans C)

Problem 3.9: A Venturi meter is shown (with a water column) in the figure. Assuming a discharge coefficient of 1.0 find the flow in the pipe.

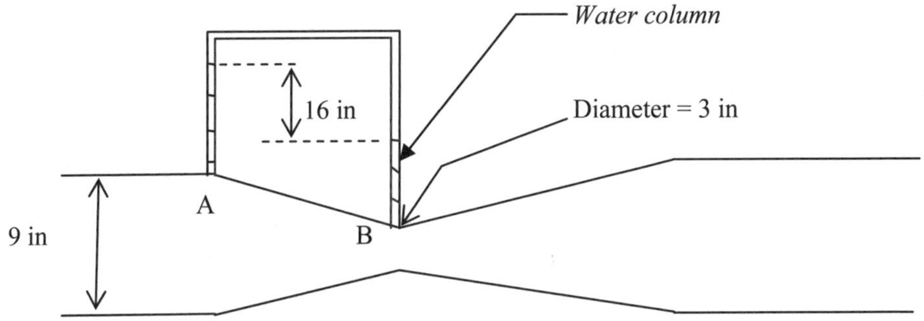

A) 0.01 cfs B) 0.45 cfs C) 0.69 cfs D) 2.1 cfs

Solution 3.9:

Area at point A is larger than area at point B. Hence velocity at point B is larger than velocity at point A. Velocity head is higher at point B compared to point A.

Pressure head at point A is larger than the pressure head at point B.

Pressure head at point A - Pressure head at point B = 16 in

$$P_A/\gamma - P_B/\gamma = 16 \text{ in} = 16/12 \text{ ft} = 1.333 \text{ ft}$$

Apply Bernoulli's equation between points A and B.

$$P_A/\gamma + V_A^2/2g + Z = P_B/\gamma + V_B^2/2g + Z$$

Gravity head (Z) is same at both points.

$$P_A/\gamma - P_B/\gamma = V_B^2/2g - V_A^2/2g$$

$$1.333 = V_B^2/2g - V_A^2/2g \text{ ------------------------------------(1)}$$

Use the continuity equation. Flow (Q) is same at both points.

Flow (Q) = $A_A \times V_A = A_B \times V_B$

A_A = Area at point A = $\pi \times (9/12)^2/4 = 0.442$ sq.ft
A_B = Area at point B = $\pi \times (3/12)^2/4 = 0.049$ sq.ft

$0.442 \times V_A = 0.049 \times V_B$
$V_A = 0.111 \times V_B$

Insert V_A to equation 1.

$1.333 = V_B^2/2g - (0.111)^2 V_B^2/2g$
$V_B = 9.32$ ft/sec

$Q = A_B \times V_B = 0.049 \times 9.32 = 0.457$ cubic ft/sec (Ans B)

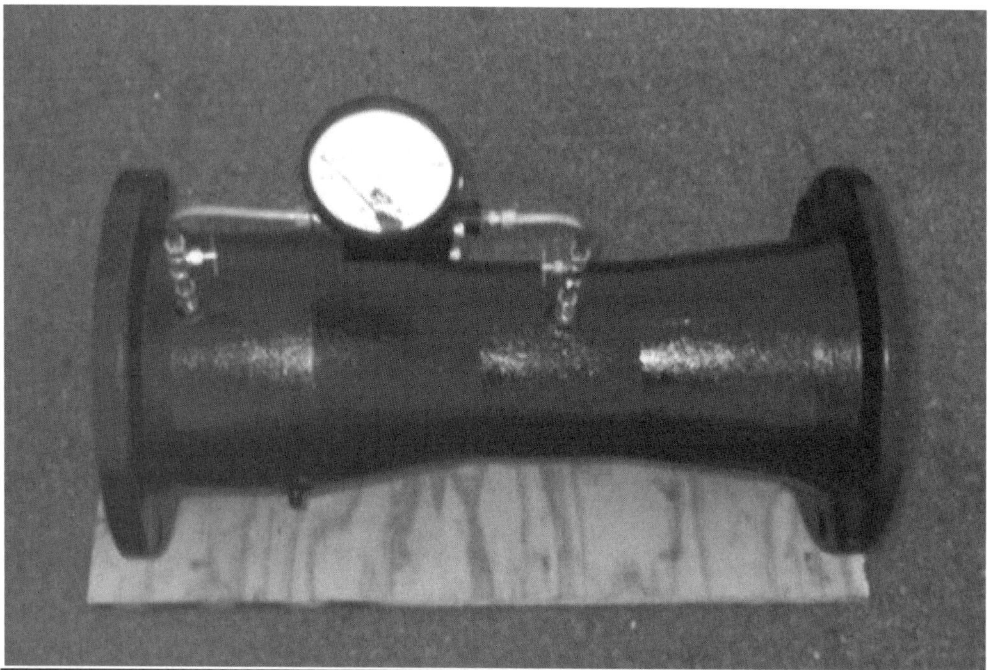

Venturi meter is shown in the figure. Two holes are made at narrow section of the pipe and wider section of the pipe. Two holes are connected with a pipe and pressure gauge is attached. The pressure gauge would measure the pressure difference

Problem 3.10) Two areas are draining into a drain pipe as shown. Information of the two areas is given below.
Area 1 = Area is 9 acres, time of concentration 31 minutes, runoff coefficient 0.51
Area 2 = Area is 3 acres, time of concentration 30 minutes, runoff coefficient 0.40

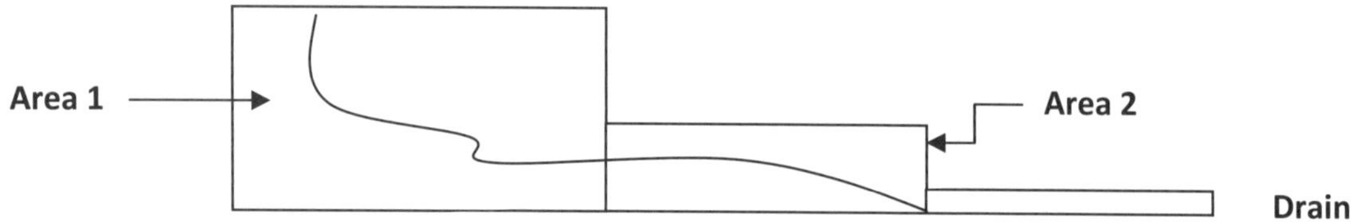

Rainfall intensity for 10 year, 15 year and 30 year rainfalls are given by following graphs.

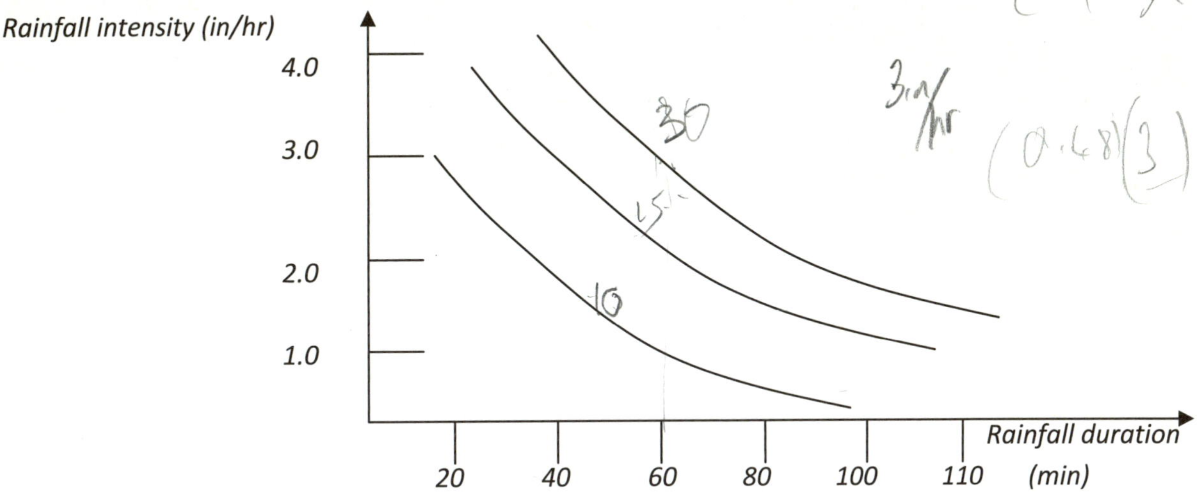

Find the maximum flow in the drain pipe for a 30 year rainfall.

A) 78.9 cu.ft/sec B) 16.84 cu.ft/sec C) 61.3 cu.ft/sec D) 23.5 cu.ft/sec

Solution 3.10): This problem has to be solved using the rational formula. Q = C. I. A.
Q = Flow; C = Runoff coefficient; I = Intensity of rainfall (in/hr); A = Area

You need to find C, I and A to find Q.
Since there are two areas, composite C needs to be found.

STEP 1: Find the composite C

$$C_{composite} = \frac{C_1A_1 + C_2A_2}{A_1 + A_2}$$

$$C_{composite} = \frac{0.51 \times 9 + 0.40 \times 3}{9 + 3} = 0.48$$

STEP 2: Find the relevant rainfall intensity (I) :
Time of concentration is the time that water from most distant point to reach the culvert or the drain pipe. Time of concentration for the area 1 is 31 minutes. Time of concentration for area 2 is 30 minutes.
Total time of concentration = 31 + 30 = 61 minutes.

Since time of concentration is 61 minutes, we should design the drain pipes to a rainfall of 61 minutes.
Three graphs are given between rainfall intensity and rainfall duration. Graphs are not named. 10 year graph should be the lowest and 30 year graph should be the highest.

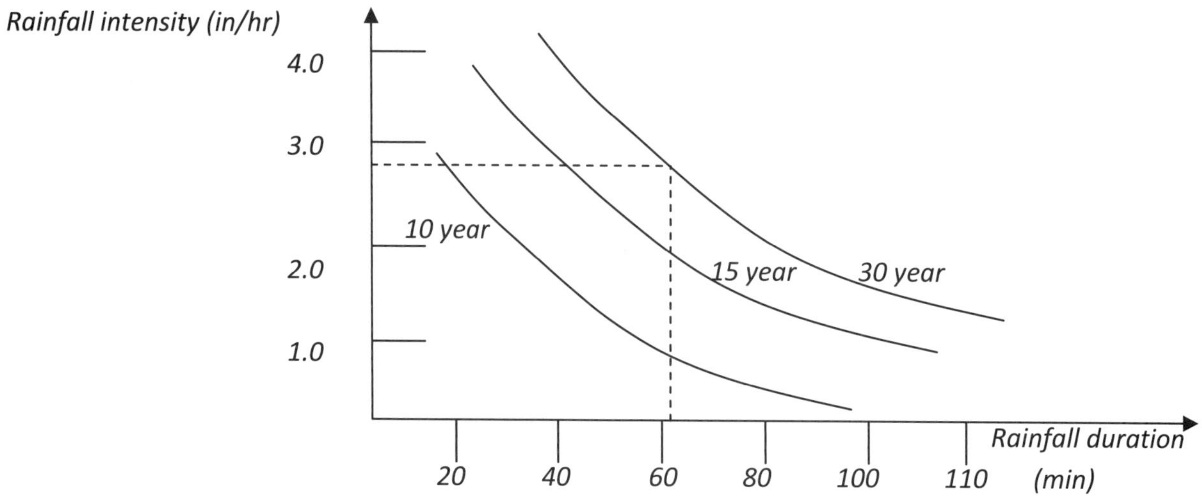

Maximum rainfall intensity (I) for a 30 year rainfall with 61 minute duration is 2.9 in/hr.

STEP 3: Apply the rational formula to obtain the maximum flow (Q):

Q = C.I.A

C = 0.48;
A = 9 + 3 = 12 acres
I = 2.9 in/hr

Q = 0.48 x 2.9 x 12 = 16.7 acre. in/hr

C is dimensionless and I is in in/hr.

The answer needs to be converted to cu.ft/sec.

16.7 acre.in/hr = 16.7/12 acre.ft/hr = 1.39 acre.ft/hr
1 Acre = 43,560 sq.ft
1.39 acre.ft/hr = 1.39 x 43,560 cu.ft/hr
= 1.39 x 43,560/(60 x 60) cu.ft/sec
= 16.84 cu.ft/sec (Ans B)

Problem 3.11) Area 1 and area 2 are discharging into a pipeline as shown. Following information is provided.

	Area	*Runoff coefficient*	*Slope*	*Distance to hydrologically most distant point*
Area 1	23 acres	0.70	8%	D1 = 0.6 miles
Area 2	35 acres	0.55	18%	D2 = 0.4 miles

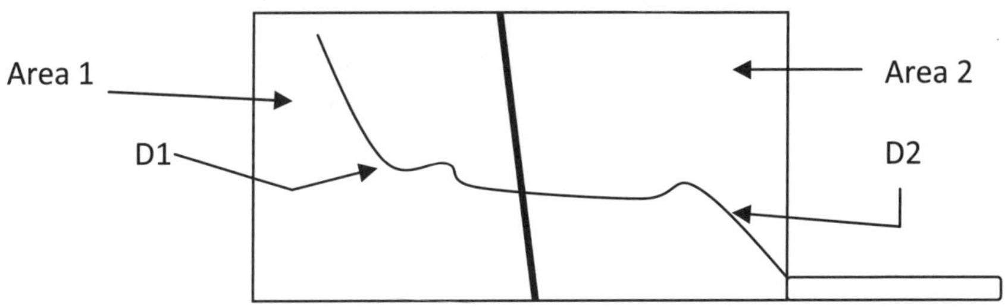

D1 = 0.6 miles
D2 = 0.4 miles

Rainfall data for the region as shown below are for 10 year, 20 year and 100 year rainfalls.

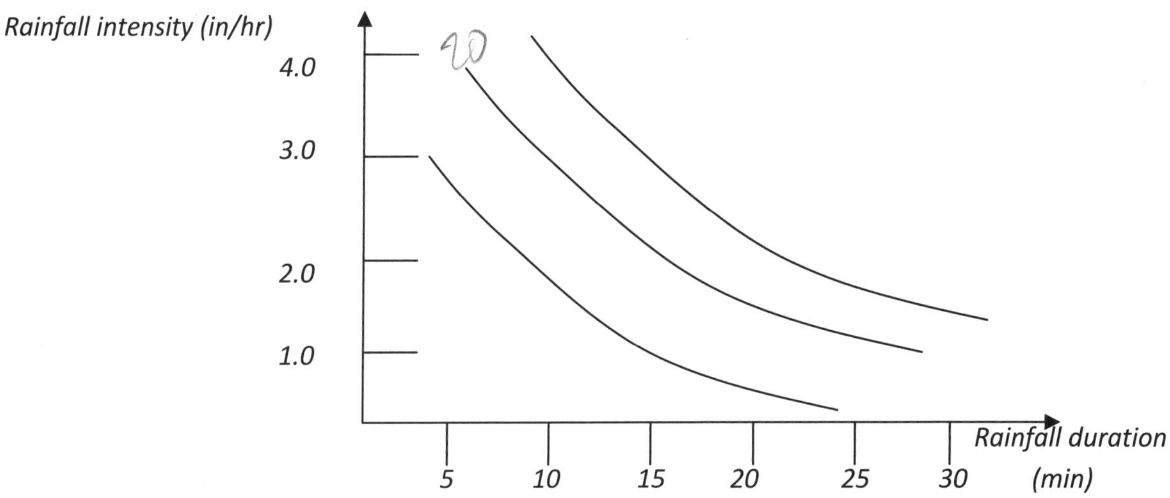

Slope	***Velocity***
5%	*2 ft/sec*
10%	*10 ft/sec*
15%	*13 ft/sec*
20%	*20 ft/sec*

Find the flow in the pipeline for a 20 year rainfall in cu.ft/sec.

A) 78.9 cu.ft/sec B) 71.6 cu.ft/sec C) 141.3 cu.ft/sec D) 110.6 cu.ft/sec

Solution 3.11): This problem has to be solved using the rational formula. Q = C. I. A.
Q = Flow; C = Runoff coefficient; I = Intensity of rainfall (in/hr); A = Area

You need to find C, I and A to find Q.
Since there are two areas, composite C needs to be found.

STEP 1: Find the composite C

$$C_{composite} = \frac{C_1A_1 + C_2A_2}{A_1 + A_2}$$

$$C_{composite} = \frac{0.70 \times 23 + 0.55 \times 35}{23 + 35} = 0.61$$

STEP 2: Find the time of concentration for each area:

Find the time of concentration in Area 1;

Distance of travel (D_1) = 0.6 miles;
Slope = 8%
A table is given between slope and the velocity of travel.

5% slope	→	2 ft/sec
8% slope	→	V_1 ft/sec
10% slope	→	10 ft/sec

We need to find the velocity for 8% slope.

Use interpolation. Assume the velocity of travel in area 1 to be V_1;

$$\frac{V_1 - 2}{8 - 5} = \frac{10 - 2}{10 - 5}$$

V_1 = 6.8 ft/sec
If velocity is 6.8 ft/sec, we can find the time of concentration in area 1.

Velocity = Distance/Time
6.8 ft/sec = D_1/time
D_1 is given to be 0.6 miles = 0.6 x 5,280 ft = 3,168 ft
6.8ft/sec = 3,168/time

Time of concentration in area 1 = 3,168/6.8 = 465.9 seconds = 7.76 minutes

Find the time of concentration in Area 2;

Distance of travel (D_2) = 0.4 miles;
Slope = 18%

A table is given between slope and the velocity of travel.

15% slope → 13 ft/sec
18% slope → V_2 ft/sec
20% slope → 20 ft/sec

We need to find the velocity for 18% slope.

Use interpolation. Assume the velocity of travel in area 2 to be V_2;

$$\frac{V_2 - 13}{18 - 15} = \frac{20 - 13}{20 - 15}$$

$V_2 = 17.2$ ft/sec

If velocity is 17.2 ft/sec, we can find the time of concentration in area 2.

Velocity = Distance/Time

17.2 ft/sec = D_2/time

D_2 is given to be 0.4 miles = 0.4 x 5,280 ft = 2,112 ft

17.2 = 2,112/time;

Time = 122.8 seconds

t_2 = 2.05 minutes

Total time of concentration = 7.76 + 2.05 = 9.81 minutes

STEP 3: Find the relevant rainfall intensity (I) :

Time of concentration is the time that water from most distant point to reach the culvert or the drain pipe. Time of concentration for the area 1 is 7.76 minutes. Time of concentration for area 2 is 2.05 minutes.
Total time of concentration is 9.81 minutes.

We need to find the 20 year rainfall with duration of 9.81 minutes.

Rainfall data for the region as shown below are for 10 year, 20 year and 100 year rainfalls. Graphs are not named. 10 year rainfall graph should be the lowest and 100 year rainfall graph should be the highest.

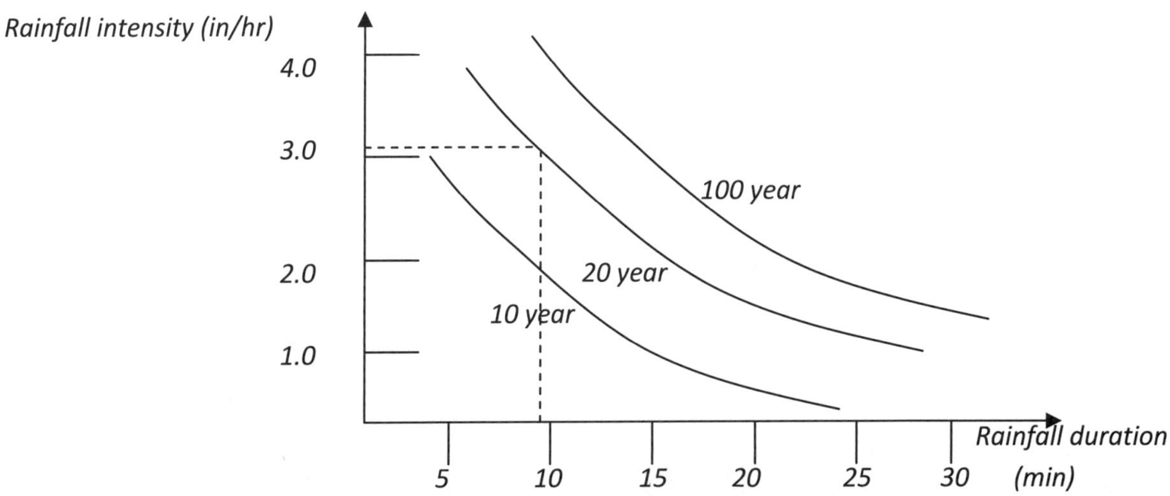

Maximum intensity for 20 year rainfall with duration 9.81 minutes = 3.1 in/hr

STEP 4: Apply the rational formula to obtain the maximum flow (Q):

Q = C.I.A

C = 0.61
I = 3.1 in/hr
A = 23 + 35 = 58 acres
Q = 0.61 x 3.1 x 58 = 109.7 acre. in/hr
C is dimensionless and I is in in/hr. A (Area) is in acres.

The answer needs to be converted to cu.ft/sec.
Q = 109.7 acre.in/hr = 109.7/12 acre.ft/hr = 9.14 acre.ft/hr
1 Acre = 43,560 sq.ft
Q = 9.14 acre.ft/hr = 9.14 x 43,560 cu.ft/hr
Q = 9.14 x 43,560/(60 x 60) cu.ft/sec
Q = 110.6 cu.ft/sec (Ans D)

WASTEWATER:

Problem 3.12) Wastewater engineer designed a circular primary sedimentation tank with a detention time of 2.5 hrs. Height of the tank is 5.6 ft. What is the diameter of the tank if the flow is 1.2 MGD?

A) 12.3 ft B) 98.7 ft C) 61.6 ft D) 24.8 ft

Solution 3.12):

Detention time = Volume of the tank/Flow

Volume of the tank = π x D^2/4 x h = π x D^2/4 x 5.6 = 4.4D^2 cu.ft
Flow = 1.2 MGD
MGD = Million gallons per day
Flow = 1.2 x 10^6 gallons per day = 1.2 x 10^6/7.48 cu.ft per day = 160,427.8 cu.ft/day = 6,684 cu.ft/hr

Detention time = Volume of the tank/Flow
2.5 = 4.4D^2/6684
D = 61.6 ft Ans (C)

Discussion: Wastewater arrives from bottom and move to the top. During the process, sludge would form at the bottom and scum would form on top.

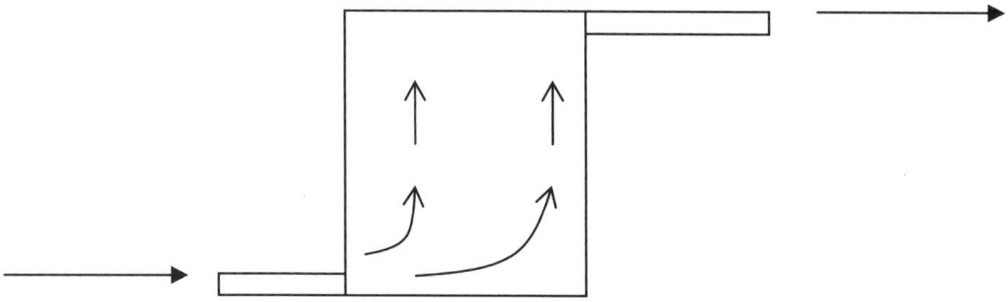

TRANSPORTATION:

Problem 3.13) Certain highway traffic consists of 8% truck traffic and 4% recreational vehicles. The terrain is considered to be rolling terrain. What is the heavy vehicle factor (f_{HV})?

Factor	Level	Type of Terrain Rolling	Mountainous
E_T (Trucks and busses)	1.5	2.5	4.5
E_R (Recreational vehicles)	1.2	2.0	4.0

HCM – Highway Capacity Manual

A) 0.764 B) 0.862 C) 0.145 D) 0.634

Solution 3.13:

One heavy vehicle creates more congestion than one passenger car. Hence heavy vehicle factor is introduced to account for extra congestion caused by heavy vehicles. Recreational vehicles create more congestion than trucks. Hence they are separated from trucks.

Heavy vehicle factor (f_{HV}) is given by following equation.

$$f_{HV} = \frac{1}{1 + P_T (E_T - 1) + P_R (E_R - 1)}.$$

HCM Equation 21-4.

f_{HV} = Heavy vehicle factor ;
P_T = Fraction of trucks and buses;
P_R = Fraction of recreational vehicles. Recreational vehicles such as mobile homes create more congestion than trucks.
E_T = Equivalent truck factor;
E_R = Equivalent recreational vehicle factor
E_T and E_R are obtained from Exhibit 21-8 of HCM

Factor	Level	Type of Terrain Rolling	Mountainous
E_T (Trucks and busses)	1.5	2.5	4.5
E_R (Recreational vehicles)	1.2	2.0	4.0

For rolling terrain; E_T = 2.5 and E_R = 2.0.
P_T = 0.08; and P_R = 0.04 (Given in the problem)

$$f_{HV} = \frac{1}{1 + 0.08\,(2.5 - 1) + 0.04\,(2 - 1)} = 0.862$$ (Ans B)

Problem 3.14) Traffic analyst have found that trips generated per household per day is to be 0.8 + 1.5P + 0.7A
P = Number of persons per household
A = Number of automobiles per household

If the area under research includes 800 households and 3.5 persons per household and 1.8 autos per household what is the number of trips generated per day.

A) 200 B) 10,065 C) 5,848 D) 4002

Solution 3.14): Find the number of trips generated per household and multiply by the number of households.

STEP 1: Find the number of trips generated per household
Number of trips generated per household = 0.8 + 1.5P + 0.7A

P = Number of persons per household = 3.5
A = Number of automobiles per household = 1.8
Number of trips generated per household = 0.8 + 1.5 x 3.5 + 0.7 x 1.8 = 7.31
Number of households = 800
Number of trips generated per day = 800 x 7.31 = 5,848 trips (Ans C)

Problem 3.15) Four lane highway has an AADT of 120,000 and a k factor of 0.9 and a directional factor of 0.8. If the rate of increase of traffic per year is 6%, what is the DDHV per lane, 5 years from now.

A) 121,345 B) 115,623 C) 45,908 D) 234,900

Solution 3.15):
K factor is defined as follows.

K = DHV/AADT

DHV = Design Hourly Volume
DDHV = Directional design hourly volume.

DDHV = D x DHV
D = Directional factor

DHV = K. AADT = 0.9 x 120,000 = 108,000
DDHV = D. DHV = 0.8 x 108,000 = 86,400

This is the DDHV at the present time.

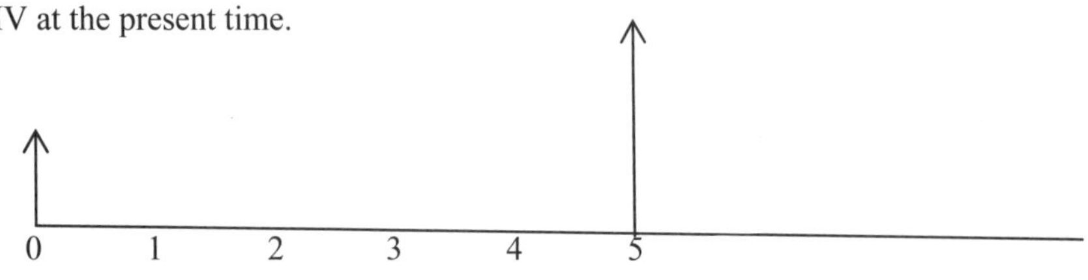

Find future value. (Present value is given).

From interest tables, $n = 5$, $i = 6\%$, $F/P = 1.3382$

Note that interest tables are not provided in this book but readily available in other books.

Present DDHV = 86,400

DDHV five years from now = 1.3382 x 86,400 = 115,623 Ans (B)

Instead of using interest tables, you can also use equations..

$F = P \times (1 + i)^n$

F = Future value; P = Present value; n = Number of years; i = Interest rate

$F = 86{,}400 \times (1 + 0.06)^5 = 115{,}623$

Problem 3.16) Design traffic volume in a highway is 1,400 vehicles per hour per lane. It is reported that this highway has 3% truck traffic and 2% RVs (Recreational vehicles). The road is in level ground. PHF (Peak hour factor) is 0.85. The traffic is mostly office commuters.

E_T = Truck equivalent factor = 1.5 for level ground

E_R = RV equivalent factor = 1.2 for level ground

(E_T and E_R obtained from table 21-8 (HCM)

f_P = Driver population factor = 1.0

If the free flow speed is 65 mph, what is the level of service of the road?

Density vs LOS table from HCM;

	Level of Service (LOS)				
	A	B	C	D	E
Density (veh/mile)	≤ 11	> 11	> 18	> 26	> 35

HCM – Highway Capacity Manual

A) LOS A B) LOS B C) LOS C D) LOS D

Solution 3.16):

Passenger car equivalent service flow rate is given by following equation. (Equation 21-3 in HCM)

$$v_p = \frac{V}{PHF \times N \times f_{HV} \times f_P}$$

v_p = 15 min passenger car equivalent flow rate (pc/h/ln)

V = Hourly volume (veh/hr); This includes all vehicles including trucks, busses and RVs.

PHF = Peak hour factor;

N = Number of lanes

f_{HV} = Heavy vehicle factor (Heavy vehicle factor converts all vehicles to passenger car equivalent)
f_P= Driver population factor; This factor ranges from 0.85 to 1.0. For weekday office commuters this factor is 1.0. Office commuters are assumed to know where they are going and familiar with the road. During weekend, there would be more drivers who are driving for leisure and lower factor such as 0.85 should be used.

STEP 1: Find f_{HV};

$$f_{HV} = \frac{1}{1 + P_T (E_T - 1) + P_R (E_R - 1)}$$

$$f_{HV} = \frac{1}{1 + 0.03 (1.5 - 1) + 0.02 (1.2 - 1)} = 0.981$$

P_T = Fraction of trucks and busses = 0.03
P_R = Fraction of recreational vehicles = 0.02
E_T = Truck equivalent factor = 1.5 for level ground
E_R = RV equivalent factor (obtained from table 21-8 (HCM) = 1.2 for level ground

STEP 2: Find v_p;
PHF = 0.85; f_P = 1.0 (Weekday commuters)

$$v_p = \frac{V}{PHF \times N \times f_{HV} \times f_P} = \frac{1400}{0.85 \times 1 \times 0.981 \times 1.0} = 1{,}679 \text{ pc/h/ln}$$

Note that V is given to be 1,400 vehicles/hr/ln. Hence number of lanes is 1.0.

STEP 3: Speed x Density = Flow

Density = Flow/Speed
Density = 1,679/65 = 25.8

Use the level of service vs. density table given by HCM.

25.8 is less than 26. Hence level of service is C. (Ans C)

LOS A: If the density is equal to 11 or less than 11
LOS B: If the density is greater than 11 and less than or equal to 18
LOS C: If the density is greater than 18 and less than or equal to 26
LOS D: If the density is greater than 26 and less than or equal to 35
LOS E: If the density is greater than 35

Problem 3.17) Station at point of intersection (PI) is given to be 112 +70 and the degree of curve (arc method) is given to be 2.3^0. Find the station at PT. (Angle of intersection is 43^0 13' 12").

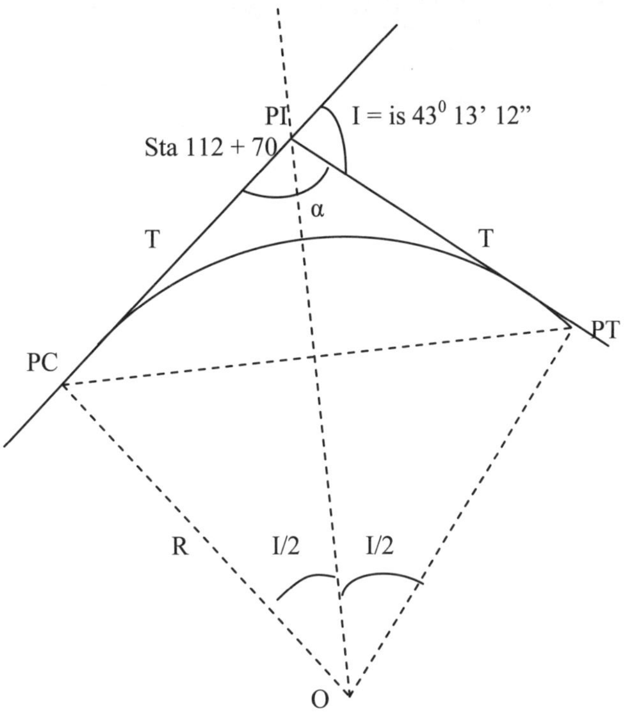

A) 121+ 62 B) 134+75 C) 220+67 D) 234+81

Solution 3.17): In this problem, station at PT is required.
Station to PT is always measured **along** the curve from PC.
Never measure the station of PT from PI.
As you could see radius of the circle is not given. Instead of giving the radius, degree of curve is provided.
There are two methods in highway surveying.

- Degree of curve (arc method)
- Degree of curve (chord method)

Degree of curve (arc method): In this method, angle subtended at the center of a circle with an arc of 100 ft is known as degree of curve (D).

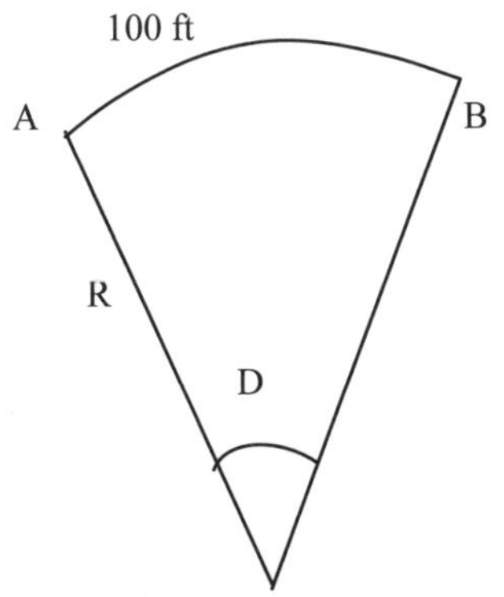

Arc Method (Arc AB = 100 ft)

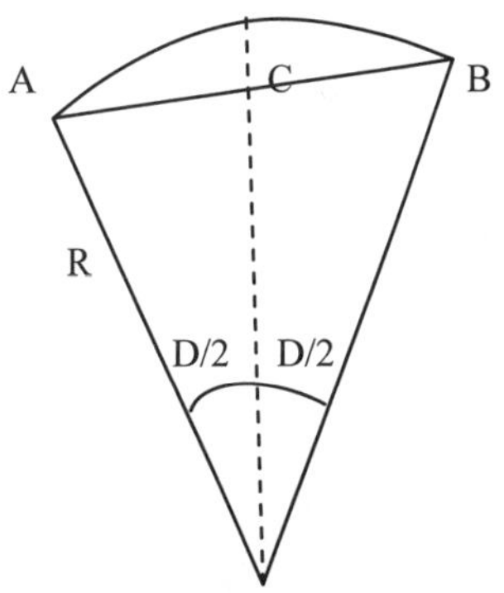

Chord Method (Chord AB = 100 ft)

Arc Method;

As per definition of degree of curve, D degrees will produce an arc of 100 ft.

Full circle is 360 degrees. Full circle produces $2\pi R$.

360 degrees ----→ $2\pi R$

1 degree --------→ $2\pi R/360$

D degrees ------→ $2\pi R.D/360$

As per definition, D degrees would produce an arc of 100 ft.
D degrees ------→ $2\pi R.D/360 = 100$ ft

$$D = \frac{100 \times 360}{2\pi R} = \frac{5{,}729.5}{R}$$

$$R = \frac{5{,}729.5}{D} \quad \text{------------------------------------(1)}$$

If "D" degrees is known, R can be computed.

Chord Method:

Chord AB (Straight line) = 100 ft

AC = R sin D/2
AB = 2 AC = 2 R Sin D/2

In chord method 100 ft chord is used.
Hence

$$100 = 2 R \sin D/2$$

$$R = 50/\sin D/2 \quad \text{--(2)}$$

Use equation 1 for arc method and equation 2 for chord method to find the radius of a horizontal curve.

Note: Instead of giving the radius of a circle, the problem may give the degree of curve. Use above equations 1 and 2 to find the radius.

In this problem, degree of curve (D) is given to be 2.3^0. Note that the degree of curve is in arc method.

$$2.3 = \frac{5{,}729.5}{R}$$

R = 2,491 ft

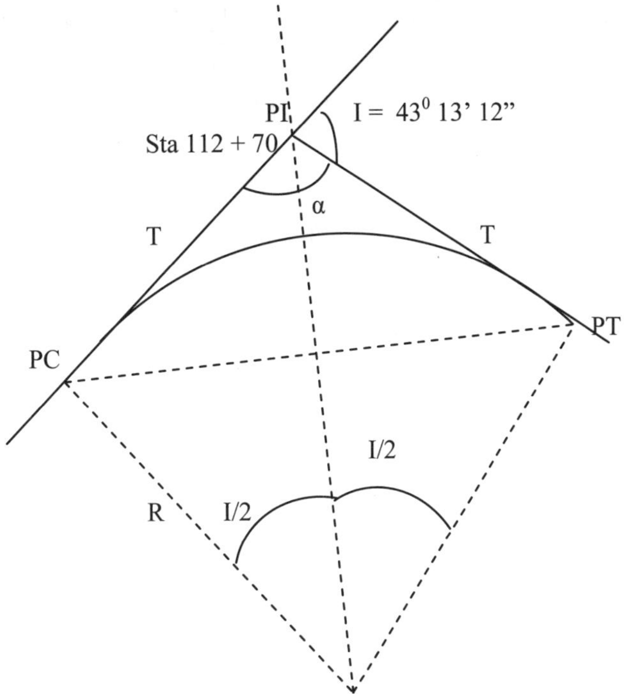

STEP 1: Find the distance T

From trigonometry, T = R tan I/2 = 2,491 x tan [(43^0 13' 12")/2] = 986.8 ft

Station at PC = Station at PI – T
Station at PI = 112 + 70 = 11,270 ft
Station at PC = 11,270 – 986.8 = 10,283.2 = Sta. 102 +83.2

STEP 1: Find the curve length from PC to PT:

Curve length from PC to PT = R. θ

R = Radius
θ = Angle at the center of the circle between PC and PT = I
θ should be measured in radians.
I = 43^0 13' 12"

Following equation can be used to convert degrees to radians.

$I_{radians} = I_{degrees} \times \pi/180$

$I_{radians}$ = 43^0 13' 12" x π/180
= 0.7543 radians
Length along the curve from PC to PT = R. θ = 2,491 x 0.7543 = 1,878.9 ft

Station at PT = Station at PC + Length along the curve from PC to PT
Station at PT = 102 + 83.2 + 1,878.9 = 10,283.2 + 1,878.9 = 12,162 = Station (121 + 62.1)
(Ans A)

Problem 3.18): Sag vertical curve of a roadway is shown in the figure. Find the elevation at point P.

Station of BVC = 105 + 45,
Elevation of BVC = 124.56 ft
Station of PVI = 109 + 34
Station at point P = 113 + 08

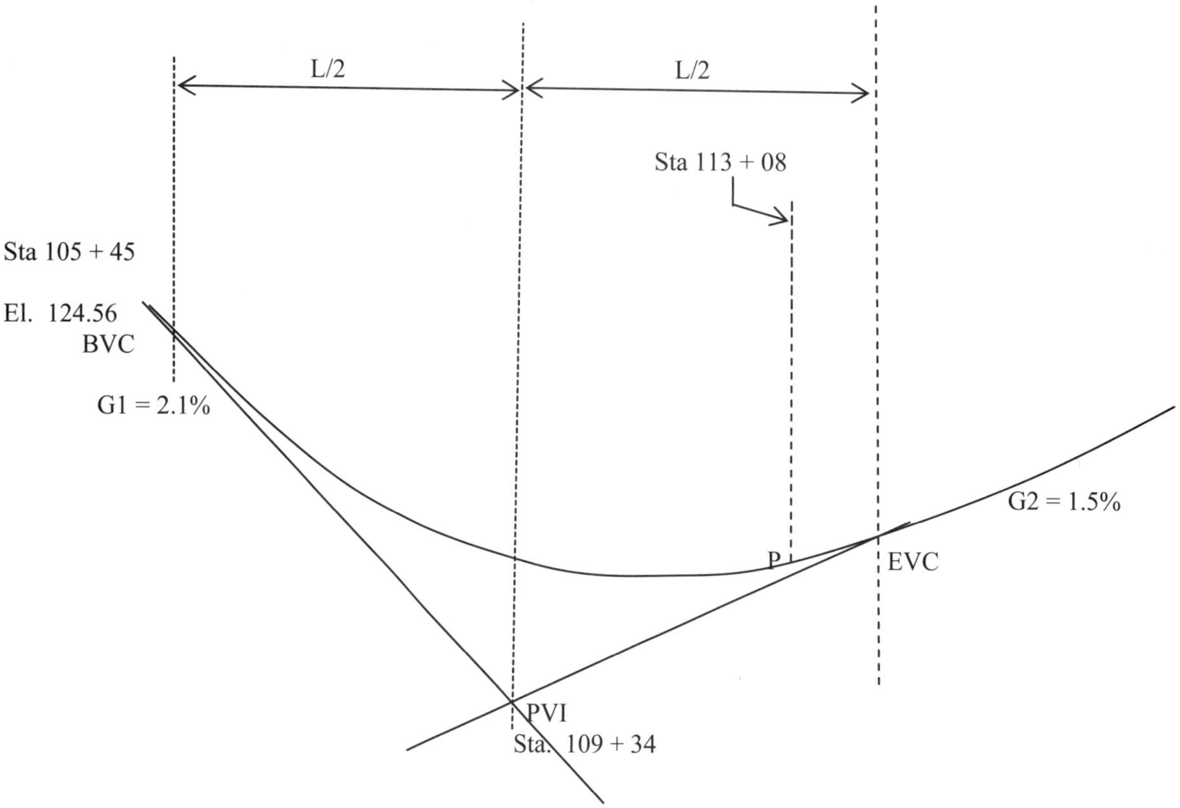

A) 124.5 ft B) 120.7 ft C) 122.01 ft D) 108.9 ft

Solution 3.18):

BVC = Beginning of vertical curve
EVC = End of vertical curve

Vertical Curve Equation;

$$y = \frac{(G2 - G1)}{2L} \cdot x^2 + G1 \cdot x$$

y = Vertical height to point in the curve from BVC in ft
x = Horizontal distance between two points
G1 and G2 are gradients.
Upward gradient is taken to be positive (+) ve.
Downward gradient is taken to be (-) ve.
G1 = -0.021;

G2 = + 0.015
Gradients should be in decimals.
L = Horizontal distance between BVC and EVC in ft.

STEP 1: Write down the given parameters.
BVC station = 105 + 45
PVI station = 109 + 34
The horizontal distance between BVC and PVI = Sta 109 + 34 – Sta 105 + 45
= 10,934 - 10,545 = 389 ft

Note that PVI station is located midway between BVC and EVC.

Hence L = 2 x 389 = 778 ft

G1 = -0.021 (Downward gradient is considered to be negative)
G2 = 0.015 (Upward gradient is considered to be positive)
We need to find the horizontal distance between BVC and point P
Horizontal distance between BVC and point P;
x = Sta. (113 + 08) – Sta. (105 + 45)
x = 11,308 – 10,545 = 763 ft

STEP 2: Find the elevation at point P:

$$y = \frac{(G2 - G1)}{2L} \cdot x^2 + G1\,x$$

$$y = \frac{(0.015 - -0.021)}{2 \times 778} \cdot 763^2 + -0.021 \times 763$$

y = - 2.55

Note that "y" is measured from BVC.

Elevation at point P = 124.56 – 2.55 = 122.01 (Ans C)

Use the data given in 3.19 to solve 3.20.

Problem 3.19) Engineers are designing a horizontal curve with a superelevation. Following information is provided regarding the highway.
Number of lanes = 3,
Width of one lane = 12 ft
Lane adjustment factor = 0.67
Superelevation = 4%
Relative gradient = 0.5%
Find the runoff length required prior to achieving the full superelevation.

A) 123 ft B) 231 ft C) 193 ft D) 351 ft

Solution 3.19) Runoff length is given by the following equation. *(AASHTO Geometric Design of Highways, 2004).* Runoff length and tangent runout were explained in sample exam 2.

$$L_r = w.\ n_1\ .\ e_d.\ b_w/\Delta$$

L_r = Runoff length (ft)
w = Width of one lane (ft) (Typically 12 ft)
n_l = Number of lanes
e_d = Superelevation (percent)
b_w = Lane adjustment factor for number of lanes
Δ = Relative gradient (percent). (The gradient difference between axis of rotation and the edge of the pavement).

Δ (Relative gradient) need to be explained further;

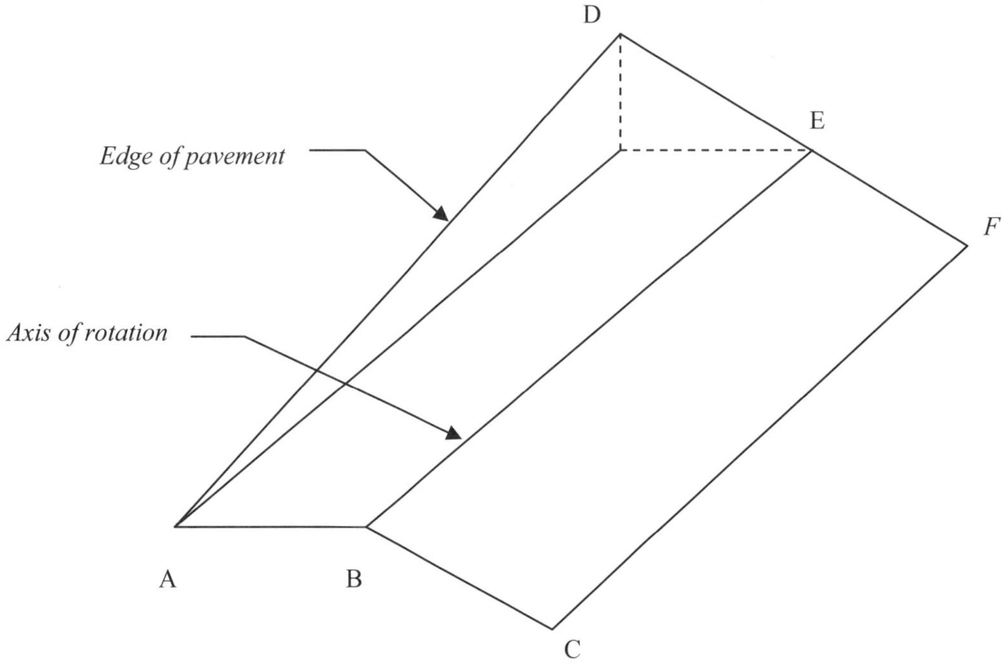

BE is the axis of rotation and AD is the edge of pavement.
At section ABC, crown is already been removed. Then edge of pavement need to be raised gradually to attain the total superelevation. At DEF, total superelevation is achieved. Δ is the gradient difference between BE and AD.

$L_r = w \cdot n_l \cdot e_d \cdot b_w/\Delta$

$L_r = 12 \times 3 \times 4 \times 0.67/0.5 = 192.96$ ft (Ans C)

Problem 3.20): Find the tangent runout for the horizontal curve given in the previous problem.
Normal cross slope = 2%
Design superlevation = 4%.

A) 102.3 ft B) 96.5 ft C) 234.8 ft D) 78.9 ft

Solution 3.20):

Tangent runout is given by the following equation; *(AASHTO Geometric Design of Highways, 2004)*

$$L_t = e_{nc}. L_r/e_d$$

L_t = Tangent runout (ft)
e_{nc} = Normal cross slope (percent) = 2%
L_r = Runoff length (ft)
Runoff length (L_r) was found in the previous problem to be 192.96 ft.
e_d = Superelevation = 4%

$L_t = e_{nc}. L_r/e_d$
L_t = 2 x 192.96/4 = 96.5 ft Ans (B)

CONSTRUCTION:

Problem 3.21) A portion of surveyor's log book is shown below. What is the elevation of point D

Location	Elevation	Backsight Reading	Line of sight El.	Foresight Reading
Benchmark X	201.67	4.56		
Point A				3.89
Point B				4.34
Point C		6.89		8.45
Point D				4.21

A) 200.46 B) 106.05 C) 103.56 D) 199.87

Solution 3.21)

In surveying, backsight reading is taken to a known benchmark and elevation of line of sight is established.

STEP 1) Take a backsight reading to a known benchmark:
Since the elevation of the benchmark is known, elevation of line of sight can be found.

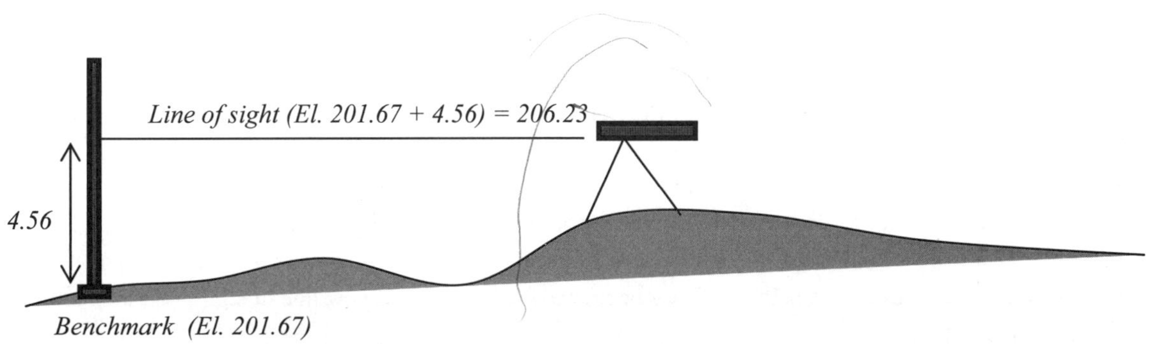

Elevation of line of sight = 201.67 + 4.56 = 206.23 ft

STEP 2) Take a reading to point A:

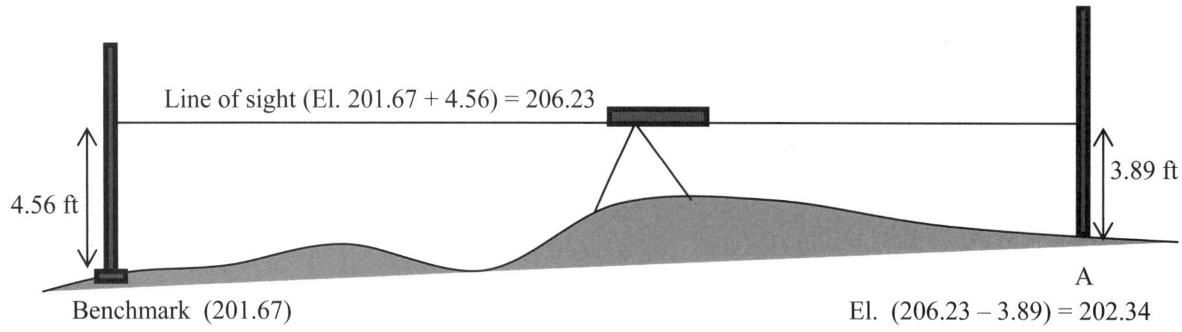

Line of sight elevation is found to be 206.23.
Hence Elevation of point A = 206.23 – 3.89 = 202.34
Elevation of point B = 206.23 – 4.34 = 201.89
Elevation of point C = 206.23 – 8.45 = 197.78

Input the known values in the table.

Location	**Elevation**	**Backsight Reading**	**Line of sight El.**	**Foresight Reading**
Benchmark X	201.67	4.56	206.23.	
Point A	202.34		206.23	3.89
Point B	201.89		206.23	4.34
Point C	197.78	6.89		8.45
Point D				4.21

Now there is a backsight reading also given to point C. The surveyor has moved the instrument to a new location after taking a reading at point C.

Let's see what he did step by step.

- He took a backsight reading to the benchmark and established the elevation of line of sight.
- Then he took readings to points A, B and C. Since he knows the elevation of line of sight he can find the elevations of points A, B and C.
- Now he wants to take a reading to point D. But due to an obstruction, he is unable to obtain a reading to point D. He has to move his instrument to a new location.
- He moves his instrument to a new location where he can see point C and point D.
- He takes a backsight reading to point C. Remember he know the elevation of point C.
- Since he knows the elevation of point C, now he can establish the elevation of new line of sight.
- Then he takes a reading to point D. Since he knows the elevation of new line of sight, he can obtain the elevation of point D.

STEP 3)

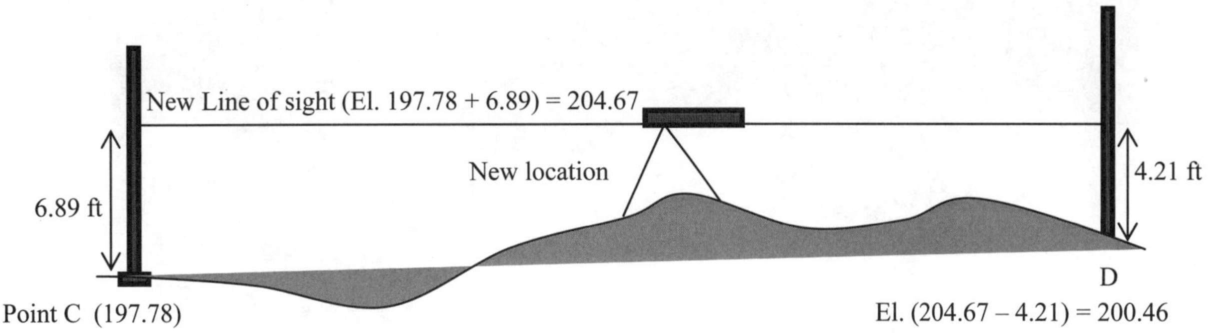

The instrument was moved to a new location and a backsight reading is taken to point C.
Elevation of point C was found in step 2 to be 197.78.
Backsight reading to point C is given to be 6.89 ft.
Hence the elevation of new line of sight can be found.
Elevation of new line of sight = 197.78 + 6.89 = 204.67

Foresight reading to point D = 4.21.
Elevation of point D = 204.67 – 4.21 = 200.46 (Ans A)

Now whole table can be filled out. (Not necessary to solve this problem).

Location	**Elevation**	**Backsight Reading**	**Line of sight El.**	**Foresight Reading**
Benchmark X	201.67	4.56	206.23.	
Point A	202.34		206.23	3.89
Point B	201.89		206.23	4.34
Point C	197.78	6.89		8.45
Point D	200.46		204.67	4.21

Note: To solve this problem, you don't need to find the elevations of points A and B.

Old survey level and a modern survey level

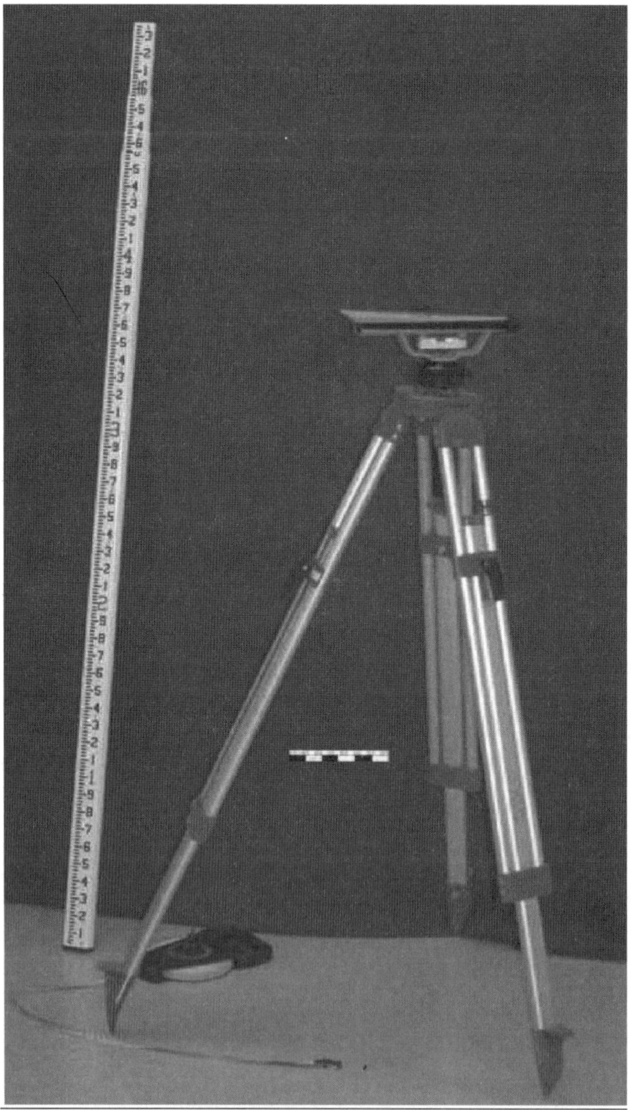

Tripod, survey level and s survey rod

Problem 3.22): Bearing of a line is given to be S 25^0 14' 20"W. Find the azimuth of the line.

A) 205^0 14' 20" B) 125^0 14' 13" C) 275^0 14' 33" D) 114^0 24' 33"

Solution 3.22): Azimuth is measured always from the North direction.
Bearing is measured either from North or South.

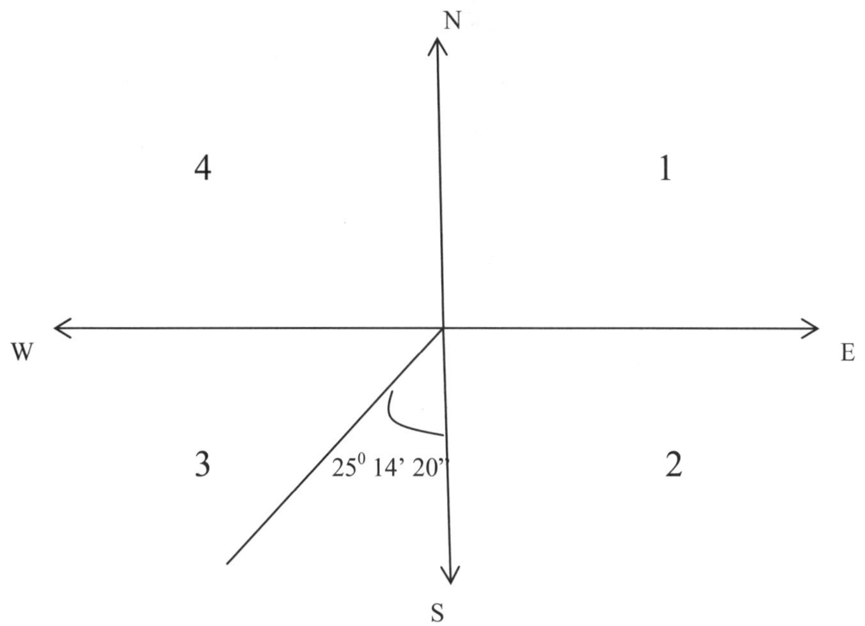

S 25^0 14' 20"W --------> This means measure 25^0 14' 20" from south to west.

For example N 60^0 12' 12" E means measure 60^0 12' 12" from North to East.
Bearing is always measured from North or South.

Azimuth is always measured from clockwise from North direction.

Azimuth = 180^0 + 25^0 14' 20" = 205^0 14' 20" (Ans A)

Problem 3.23): Following successor relationships are available for a project. Draw an activity on node diagram and find the early start of activity E?

Activity ID	Duration (days)	Successors
A	6	B
B	12	C, D, E
C	6	D,F
D	12	E
E	11	F
F	4	

A) 22 days B) 26 days C) 36 days D) 40 days

Solution 3.23):

STEP 1: Draw an activity on node diagram for the data given.

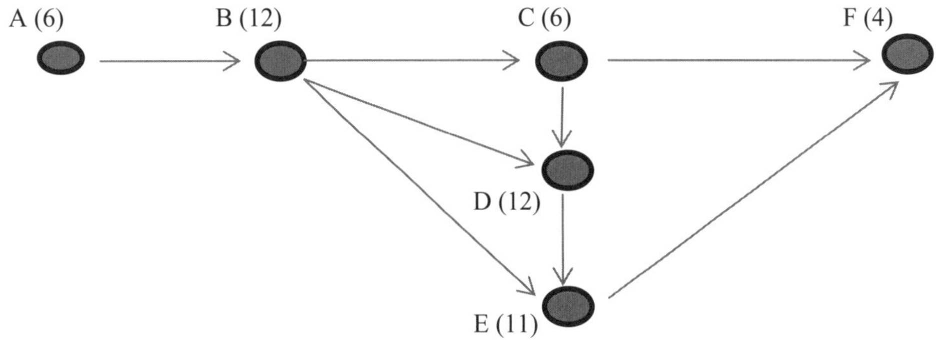

STEP 2: How to draw the activity on node diagram;
Activity is represented in the node.
First activity is activity A
There is only one successor to activity A. That is activity B.
Make a connection to B from A.
Activity B has three successors. They are activities C, D and E.
Draw C, D and E connected to B.
Activity C has two successors. They are D and F.
Make connections to D and F from C.
Activity D has one successor. That is E.
Activity E has one successor. That is F.

STEP 3: Find the early start of activity "E"
For activity E to start, activity D and B has to be completed.

Duration for path A, B, C, D = 6 + 12 + 6 + 12 = 36
Duration for path A, B, D = 6 + 12 + 12 = 30

Earliest, activity E can be started is on day 36. Note that activity E cannot be started until activity D is completed. Activity D cannot start until activity C is completed. Activity C cannot start until activity B is completed. Activities A, B, C and D has to be completed to start activity E. Hence activity E cannot be started on day 30. Earliest day activity E can be started is on day 36.
(Ans C)

Problem 3.24: Shallow foundation of the building shown is 3 ft wide and 18 in high. Find the square footage of formwork needed.

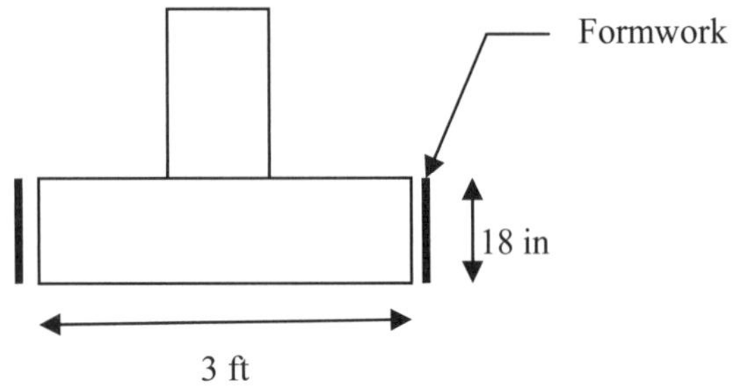

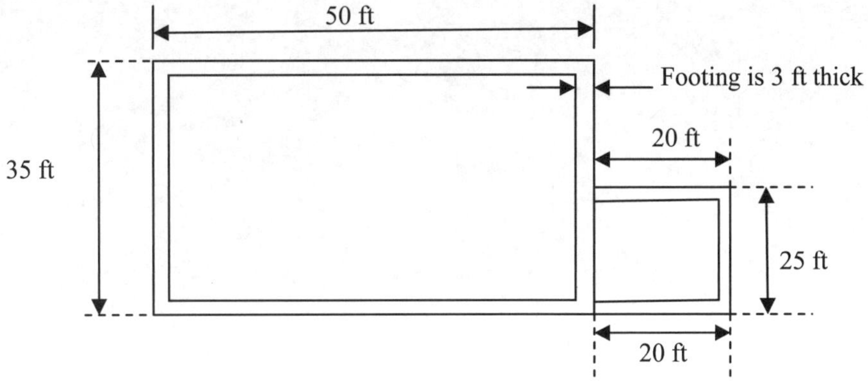

All dimensions are measured from outside.

A) 686 sq. ft B) 651 sq. ft C) 532 sq. ft D) 444 sq. ft

Solution 3.24):

Perimeter length along the center line of the footing is taken and multiplied by 2.0. Centerline distance has to be multiplied by 2 since formwork is needed on both sides of the footing.

Centerline distance measurement is obtained using **In to In and Out to Out** method.

In this process all horizontal dimensions are taken outside to outside and all vertical dimensions are taken inside to inside. In reality, all measurements are horizontal since what is shown is the plan view.

Horizontal dimensions -------------> In to In
Vertical dimensions -------------> Out to Out

STEP 1: Total horizontal length of the footing: (Out to Out):

Total horizontal length of the footing: (Out to Out) = 50 + 20 + 20 + 50 = 140 ft

STEP 2: Total vertical length of the footing: (In to In):

Total vertical length of the footing: (In to In) = (35 – 6) + (35 – 6) + (25 – 6) = 77 ft
Note that here vertical refers to the vertical length measured in the plan view. In reality all measurements are horizontal.

STEP 3: Find the total length of the footing along the centerline
Total length of footing along the centerline = 140 + 77 = 217 ft

STEP 4: Formwork is needed for both sides of the footing.
Total length of formwork = 2 x 217 ft = 434 ft

STEP 5: Find the area of formwork;

Area of formwork = Length x Height = 434 x 18 inches = 434 x 1.5 = 651 sq. ft (Ans B)

Left: Formwork for a footing; Right: Footing is poured

Problem 3.25): A company has 20 painters, 15 metal workers and 40 laborers. Painters and metal workers work 40 hours a week and laborers work 50 hours a week. All workers worked 50 weeks in a given year. There were nine injuries reported by painters, 2 by metal workers and 13 by laborers. What is the incidence rate?

A) 32.4 B) 28.2 C) 86.6 D) 12.3

Solution 3.25):

Incidence rate means number of injuries or lost work days per 100 full time workers per year.
One worker works 40 hours a week. Assuming he works 50 weeks, one worker would work 2,000 hours per year.

100 workers would work 200,000 hours per year. (2,000 x 100)

$$\text{Incidence Rate} = N \times 200{,}000/EH$$

N = Number of injuries
EH = Number of work hours per year

Total number of work hours for painters per year = 20 x 40 x 50 = 40,000 hrs
Total number of work hours for metal workers per year = 15 x 40 x 50 = 30,000 hrs
Total number of work hours for laborers per year = 40 x 50 x 50 = 100,000 hrs

Total number of work hours = 40,000 + 30,000+ 100,000 = 170,000
Total injuries reported = 9 + 2 + 13 = 24

Incidence Rate = N x 200,000/EH = 24 x 200,000/170,000 = 28.2 (Ans B)

Problem 3.26) Following information was obtained from a traverse survey. Find the latitude of CA.

Leg	Latitude (ft)	Departure (ft)
AB	20	25
BC	-15	30

Note: Positive latitude is measured from South to North and positive departure is measured from West to East.

A) 5 B) -5 C) -20 D) 20

Solution 3.26): Draw the traverse using available information.

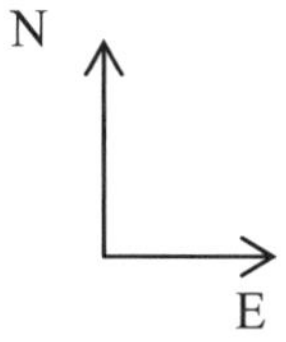

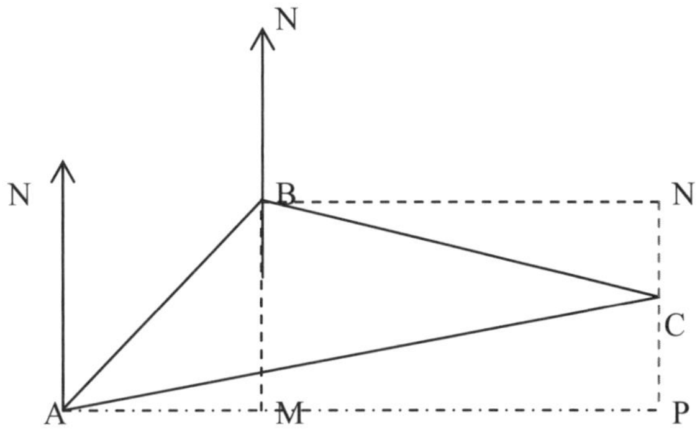

Positive departure is measured from west to east.
Positive latitude is measured from south to north.

Departure of AB = AM = 25
Latitude of AB = MB = 20

Departure of BC = BN = 30
Latitude of BC = NC = -15 (Minus latitude is measured from North to South direction)

Latitude of CA = CP

CP is measured from North to south. Hence it should be negative.
CP = NP – NC
NP = MB = 20
NC = 15
Hence CP = -5. Since CP is measured downwards or North to south direction it should be negative.
Departure values given in this problem are not needed for the solution. (Ans B)

Problem 3.27): Activity on node network is shown below. Durations are as shown.
Find the late start times of activities C and D.

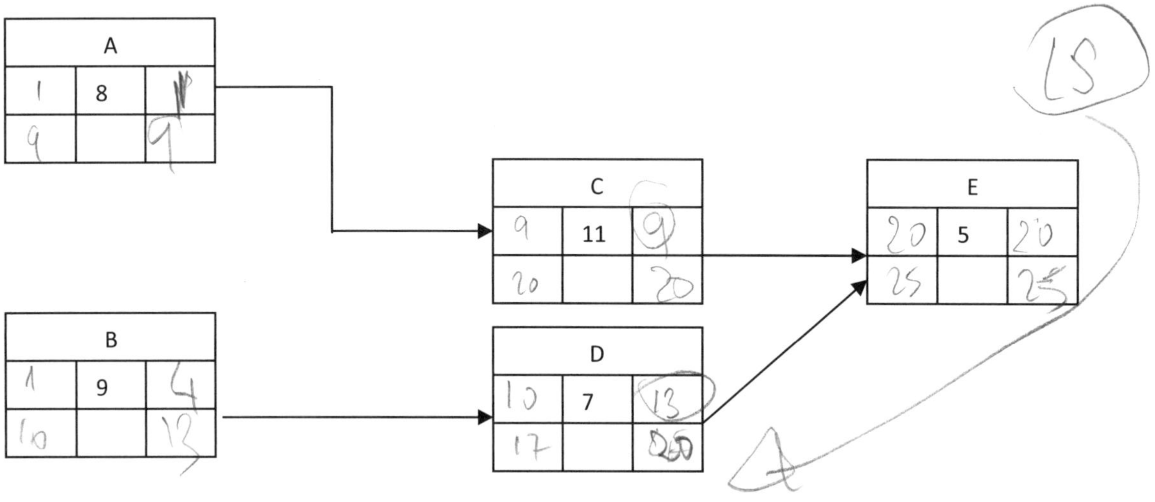

A) LS of activity C = 11, LS of activity D = 15
B) LS of activity C = 9, LS of activity D = 13
C) LS of activity C = 9, LS of activity D = 12
D) LS of activity C = 10, LS of activity D = 12

Key:

Activity name		
ES	Duration	LS
EF		LF

Solution 3.27): When completing activity on node networks, following rules should be followed. I have compiled these rules based on examples given in NCEES booklet.

- NCEES books start the network with 1 (one). Some books start the network with zero. You should start with 1, in the exam.
- One (1) is not added to the finish time of an activity to get the start time of the next activity.

Complete the forward pass: During the forward pass early start times and early finish times are found.

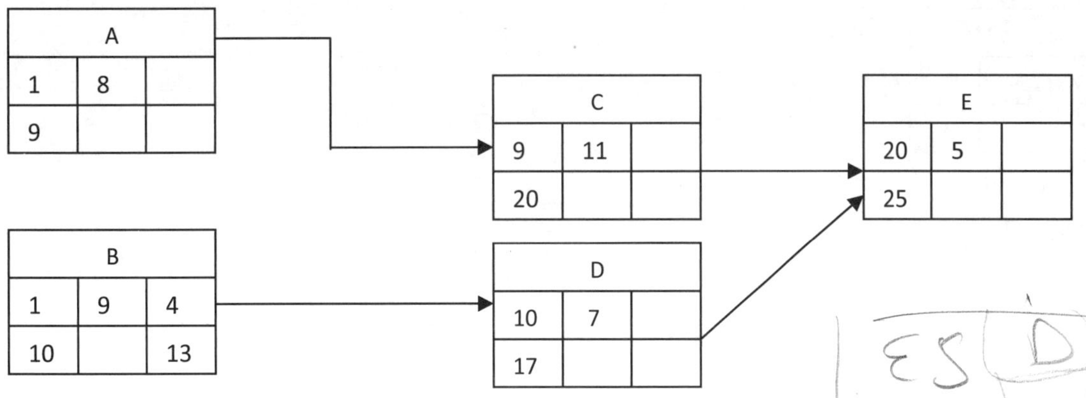

Complete the backward pass:

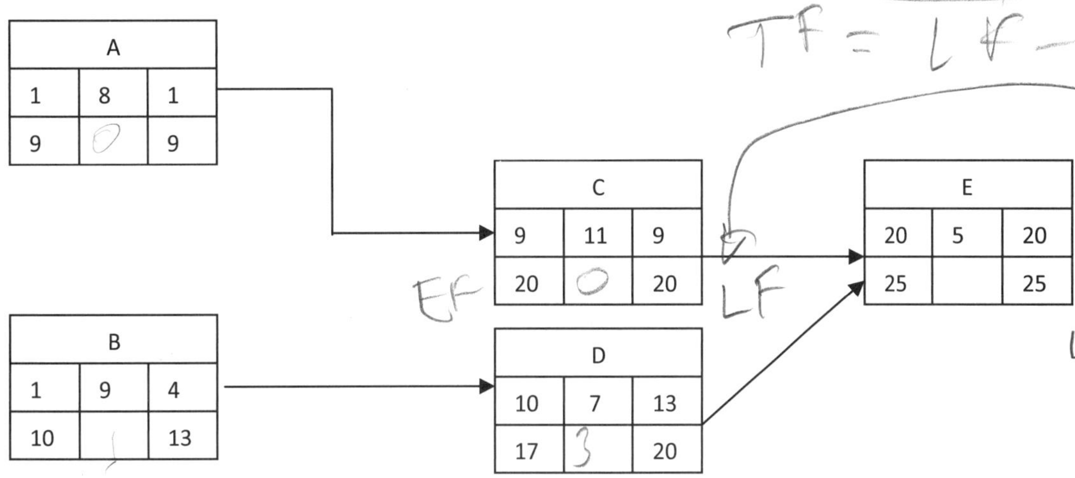

Late start of activity C = 9
Late start of activity D = 13
(Ans B)

Problem 3.28): Find the total float of activities C and D in the previous problem;

A) Total float of a activity C = 1, Total float of a activity D = 3
B) Total float of a activity C = 0, Total float of a activity D = 3
C) Total float of a activity C = 0, Total float of a activity D = 2
D) Total float of a activity C = 1, Total float of a activity D = 1

Solution 3.28): Total Float:

ES = Early Start,
LS = Late Start
EF = Early Finish,
LF = Late Finish,
D = Duration

Activity No:		
ES	D	LS
EF		LF

Total Float:

Total float of an activity is given by the following equation;

$$\text{Total Float} = LF - EF$$

LF – EF is same as LS – ES.

Total float of activity C = 20 – 20 = 0
Total float of activity D = 20 – 17 = 3
(Ans B)

Problem 3.29): Find the free float of activities C and D of the previous problem;

A) Free float of a activity C = 1, Free float of a activity D = 3
B) Free float of a activity C = 0, Free float of a activity D = 3
C) Free float of a activity C = 0, Free float of a activity D = 2
D) Free float of a activity C = 1, Free float of a activity D = 1

Solution 3.29): Free Float:

ES = Early Start,
LS = Late Start
EF = Early Finish,
LF = Late Finish,
D = Duration

Activity No:		
ES	D	LS
EF		LF

Free float is given by the following equation;

$$\text{Free Float of Activity A} = ES_{\text{successor}} - \text{EF of activity A}$$

Step by step procedure to find the free float;

STEP 1: Find the free float of activity C.

Find the successor of activity C.

In this case, successor of activity C is activity E.

STEP 2: Write down the early start time of successor.
$(ES)_{activity\ E}$

STEP 3: Write down the early finish of activity that you need to find the free float.
In this case we need to find the free float of activity C.
Hence write down the early finish of activity C.
$(EF)_{activity\ C}$

STEP 4: Find the free float using the following equation.

Free Float of activity C = $ES_{successor}$ – EF of Activity C

Now find the free float of activity C;

Free Float of activity C = $ES_{successor}$ – EF of Activity C

Successor of activity C for the given network is activity E.
$(ES)_{activity\ E} = 20$;
$(EF)_{activity\ C} = 20$

Free float of activity C = $(ES)_{activity\ E} - (EF)_{activity\ C} = 20 - 20 = 0$

STEP 5: Find the free float of activity D;

Free Float of activity D = $ES_{successor}$ – EF of Activity D
Successor of D is E.
Hence $ES_{activity\ E} = 20$
EF of Activity D = 17
Free Float activity D = 20 – 17 = 3 (Ans B)

STRUCTURAL:

Problem 3.30) A beam is loaded as shown. A distributed load of 60 lbs/ft is applied and a concentrated load of 570 lbs applied at 5 ft from point A. Find the larger shear force at point M.

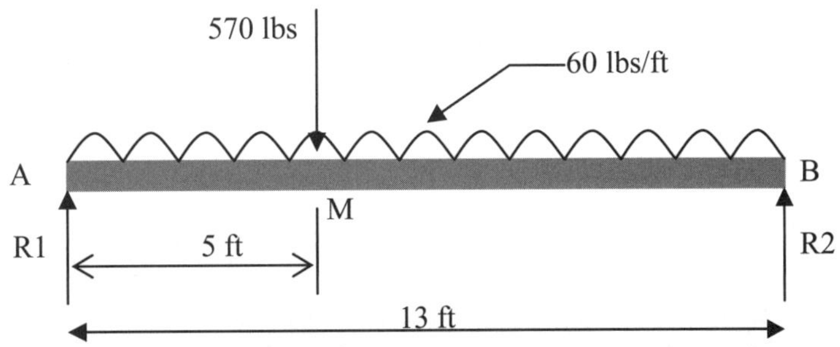

A) 440 lbs B) 590 lbs C) 780 lbs D) 800 lbs

Solution 3.30):

STEP 1: Find the support reactions R1 and R2.

R1 + R2 = 60 x 13 lbs + 570 = 1,350 lbs -----------------------------(1)

Obtain moments around point A;

R2 x 13 = (570 x 5) + (60 x 13) x 13/2----------------------------(2)

Above (570 x 5) is the moment around point A due to concentrated load of 570 lbs.
Above (60 x 13) x 13/2 is the moment due to distributed load. (60 x 13) is the total load and 13/2 is the distance to center of gravity of the distributed load.

Above equation (2) can be solved to obtain R2 = 609.2 lbs

From (1), R1 = 1,350 – 609.2 = 740.8 lbs

STEP 2: Draw the shear force diagram

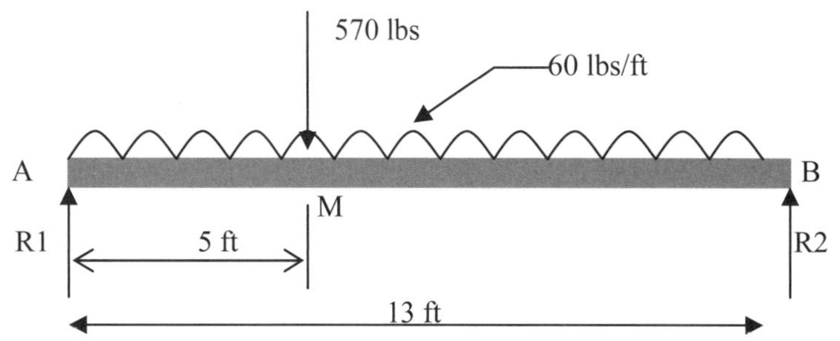

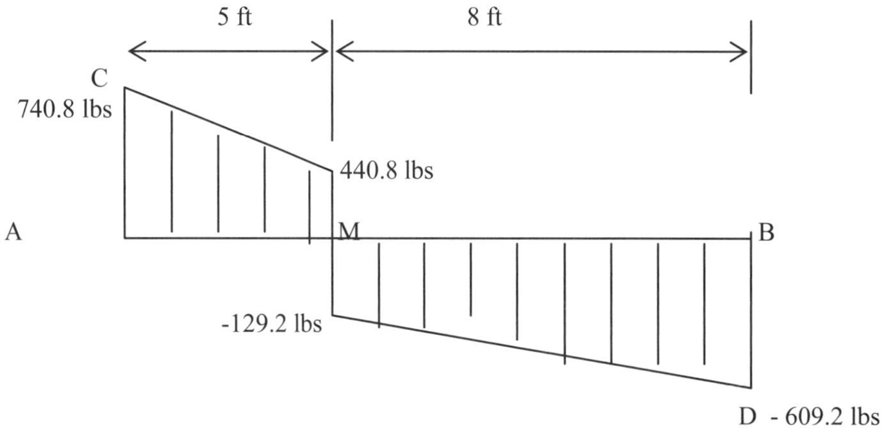

How to draw the shear force diagram?

Assume upward shear is positive and downward shear is negative.

At support A , reaction R1 is upward. Hence draw a line 740.8 lbs upwards. (Line AC)

Then due to distributed load, line drops down since distributed load is downward.
At point M, shear force is 740.8 – (60 x 5) = 440.8 lbs
Then at point M, concentrated load of 570 lbs acts downward. Hence shear force drops down to 440.8 – 570 = -129.2 lbs

Then it keeps dropping down to the end of the beam due to the distributed load. Shear drop from point M to B is (60 x 8).

Shear force at point D = -129.2 – (60 x 8) = -609.2
At point B, reaction is upward again. Reaction R2 is 609.2. (Found earlier)

The question is to find the shear force at point M. Its 440.8 lbs (ans A). -129.2 is no consequence since it is less than 440.8. In the case of shear, the sign is not important. What is important is the value.

Problem 3.31): Find the maximum bending moment of the beam given in the previous problem.

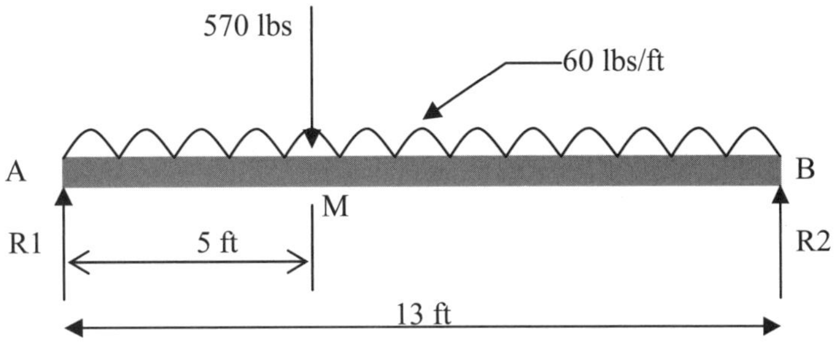

A) 2440 lbs B) 2954 lbs C) 2780 lbs D) 3800 lbs

Solution 3.31):

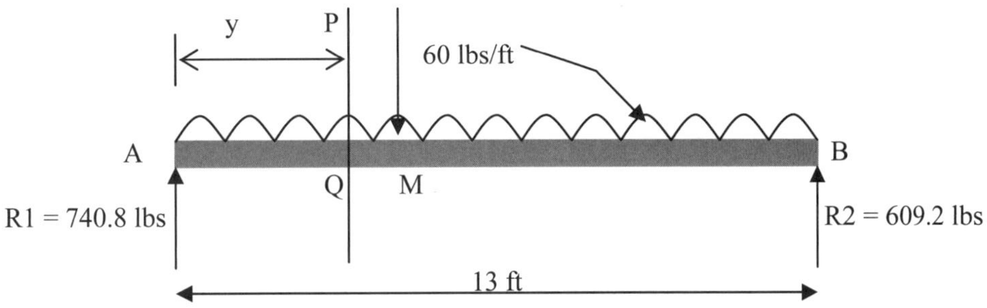

Maximum bending moment occurs at zero shear

STEP 1: Maximum bending moment occurs at locations of zero shear.
Zero shear occurs at point M.

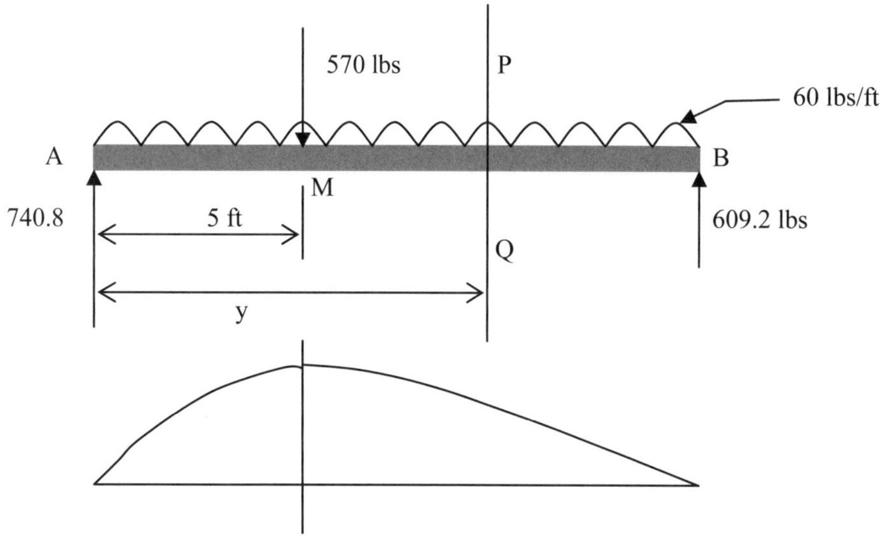

Maximum bending moment occurs at point M. (y = 5 ft)
Obtain moments at point M;

Bending moment at point M = 740.8 x 5 – 60y x (y/2)---(1)
M_{max} = 740.8 x 5 – 60 x 5 x (5/2).---(1)
M_{max} = 2,954 bs. ft. (Ans B)

Note that R1 and R2 reactions were found in the previous problem.

Problem 3.32): Find the distance (y) to centroid of the T-beam shown.

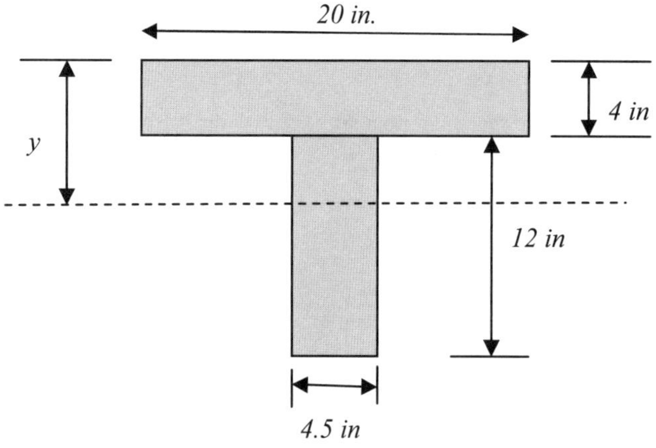

A) 5.22 in B) 3.78 in C) 4.12 in D) 4.81 in

Solution 3:32:

STEP 1: Divide the T-beam into two sections, A1 and A2.

A1 is the flange and A2 is the web.

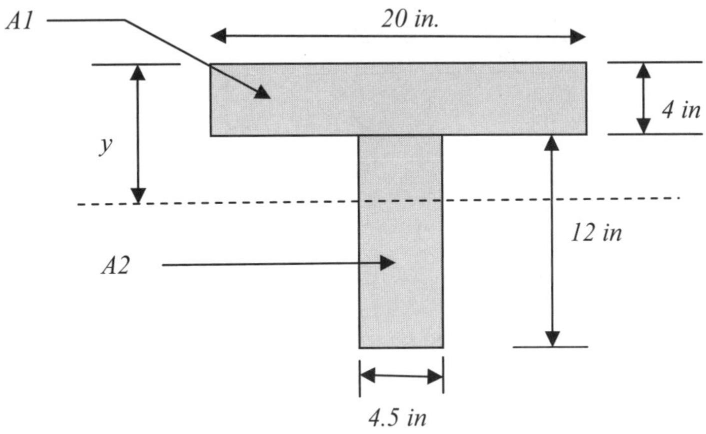

STEP 2: Find the distance to center of gravity of A1 and A2 from top of the beam.

Distance to center of gravity of A1 from top = 2 inches.
Distance to center of gravity of A2 from top = (4 + 6) inches. = 10 inches.

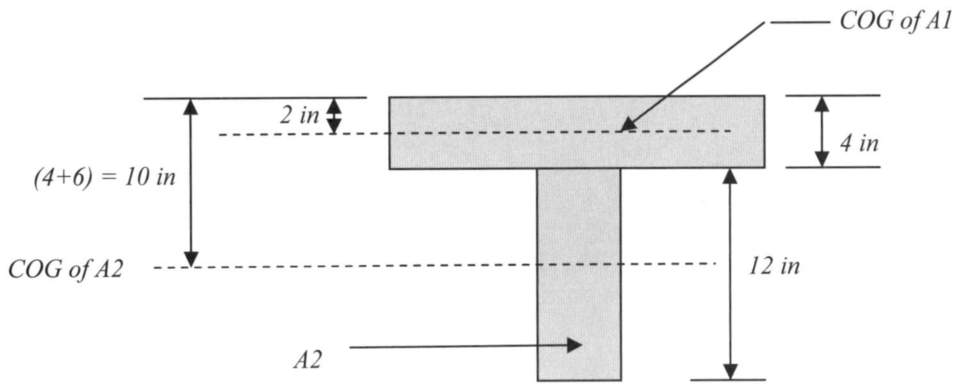

COG = Center of gravity

COG of area A1 is 2 inches below the top surface of the beam.
COG of area A2 is (4 + 6) inches below the top surface of the beam.

STEP 3: Find the center of gravity of T-beam (y):

Following equation can be used to find "y";

$(A1 + A2) \times y = (A1 \times 2) + (A2 \times 10)$

$A1 = 20 \times 4 = 80$ sq. in
$A2 = 4.5 \times 12 = 54$ sq. in

(80 + 54) x y = (80 x 2) + (54 x 10)

y = 5.22 in. (Ans A)

Problem 3.33): Find the nominal moment of the T-beam given.

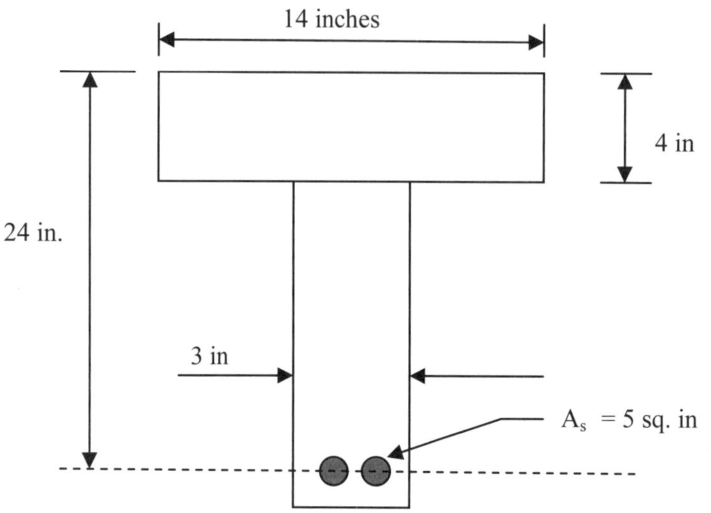

Steel yield stress = 60,000 psi
Concrete compressive strength = 4,000 psi
Steel area = 2 sq. in

A) 483 kip. ft B) 227 kip. ft C) 142 kip. ft D) 331 kip. ft

Solution 3.33):

STEP 1: Write the force balance equation.

Force in concrete = $0.85. f_c' \, A_c$

f_c' = Concrete compressive strength = 4,000 psi
A_c = Compressed are of concrete
(A_c is not known. This needs to be found).

Force in steel = $f_y. A_s$

f_y. = Steel yield stress = 60,000 psi
A_s = Area of steel = 5 sq. in

Force in concrete = Force in steel

$$0.85. f_c' \, A_c = f_y. A_s$$

$0.85. f_c' \, A_c = f_y. A_s$

0.85 x 4,000 x A_c = 60,000 x 2

$A_c = 35.3$ sq. in

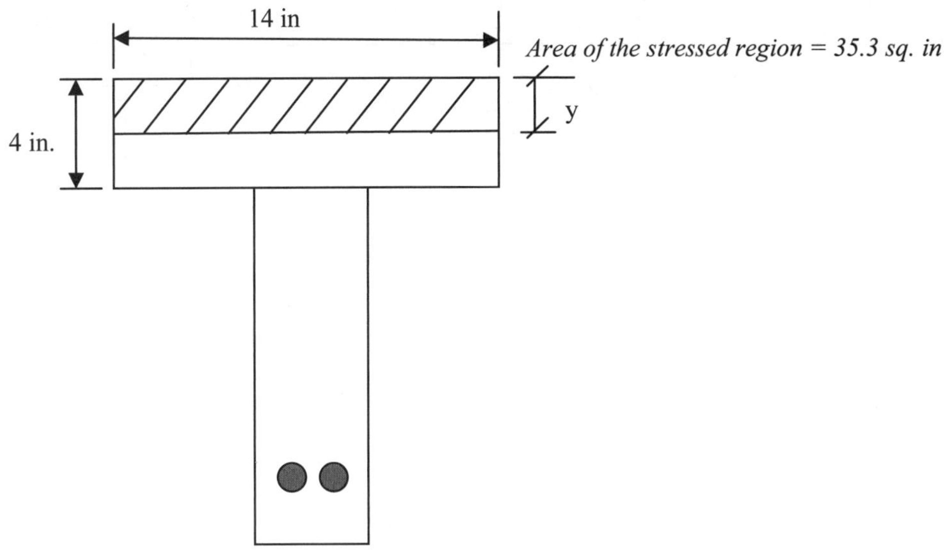

14. y = 35.3

Hence y = 2.52 in.

STEP 3: Find the center of gravity of the stressed section in concrete;

Assume the distance to center of gravity of the stressed section is "C" from the top.

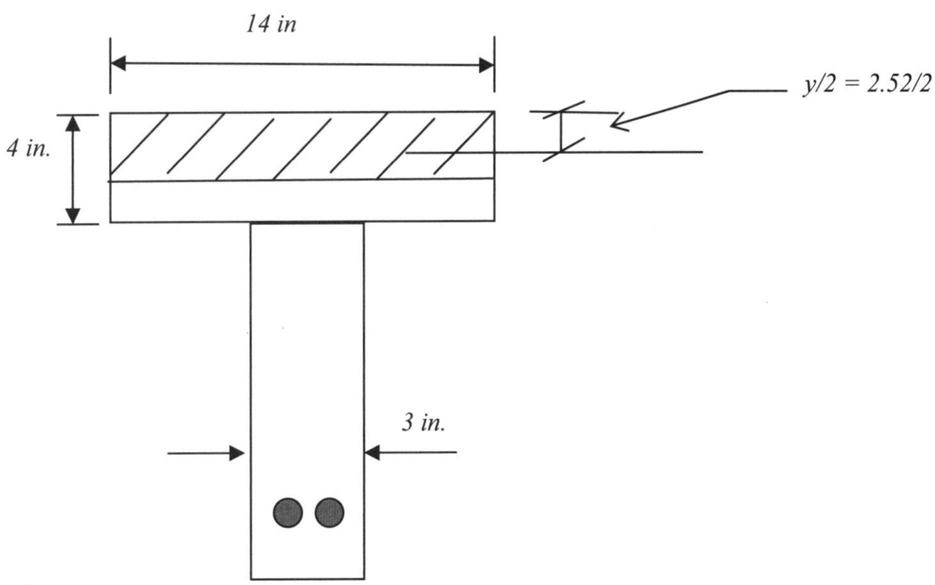

Center of gravity is at the center of the stressed section.

STEP 4: Find the moment arm (Z);
Moment arm is the distance between center of gravity of concrete stressed area and steel.

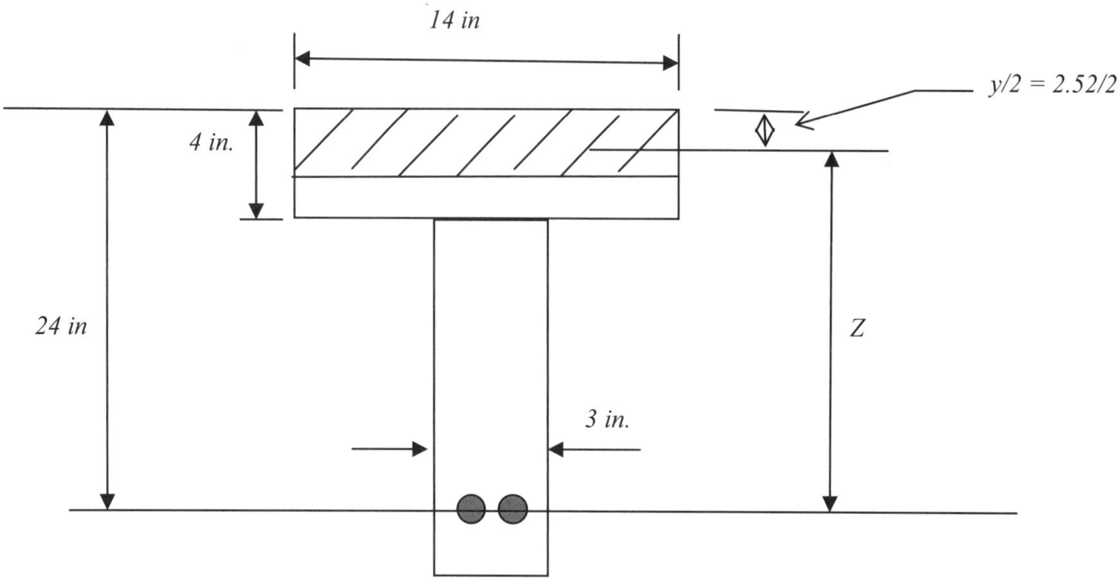

$Z = 24 - y/2 = 24 - 2.52/2$
$Z = 22.74$ in.

STEP 5: Find the nominal moment of the section;
Force in steel = $f_y \cdot A_s$

Since force in steel is equal to force in concrete one can use the force in concrete for the above equation.

Nominal moment = Force in steel x Moment arm

Nominal moment = $f_y \cdot A_s \times Z$

Nominal moment = 60,000 x 2 x 22.74
= 2,728,800 lbs. in
= 2,728,800/(12 x 1,000) kips.ft
= 227.4 kips. ft (Ans B)

Problem 3.34): Find the total force in the reinforcements in the beam shown. $f_y = 60{,}000$ psi.

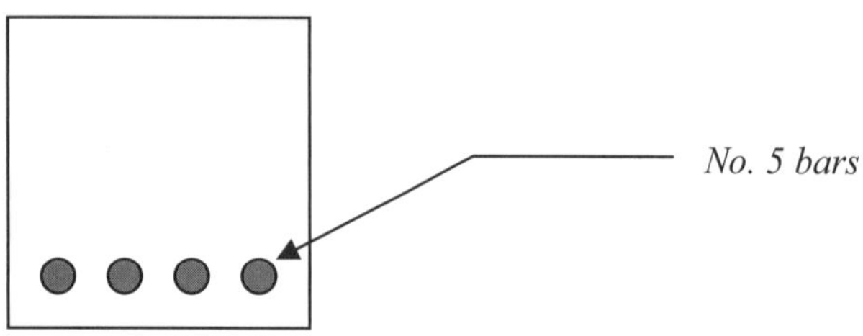

A) 23.8 tons B) 45.7 tons C) 36.8 tons D) 76.9 tons

Solution 3.34):

STEP 1: Find the total steel area

There are four No. 5 bars. No. 5 bar has a diameter of 5/8 in.

Area of one bar = $\pi D^2/4 = \pi \times (5/8)^2/4 = 0.307$ sq. in
Area of four bars = 4 x 0.307 sq. in = 1.228 sq. in

STEP 2: Find the total force:

Total force = $f_y \times A_s$ (f_y = yield stress A_s = steel area)

Total force = 60,000 x 1.228/2000 tons = 36.8 tons (Ans C)ble

GEOTECHNICAL:

Problem 3.35): Specific gravity of a soil sample found to be 2.65. Water content of the sample and the degree of saturation was measured to be 0.22 and 0.90 respectively. What is the dry density of the soil sample.?

A) 100.3 lbs/cu.ft
B) 121.5 lbs/cu.ft
C) 96.9 lbs/cu.ft
D) 116.9 lbs/cu.ft

Solution 3.35):

Followi equation can be used to solve the problem. (See appendix)

$$\gamma_d = \gamma_w . G_s/[1 + (w/S)G_s]$$

γ_d = Dry c
γ_w.= Den .4 lbs/cu.ft
G_s = Spe = 2.65 (Given)
w = Wa ven)
S = De) (Given)
γ_d = D ft
(Ans

Pro ıre used to;

A) soils
B dy soils
C ty soils
D) c

Solution 3.36): Ans (A)

Shelby tubes are used to obtain soil samples in clay soils. Some times it may be possible to obtain Shelby tube samples from silty cohesive soils as well.

Shelby tube is a tube with diameter of 3 inches. The tube is pressed into the soil stratum and a sample is obtained.

Shelby tubes. (These tubes are pushed into soil to obtain a soil sample).

Shelby tube is pushed into the ground to obtain a soil sample

Problem 3.37): Rock core is shown in the figure. What is the RQD of the core if the core length is 5 ft?

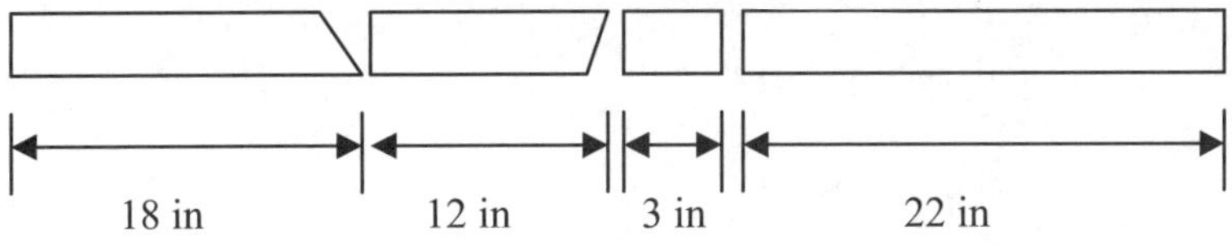

A) 92% B) 87% C) 65% D) 21%

Solution 3.37): RQD (Rock quality designation) is obtained by using following procedure;.

STEP 1: Obtain a rock core. Rock cores are obtained using rock coring rigs. Rock core is typically 60 inch long.
STEP 2: Measure the lengths of all rock pieces greater than 4 inches.
STEP 3: Obtain the RQD using the following equation;

$$RQD = \frac{\text{Length of all pieces greater than 4 inches}}{\text{Length of the core}}$$

Length of all pieces greater than 4 inches = 18 + 12 + 22 = 52 inches
Length of the core = 5 ft = 60 inches

RQD = 52/60 = 87% (Ans B)

Rock cores. (Most rock pieces are greater than 4 inches in this case. High RQD).

Most rock pieces are less than 4 inches. Low RQD.

Problem 3.38): Sieve analysis curve is shown below. What is the D_{30} value of the soil.

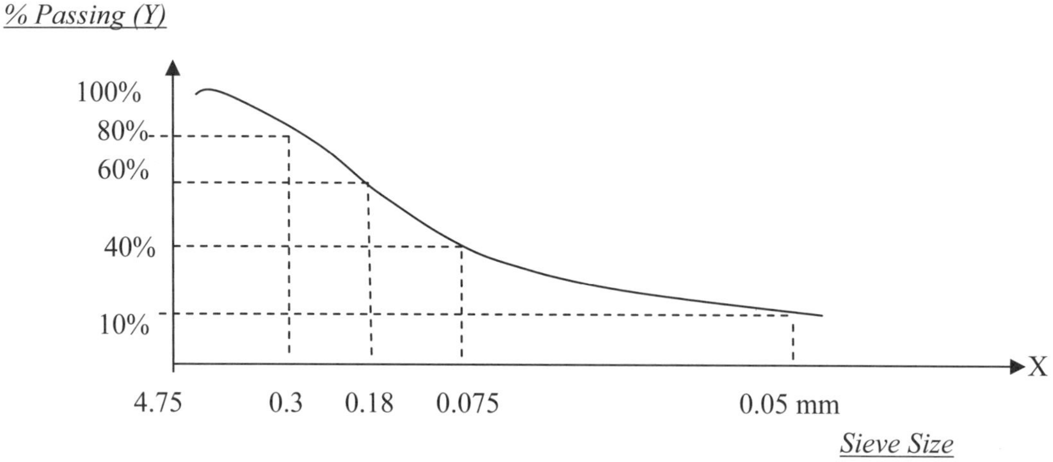

X – axis - Sieve size.
Y - axis - Percent passing.

A) 0.075 mm B) 0.18 mm C) 0.07 mm D) 0.40 mm

Solution 3.38):

To find the D_{30} value, draw a line thru 30% passing line.

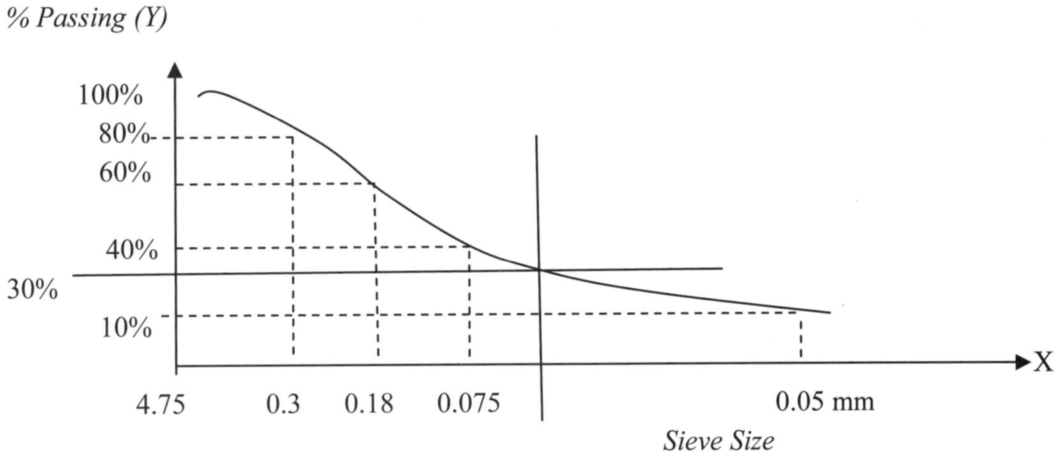

X – axis - Sieve size.
Y - axis - Percent passing.

D_{30} value means the diameter of a sieve that would have 30% passing.
You could see that D_{30} value should be between 0.075 and 0.05.

Hence answer is 0.07 mm. (Ans C)

Problem 3.39): What is the difference between normally consolidated soils and overconsolidated soils.

A) Normally consolidated soils are saturated and overconsolidated soils are unsaturated.
B) Normally consolidated soils are suitable for geotechnical work while overconsolidated soils are not suitable for geotechnical work.
C) Normally consolidated soils have been subjected to large stresses in the past while overconsolidated soils are not subjected to large stresses in the past.
D) Normally consolidated soils have not been subjected to large stresses in the past while overconsolidated soils have been subjected to large stresses in the past.

Solution 3.39): Ans D

Consider a lake bed. Over millions of years lake bed would be deposited with soil particles.

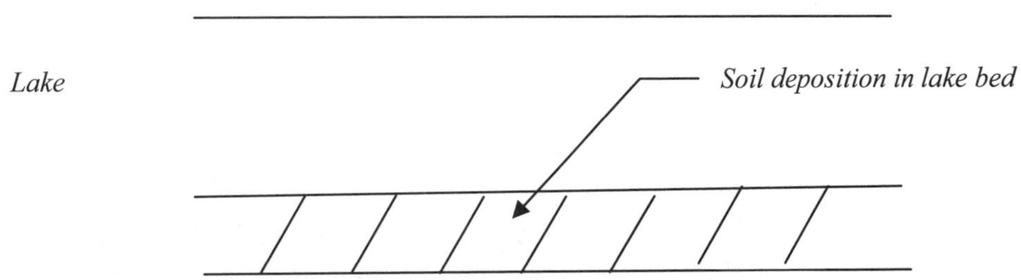

Soil in lake bed is normally consolidated.

Now consider the lake gets dried out and a glacier comes into the region.

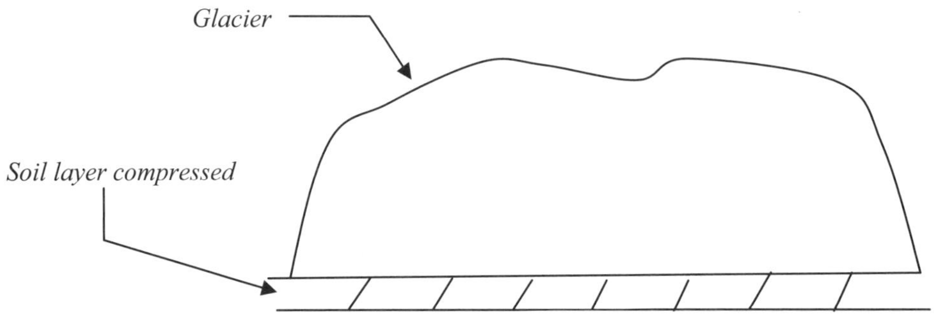

Soil layer would be compressed due to the weight of the glacier. Glaciers do not last forever. Glaciers come and go. Lets assume the glacier is melted away over time.

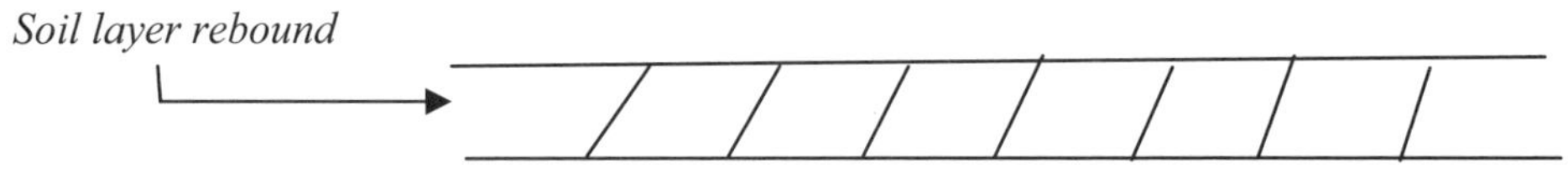

When the glacier is gone, soil would rebound. Thickness of the soil layer would increase. Now this soil layer is overconsolidated. It has been subjected to larger stresses in the past.

Problem 3.40): Road project needs a gravel base. Which gravel do you recommend for the project?

A) Well graded gravel since gravels are of same size and easily compacted.
B) Poorly graded gravel since gravels are of same size and easily compacted.
C) Poorly graded gravel since gravels are of different sizes and easily compacted.
D) Well graded gravel since gravels are of different sizes and easily compacted.

Solution 3.40):

- Gravel particles in poorly graded gravel are of same size.
- Gravel particles in well graded gravel are of different sizes.

Well graded gravel consists of small and large particles. During compaction, voids are filled with smaller particles. (Ans D)

SAMPLE EXAM 4

Problem 4.1): Hydraulic jump is observed in a rectangular channel with an upstream depth of 3 ft. Froude number of the channel upstream section is 5.2. What is the downstream depth of the channel after the jump.

A) 12.34 ft B) 20.61 ft C) 5.67 ft D) 9.08 ft

Problem 4.2) Broad crested weir has a crest length of 300 ft and a discharge of 646 MGD. What is the height of water at the weir if the weir coefficient is 0.8.

A) 2.59 ft B) 5.45 ft C) 12.78 ft D) 4.12 ft

Problem 4.3) A circular channel has a diameter of 12 ft. The channel depth is 8.5 ft as shown. Find the flow in the channel if the slope is 0.5% and Manning coefficient is 0.02.

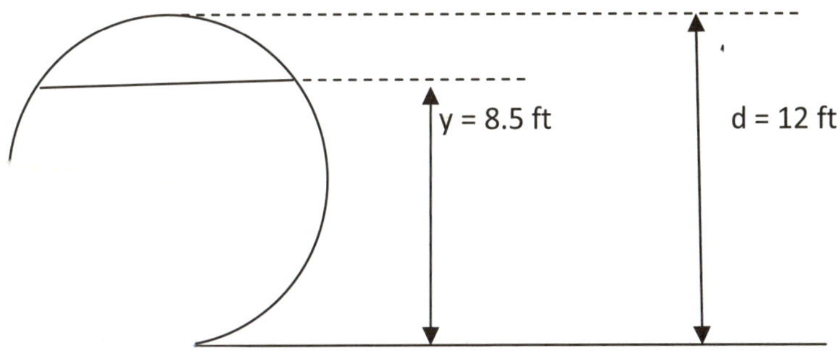

A) 1 ... sec C) 672 cu.ft/sec D) 418 cu.ft/sec

Pro ... shown below. The flow rate was found to be 1,200 gallons/minute. Mar ... 3 and slope was found to be 0.0025. Find the depth of the channel.

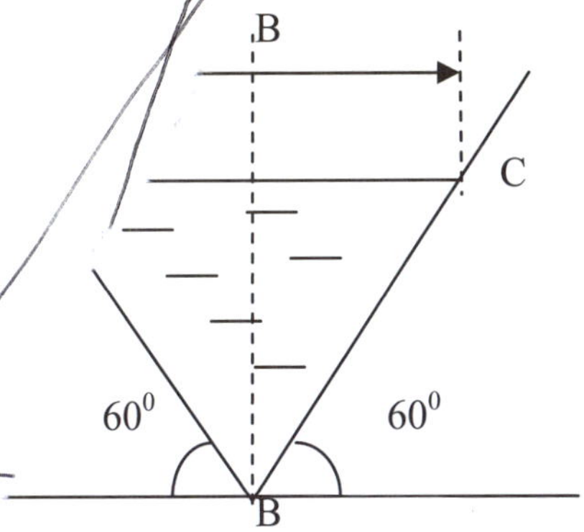

A ... B) 1.28 ft C) 0.82 ft D) 0.55 ft

Problem 4.5): Find the total head added by the pump in the figure below.
Assume the following;

- Friction head loss in pipes is zero.
- Pressure at point B is 3 psi
- Velocity head at point B is negligible.

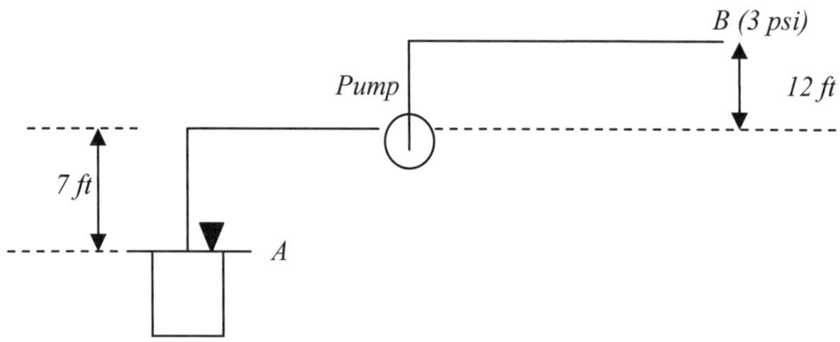

A) 5.0 ft B) 25.9 ft C) 7.1 ft D) 12.2 ft

Problem 4.6): Parallal pipe system is shown in the figure.

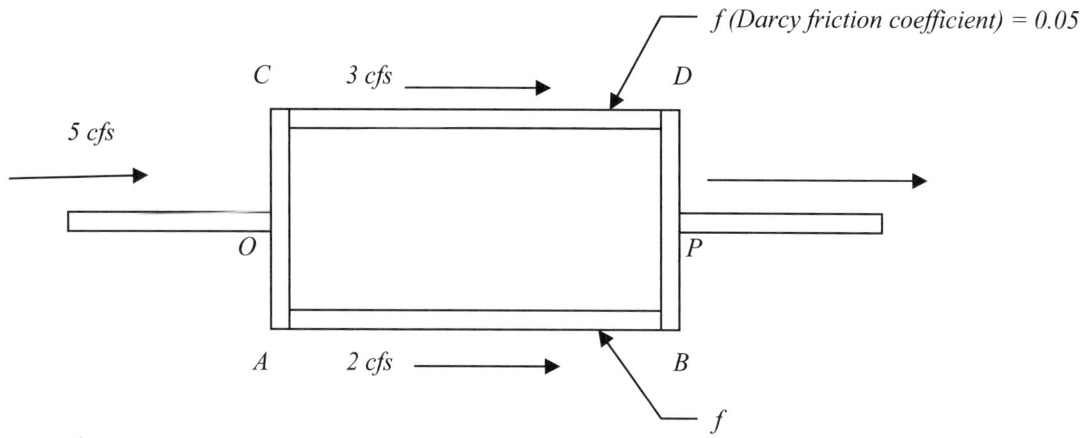

Length of pipe along OABP = 120 ft
Length of pipe along OCDP = 170 ft
Diameter of all pipes are 6 inches.
Find the Darcy friction coefficient of pipe AB.

A) 0.022 B) 0.114 C) 0.003 D) 0.159

Problem 4.7): Suction head of a pump is 8 ft and the discharge head is 9 ft. The flow is 300 gal/min. Find the horse power of the pump. Ignore the head loss due to friction and velocity head in pipes.

A) 5.3 HP B) 1.2 HP C) 0.3 HP D) 4.1 HP

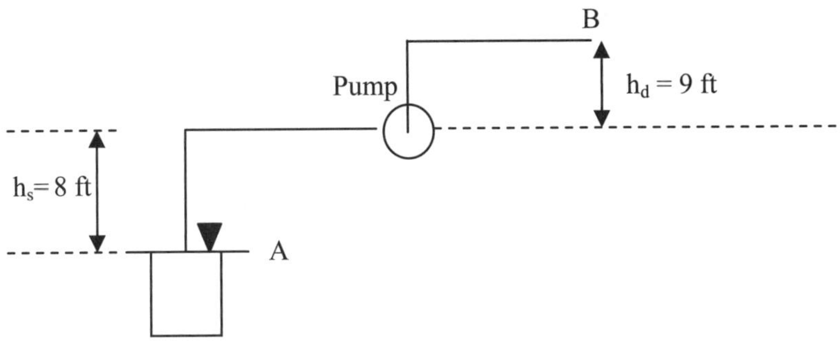

Problem 4.8): A 6 in pipe with Tee runs, elbows and valves are shown below. There are two valves, one Tee run and five elbows. Energy loss at elbows, Tee runs and valves are given below. Flow is 200 gallons per minute and total length of pipes is 4,400 ft.

Energy losses:

Elbow = 2.1 ft;
Tee run = 2.8 ft;
Valve = 1.3 ft

Find the Darcy coefficient of the pipe line.

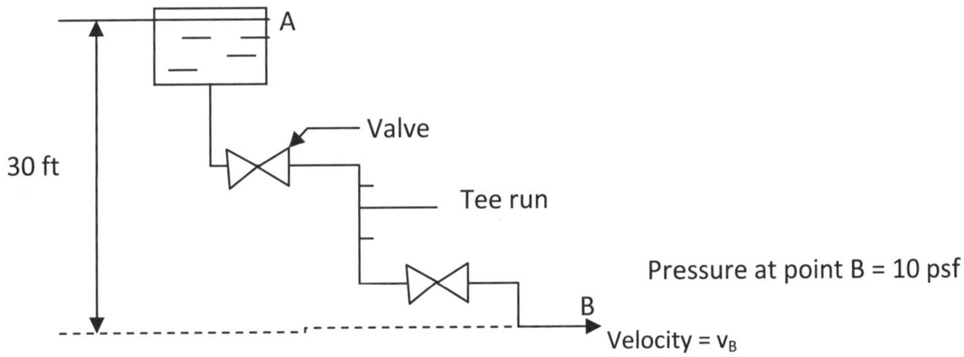

A) 0.020 B) 0.080 C) 0.002 D) 0.110

Problem 4.9): Approximate hydrograph for a stream is shown below.

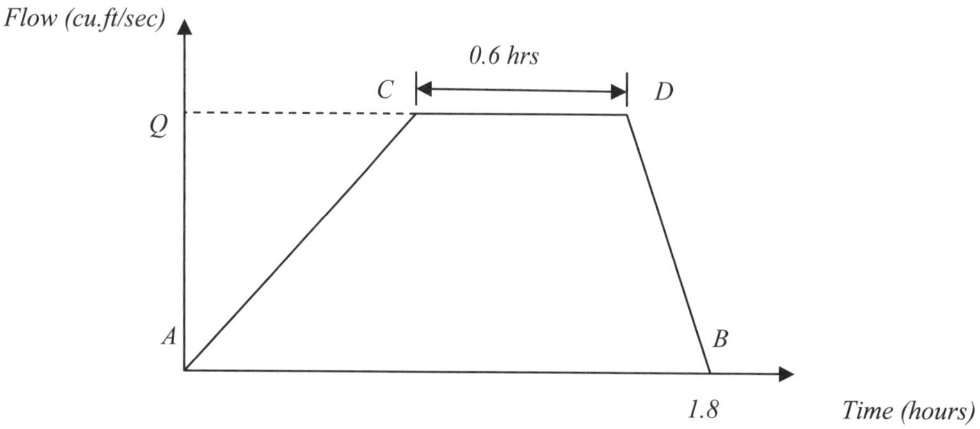

Area of the drainage basin was found to be 15 acres and the net precipitation was 1.3 inches.

Find the maximum flow (Q) in cu.ft/sec.

A) 16.4 cu.ft/sec B) 22.9 cu.ft/sec C) 76.7 cu.ft/sec D) 42.2 cu.ft/sec

Problem 4.10): Find the runoff for a unit hydrograph for the data given in the previous problem.

A) 54,450 cu.ft B) 64,950 cu.ft C) 12,512 cu.ft D) 23,887 cu.ft

Problem 4.11) What is NOT an occupancy category with regard to seismic design

A) Buildings located in soft soils
B) Buildings that are not subjected to human occupancy.
C) Buildings with large number of occupants.
D) Buildings which are essential for post earthquake response work.

Problem 4.12) A building is located in a high groundwater area with loose sandy soil. The site class of the building in relation to seismic design is

A) Site class A or B
B) Site class C or D
C) Site class D
D) Site class F

Problem 4.13) Seismic response coefficient (C_s) of a building was found to be 0.10. The effective seismic weight of a column is 300 tons. What is the seismic shear force acting on the column base.

A) 30 tons B) 20 tons C) 3,000 tons D) 200 tons.

Problem 4.14): Find the area of steel required for a balanced section for the beam shown below.

Steel Young's modulus = 29 x 10^6 psi.
f_y= 60,000 psi
Concrete compression strength (f_c') = 4,000 psi
Concrete is assumed to crush when the strain in concrete is equal to 0.003. (ACI 318)

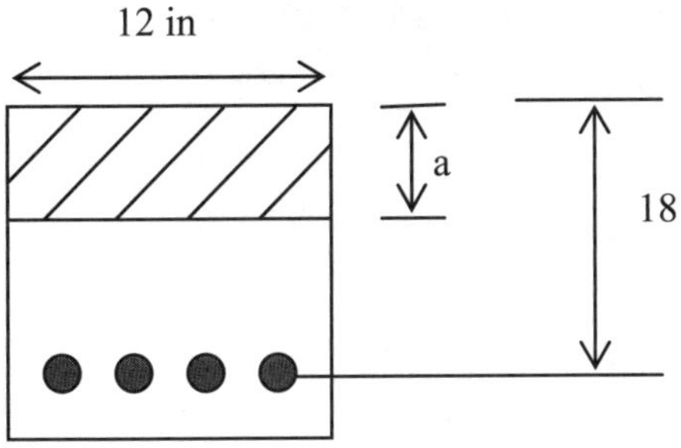

A) 7.08 sq. in B) 6.12 sq. in C) 1.24 sq. in D) 5.44 sq. in

Problem 4.15): Find the nominal moment of the T-beam given.

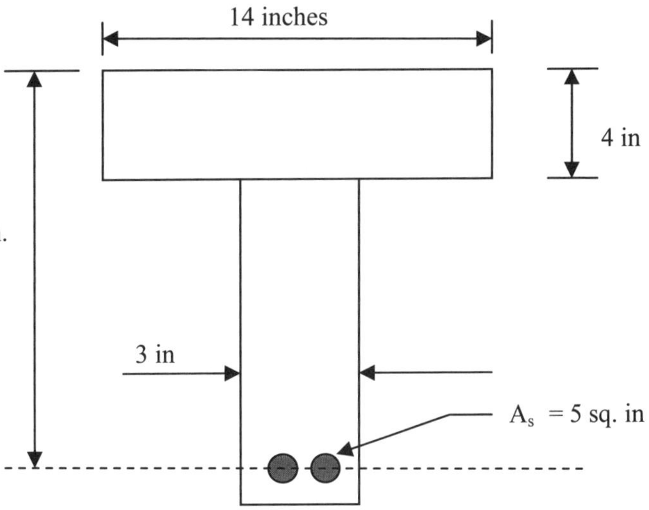

Steel yield stress = 60,000 psi
Concrete compressive strength = 4,000 psi
Steel area = 5 sq. in

A) 483 kip. ft B) 271 kip. ft C) 342 kip. ft D) 381 kip. ft

Problem 4.16): Column to beam connection is shown in the figure. It can be considered as a
A) Moment connection B) Shear connection C) Can't tell D) Fixed end

Problem 4.17): Find the moment of inertia (I) thru the centroid of the T- beam shown.

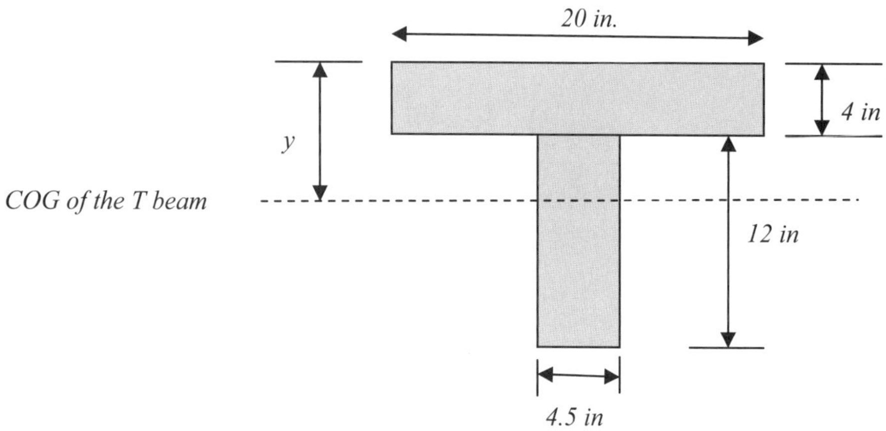

A) 2 ,143 in^4 B) 2,818 in^4 C) 986 in^4 D) 3,194 in^4

Problem 4.18): W18 x 67 column has one end pinned and other end fixed. What is the nominal compression strength of the column in kips? Length of the column is 12 ft.

	Effective length factor (K)
Both ends pinned	1.0
One end pinned, other end fixed	0.8
Both ends fixed	0.65

Portion of AISC Table 4.1 (W 18 x 67)

K.L (with respect to least radius of gyration)	Axial compression strength (kips)
7	790
8	762
9	733
10	701
11	667
12	632
13	596

A) 596 kips B) 632 kips C) 667 kips D) 701 kips

Problem 4.19): Station at point of intersection (PI) is given to be 112 +70 and the degree of curve (arc method) is given to be 2.3^0. Find the station at PT. (Angle of intersection is 28^0 15' 08").

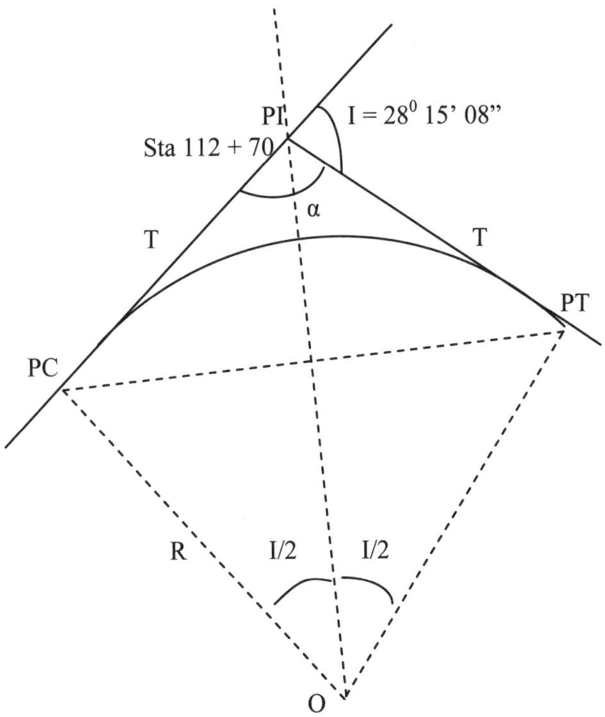

A) 118 + 71 B) 134+75 C) 220+67 D) 234+81

Problem 4.20) Radius of a horizontal curve is given to be 1,300 ft. Station at PC is given to be 187+34 and angle of intersection is 73^0. Find the deflection angle to station 190+00.

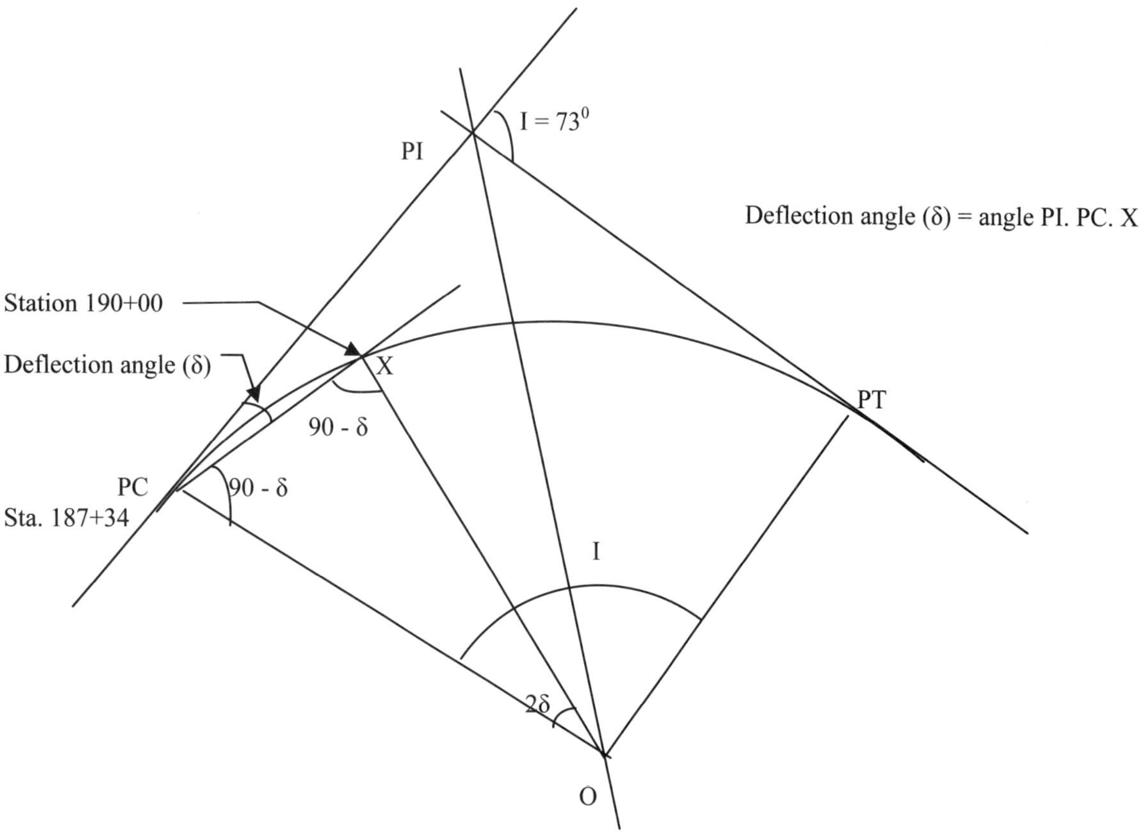

A) 3.95^0 B) 9.20^0 C) 5.86^0 D) 3.67^0

Problem 4.21) Base free flow speed of a highway was assumed to be 70 mph. Following information is provided.

Lane width = 10 ft,

Right shoulder clearance = 3 ft,

Number of lanes in one direction = 3,

Two interchanges per mile.

Find the free flow speed.

Following parameters are given;

$f_{LW} = 6.6$, $f_{LC} = 1.2$, $f_N = 3.0$, $f_{ID} = 7.5$

A) 53.5 mph B) 51.7 mph C) 60.0 mph D) 54.3 mph

Problem 4.22) Design traffic volume in a 6 lane freeway is 1,600 vehicles in both directions. It is reported that this highway has 3% truck traffic and 2% RVs (Recreational vehicles). The road is in rolling ground. Following vehicle flow rates were provided.

5:00 PM – 5:15 PM - 385
5:15 PM – 5:30 PM - 763
5:30 PM – 5:45 PM - 563
5:45 PM – 6:00 PM - 756

The traffic is mostly office commuters. Directional factor is 0.7.
Base free flow speed of the highway was assumed to be 70 mph.
Lane width = 10 ft, Right shoulder clearance = 3 ft, Number of lanes in one direction = 3,
Two interchanges per mile.
What is the level of service provided by the freeway?

$f_{LW} = 6.6$, $f_{LC} = 1.2$, $f_N = 3.0$, $f_{ID} = 7.5$
E_T = Truck equivalent factor = 2.5 for rolling ground
E_R = RV equivalent factor = 2.0 for rolling ground (obtained from table 21-8 (HCM)

Density vs LOS table from HCM;

	Level of Service (LOS)				
	A	B	C	D	E
Density (veh/mile)	11	18	26	35	45

A) LOS A B) LOS B C) LOS C D) LOS D

Problem 4.23): Sag vertical curve of a roadway is shown in the figure. The roadway goes under an overpass.
Station of BVC = 115+45,
Elevation of BVC = 134.56 ft

Station of PVI = 120 + 34
Station of overpass = 123 + 13
Elevation of overpass = 143.25 ft

What is the maximum height of trucks that can be allowed in the roadway assuming 1 ft clearance between roof of trucks and the overpass.
Note that vertical curves are constructed in a manner so that PVI station is at the center of BVC and EVC.

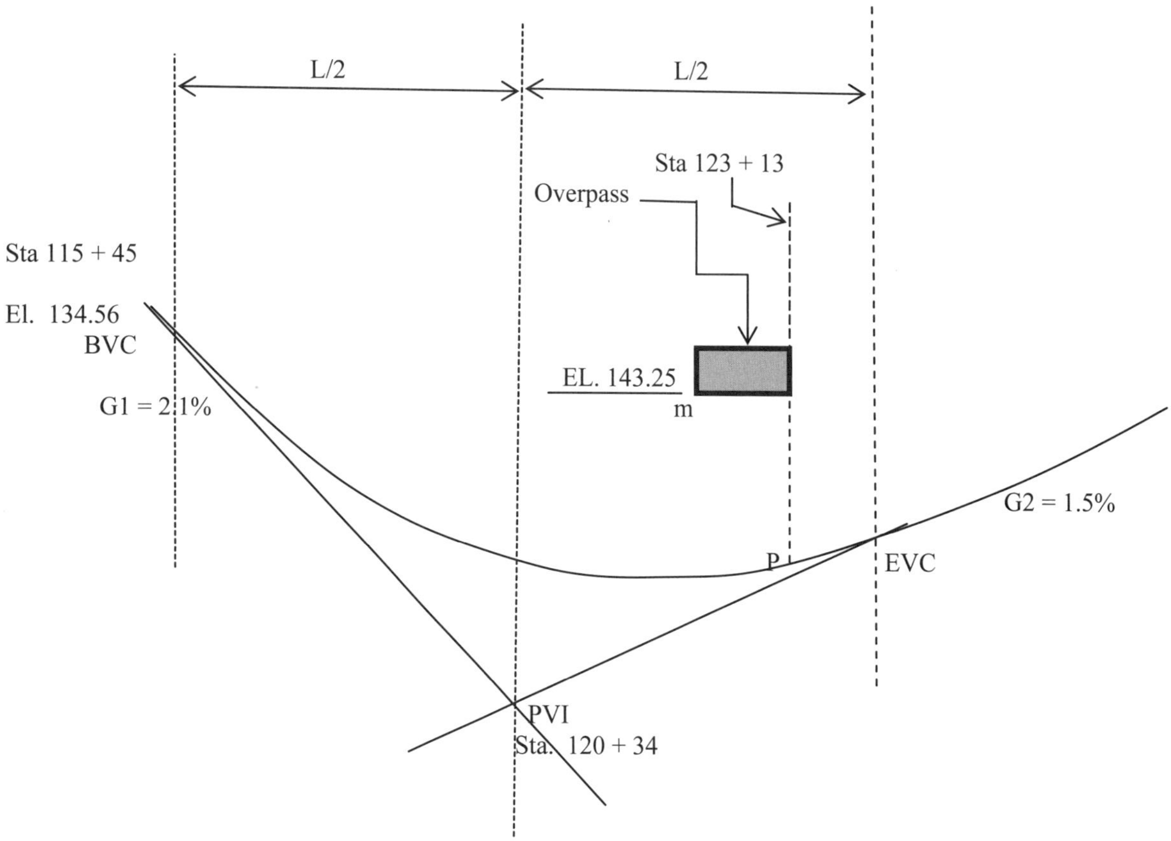

A) 13.45 ft B) 12.97 ft C) 14.56 ft D) 10.89 ft

Problem 4.24) A portion of surveyor's log book is shown below. What is the foresight reading could have been to point C.

Location	Elevation	Backsight Reading	Line of sight El.	Foresight Reading
Benchmark X	191.67	3.56		
Point A				2.81
Point B				4.34
Point C		10.89		
Point D	200.76			4.21

A) 2.46 B) 6.05 C) 13.56 D) 1.15

Problem 4.25): Bearing of line AB is given to be S 45^0 14' 13"W. Interior angle of ABC is 80^0 23' 11". Find the azimuth of line CB.

A) 265^0 14' 13" B) 144^0 51' 2" C) 175^0 14' 33" D) 134^0 24' 33"

Problem 4.26) Activity on arrow diagram is shown below. Find the duration of the project.

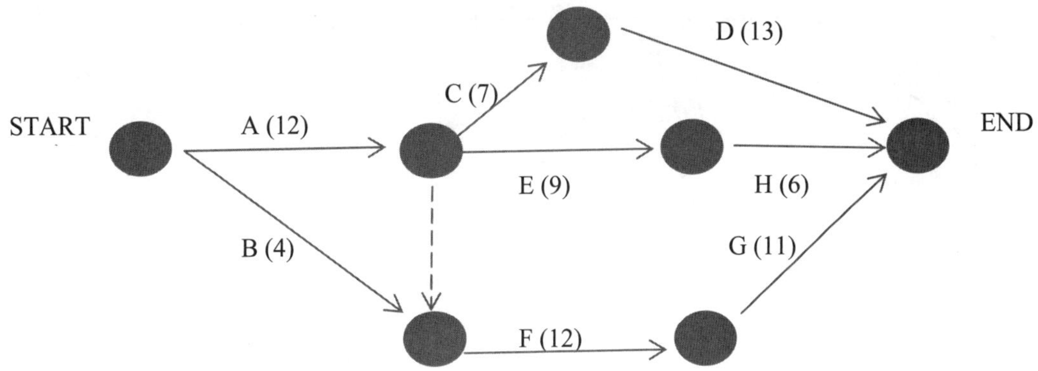

A) 32 days B) 25 days C) 27 days D) 35 days

Problem 4.27) Find the early start and late start of activity D for the project given in the previous problem.

A) ES = 12, LS = 15 B) ES = 13, LS = 22 C) ES = 19, LS = 22
D) ES = 22, LS = 24

Problem 4.28) Find the total float of activity D.

A) 1 B) 3 C) 9 D) 2

Problem 4.29) What type of cement is known as general purpose cement?

A) Type I B) Type II C) Type III D) Type IV

Problem 4.30): Concreting has to be done in cold weather. What type of cement is not desirable?

A) Type I/II B) Type III C) Type IV D) Type V

Problem 4.31): Topography of a proposed road is shown below.

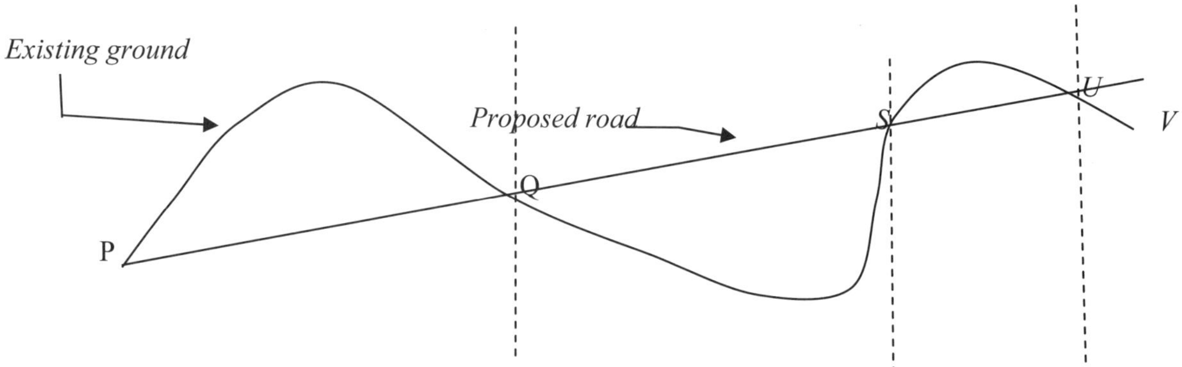

What is the correct mass-haul diagram for the above project?

A)

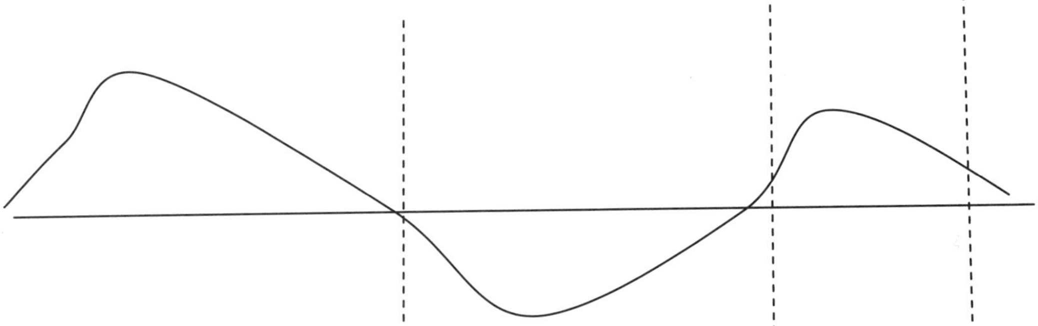

B)

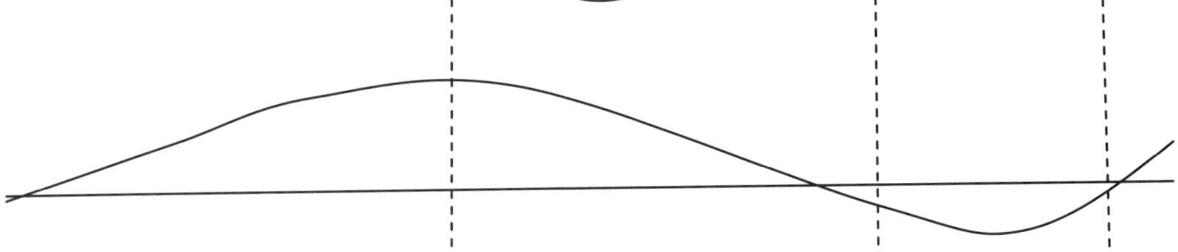

C)

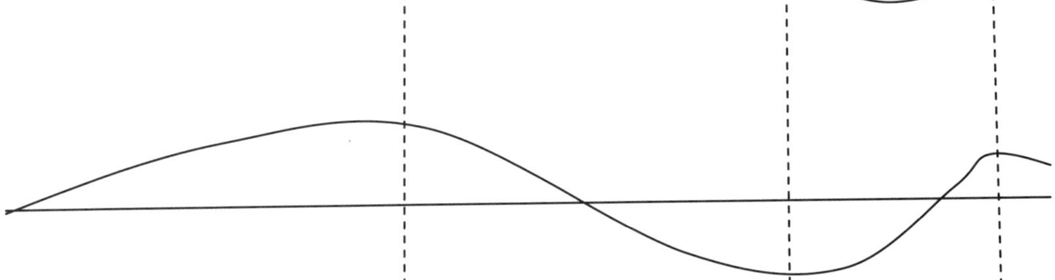

D)

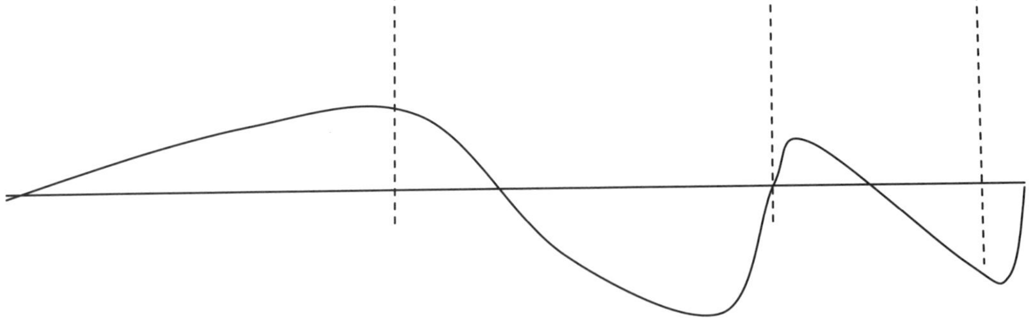

Problem 4.32): Painting crew consists of 3 painters and 2 helpers. Wages of a painter is \$60/hr and of a helper is \$40 per hour. What is the labor hour rate (LH).

A) 55 B) 52 C) 77.2 D) 34.7

Problem 4.33): Site with gray sandy silt is shown below. Original groundwater was at 5 ft below the ground surface. Due to a severe drought, groundwater level drops down to 9 ft below the ground surface.
What would happen to the effective stress at point A when groundwater level drops down to 9 ft below ground surface. Total density of soil is 110 pcf.

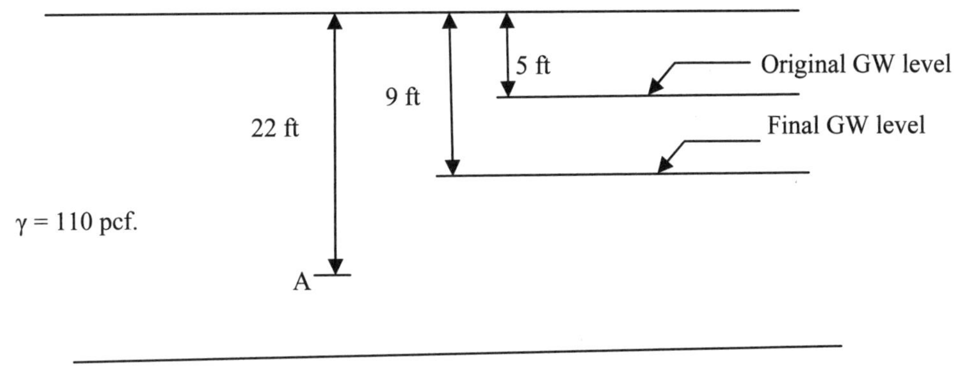

A) Vertical effective stress increases by 249.6 psf.
B) Vertical effective stress decreases by 249.6 psf.
C) Vertical effective stress increases by 169.6 psf.
D) No change

Problem 4.34): Mechanism of secondary compression of a clay soil is best described;

A) Secondary compression is caused by expulsion of air particles due to applied load.
B) Secondary compression is caused by expulsion of water particles due to applied load.
C) Secondary compression is caused by rearranging of soil particles due to applied load.
D) Due to combination of above A and B.

Problem 4.35): Clay soil is loaded in a consolidation test apparatus. The void ratio was measured during loading and plotted and shown below.

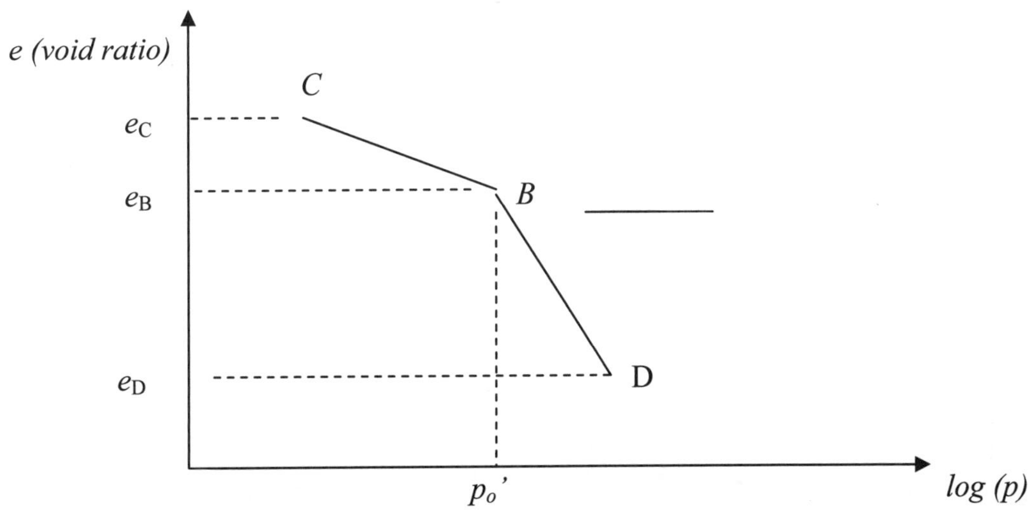

X axis represents log pressure and Y axis represents void ratio.

This clay is a
A) Normally consolidated clay
B) Overconsolidated clay
C) Silty clay
D) Organic clay

Problem 4.36): In the previous problem, point C to B is known as
A) Recompression curve
B) Virgin curve
C) Overconsolidated curve
D) None of the above

Problem 4.37): Engineers are planning to place a shallow foundation as shown.

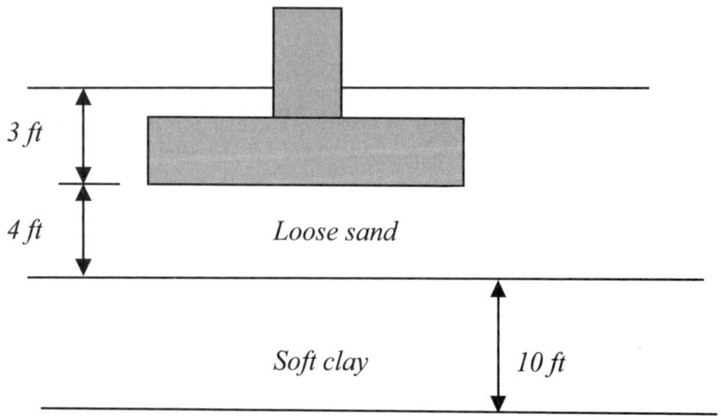

Engineers have decided that loose sand and soft clay are not capable of supporting the foundation without been treated. What is the best ground modification method for this site?
A) Conduct dynamic compaction to treat soft clay and conduct pressure grouting to treat loose sand.
B) Conduct dynamic compaction to treat loose sand and conduct surcharging to treat soft clay.
C) Conduct vibrocompaction to treat soft clay and conduct pressure grouting to treat loose sand.
D) Conduct pressure grouting to treat soft clay and conduct surcharging to treat loose sand.

Problem 4.38): Find the length of the footing required to carry a load of 120 tons with a factor of safety of 2.0.

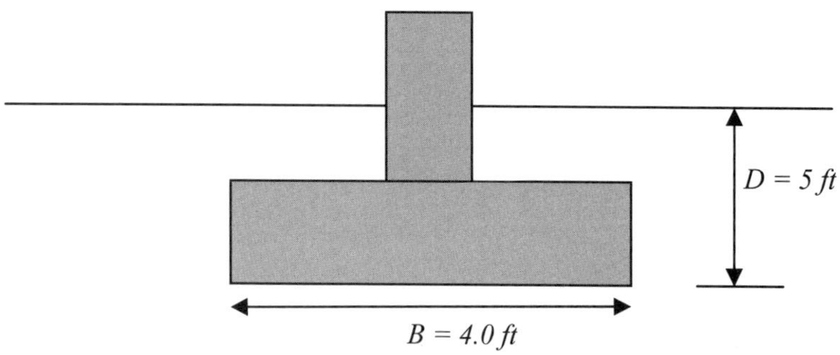

Following parameters are given;
Width of the footing = 4.0 ft
$N_c = 12.9$
$N_q = 4.4$
$N_\gamma = 2.5$

Shape factors S_c, S_q and S_γ;
$S_c = 1.2$
$S_q = 1.0$
$S_\gamma = 0.8$
Total density of soil (γ) = 120 pcf
Cohesion of the soil (c) = 400 psf
Factor of safety = 2.0

Use the Terzaghi Bearing Capacity Equation:
Bearing Capacity in psf = $c.N_c\, S_c + q.\, N_q\, S_q + \frac{1}{2}\, B.\, \gamma.\, N_\gamma\, S_\gamma$
Ignore depth factors.
Find the length of the footing required to carry a load of 120 tons with a factor of safety of 2.0.

A) 11.34 ft B) 8.27 ft C) 4.45 ft D) 13.77ft

Problem 4.39) Soil sample has a porosity of 0.55 and specific gravity of 2.65. Moisture content of the soil sample was found to be 33%. What is the degree of saturation of the soil sample?

A) 43% B) 71% C) 12% D) 34%

Problem 4.40): There is a weak soil layer 12 ft below an existing column footing (5 ft x 5 ft) as shown in the figure. The maximum additional stress that can be applied at the midpoint of the weak soil layer is 700 psf. Any stress more than 700 psf would cause excessive settlement.
What is the maximum load that can be applied on the footing?
Assume stress distribution to be 1:1.

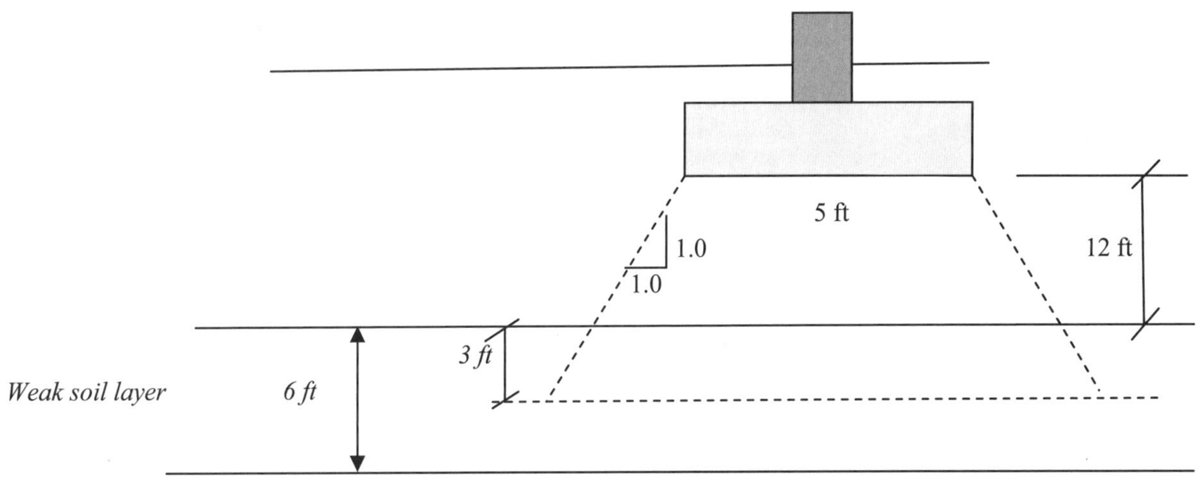

A) 341 tons B) 171 tons C) 428 tons D) 487 tons

SAMPLE EXAM 4 (SOLUTIONS)

Problem 4.1): Hydraulic jump is observed in a rectangular channel with an upstream depth of 3 ft. Froude number of the channel upstream section is 5.2. What is the downstream depth of the channel after the jump.

A) 12.34 ft B) 20.61 ft C) 5.67 ft D) 9.08 ft

Solution 4.1):

Hydraulic jump can be expressed by following two equations for rectangular channels. (Equations 1 and 2, see below).

$$y_2/y_1 = 1/2 \times [(1 + 8 Fr_1^2)^{1/2} - 1] \text{ -----------(1)}$$

y_1 = Upstream depth of the channel

y_2 = Downstream depth of the channel after the jump. Note that y_2 is always larger than y1.

Fr_1 = Froude number upstream of the channel.

Froude number is defined as follows.

$Fr = v/(gD)^{1/2}$;

D = Hydraulic dsepth = A/T;

A = Area;

T = Length of the top surface

For rectagular channels, A/T becomes equal to depth.

A = Area and T = Top surface

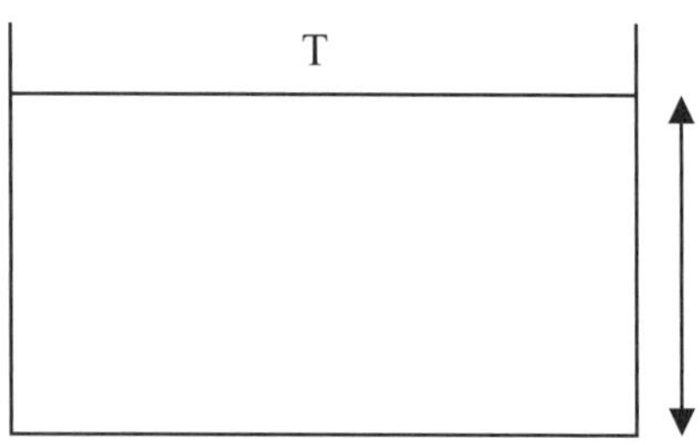

A = T. y

D = Hydraulic depth = A/T

$D = T.y/T = y$

As you could see hydraulic depth of rectangular channels are equal to depth. This works only for rectangular channels.

$$Fr_1 = v_1/(gy_1)^{1/2}$$

v_1 = Upstream velocity and y_1 = upstream depth of the channel.

Replace Fr_1 with $v_1/(gy_1)^{1/2}$ in above equation 1.
You would get equation 2.

$$y_2/y_1 = ½ \, [(1 + 8 \, v_1^2/gy_1)^{1/2} \; - 1] \text{ --------------------(2)}$$

In this problem, Froude number of the upstream section is given. Hence it is easier to use equation 1.

$y_2/y_1 = 1/2 \times [(1 + 8 \, Fr_1^2)^{1/2} - 1]$

$y_2/y_1 = 1/2 \times [(1 + 8 \times 5.2^2)^{1/2} - 1]$

$y_2/y_1 = 6.87$
$y_1 = 3$ ft (Given)
$y_2 = 3 \times 6.87$ ft
$y_2 = 20.61$ ft (Ans B)

Problem 4.2) Broad crested weir has a crest length of 300 ft and a discharge of 646 MGD. What is the height of water at the weir if the weir coefficient is 0.8.

A) 2.59 ft B) 5.45 ft C) 12.78 ft D) 4.12 ft

Solution 4.2):

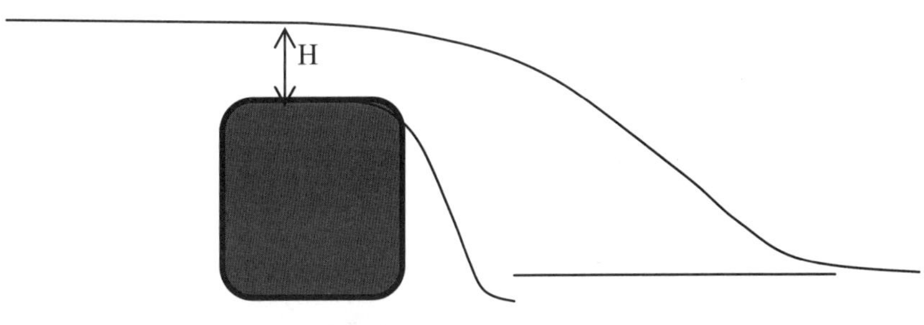

Discharge of a broad crested weir is given by the following equation;

$$Q = C. \, L. \, H^{3/2}$$

Q = Discharge (cu.ft/sec);
L = Length of the weir (ft);

H = Head at the weir (ft)

In this problem Q = 646 MGD (million gallons per day) = 7,476.85 gallons/sec
1 cu. ft = 7.4805 gallons

Q = 999.5 cu.ft/sec;
L = 300 ft;
C = 0.8
H = ?

$999.5 = 0.8 \times 300 \times H^{3/2}$

$H^{3/2} = 4.1646$
H = 2.59 ft (Ans A)

Left Photo: *2000 year old ancient stone built broad crested weir in Sri Lanka*
Right Photo: *Modern broad crested weir from New Hampshire*

Problem 4.3) A circular channel has a diameter of 12 ft. The channel depth is 8.5 ft as shown. Find the flow in the channel if the slope is 0.5% and Manning coefficient is 0.02.

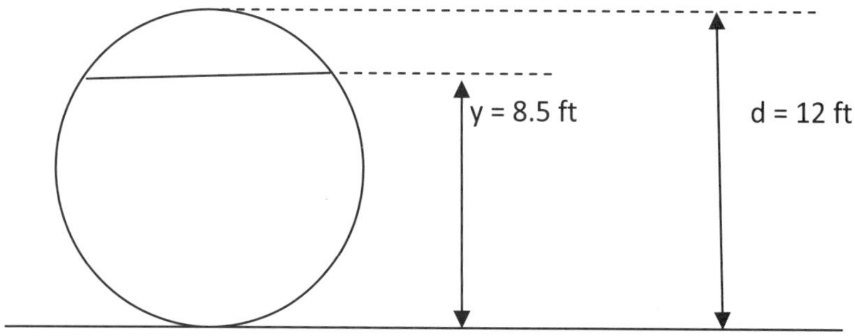

A) 1,421 cu.ft/sec B) 1,053 cu.ft/sec C) 672 cu.ft/sec D) 418 cu.ft/sec

Solution 4.3) The flow equation is

$$Q = 1.49/n \times A (R)^{2/3} S^{1/2}$$

Q = Flow in cfs;
n = Manning's coefficient;
A = Area
R = Hydraulic Radius = A/P, where A is the area and P is the wetted perimeter;
S = Slope in decimals

STEP 1: Find the area (A);

Area of a circular section is given by the following equation;

$$A = 1/8\ (\theta - \text{Sin}\ \theta) \times d^2$$

θ is the angle as shown in the figure. θ should be measured in radians.
d = Diameter
The channel depth is 8.5 ft as shown.

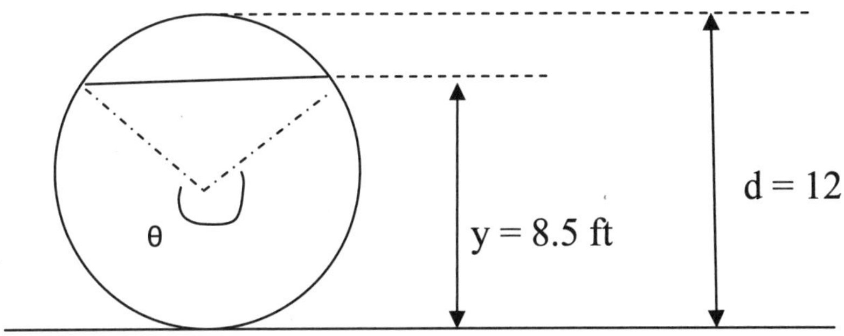

Radius is 6 ft and depth is 8.5 ft.

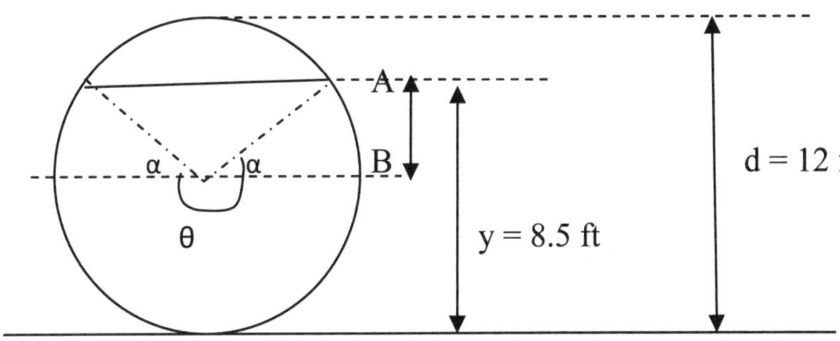

AB = 8.5 ft – 6 ft = 2.5 ft
R. Sin α = AB
R = Radius = 6 ft
6. Sin α = AB
6. Sin α = 2.5
$\alpha = 24.64^0$
$\theta = 2.\ \alpha + 180 = 2 \times 24.64 + 180 = 229.3^0$

Convert θ to radians.

$\theta = 4.00$ radians

Any scientific calculator let you convert degrees to radians. Also you can use the following method.

$180^0 = \pi$ radians

$1^0 = \pi/180$ radians

$229.3^0 = \pi/180 \times 229.3$ radians $= 4.00$ radians

Now you can find the area;

$A = 1/8\ (\theta - \text{Sin}\ \theta) \times d^2$

$A = 1/8\ (4.00 - \ \text{Sin}\ 229.3) \times 12^2$

$A = 1/8\ (4.00 - \ -0.758) \times 12^2$

Note that Sin 229.3 = -0.758

A = 85.64 sq. ft

STEP 2: Find the wetted perimeter (P);

Wetted Perimeter (P) = R. θ

θ should be measured in radians.

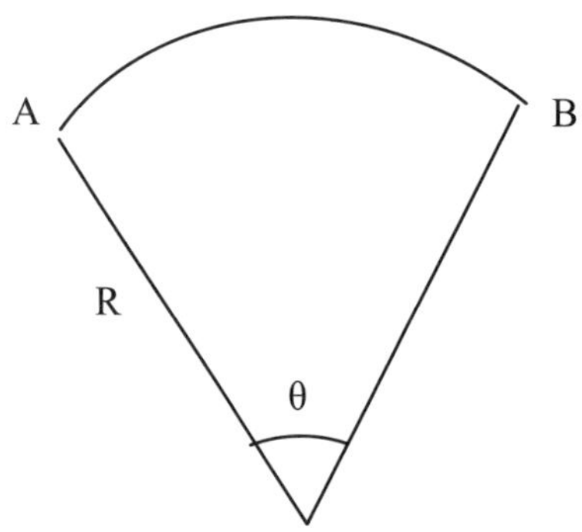

Arc AB = R. θ

(θ should be measured in radians).

Wetted Perimeter (P) = R x θ = 6 x 4.00 = 24 ft

STEP 3: Find the hydraulic radius (R);

Hydraulic radius (R) = A/P = 85.64/24 = 3.568

STEP 4: Find the flow (Q);

$Q = 1.49/n \times A\ (R)^{2/3}\ S^{1/2}$

Q = Flow in cfs;

n = Manning's coefficient; (n= 0.02)

A = Area (A = 85.64 sq. ft)

R = Hydraulic Radius = A/P, Where A is the area and P is the wetted perimeter; (R = 3.568)

S = Slope in decimals; (S = 0.5% = 0.005)

$Q = 1.49/0.02 \times 85.64 \times\ (3.568)^{2/3}\ 0.005^{1/2}$

Q = 1,053.3 cu.ft/sec (Ans B)

Problem 4.4) Triangular channel is shown below. The flow rate was found to be 1,200 gallons/minute. Manning coefficient was found to be 0.003 and slope was found to be 0.0025. Find the depth of the channel.

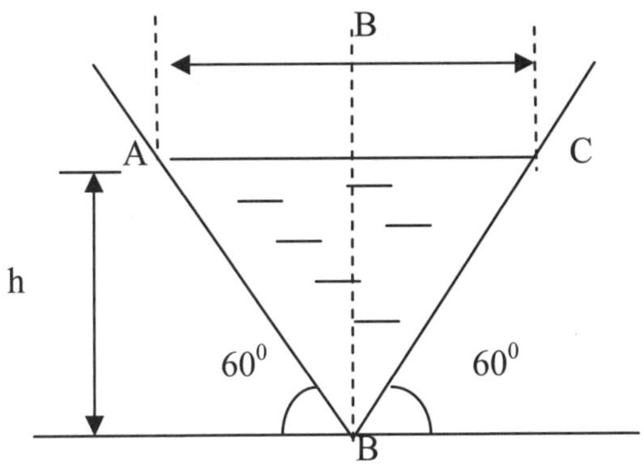

A) 0.75 ft B) 1.28 ft C) 0.82 ft D) 0.55 ft

Solution 4.4):

STEP 1: Gather all the information given.

Flow rate is given to be 1,200 gallons per minute. Flow rate needs to be converted to cu. ft/sec.

Q = 1,200 gallons/min = 20 gal/sec
1 cu. ft = 7.48 gallons

Q = 20/7.48 cu.ft/sec = 2.67 cu.ft/sec

STEP 2: Apply the Manning equation:

$$v = \frac{1.49 \times R^{2/3} S^{1/2}}{n}$$

v = Velocity = Q/Area

Q = 2.67 cu.ft/sec (Found in step 1)
A = Area = B x h/2

From trigonometry; $B/2 = h \times \tan 30^0$.

$B/2 = h \times 0.577$
$B = 1.154\ h$

A = Area = B x h/2 = (1.154 h) x h/2 = $0.577\ h^2$.

Velocity = Q/Area = $2.67/(0.577\ h^2) = 4.63/h^2$.

STEP 3: Find the hydraulic radius (R):

R = A/P

P = Wetted perimeter.
P = AB + BC

From trigonometry; AB Sin 60 = h

$AB = h/\sin 60$
$AB = h/0.866 = 1.155h$

$P = AB + BC = 2 \times 1.155\ h = 2.31\ h$

$R = A/P = (0.577h^2)/(2.31h) = 0.25\ h$

Slope (S) is given to be 0.0025.
Now you are ready to apply the Manning's equation.

$$v = \frac{1.49 \times R^{2/3}\ S^{1/2}}{n} = \frac{1.49 \times (0.25h)^{2/3} \times (0.0025)^{1/2}}{0.003} = 9.85\ h^{2/3}$$

Velocity was found to be $4.63/h^2$ (See Step 2)

$4.63/h^2 = 9.85\ h^{2/3}$
$0.47 = h^2 \times h^{2/3} = h^{(2+2/3)} = h^{2.66}$
$h = 0.753$ ft (Ans A)

Problem 4.5): Find the total head added by the pump in the figure below.
Assume the following;

- Friction head loss in pipes is zero.
- Pressure at point B is 3 psi
- Velocity head at point B is negligible.

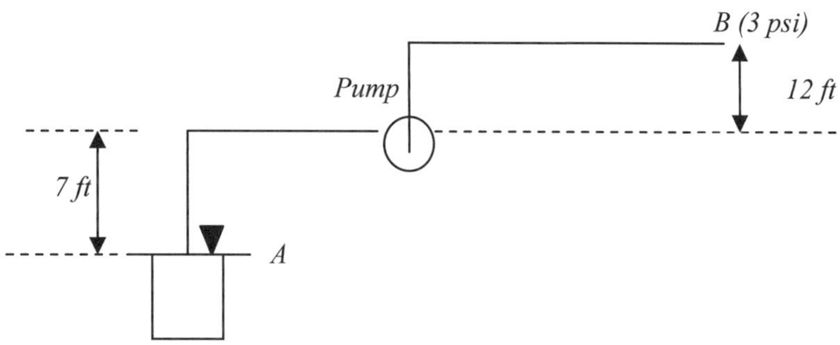

A) 5.0 ft B) 25.9 ft C) 7.1 ft D) 12.2 ft

Solution 4.5):

Total head added by the pump = Head at point B – Head at point A + Friction head loss

Head at point A = $Z_A + V_A^2/2g + P_A/\gamma$
$Z_A = 0$
$V_A = 0$ (Velocity of a reservoir surface is taken to be zero)
P_A = Pressure at open reservoir is zero.

Head at point B = $Z_B + V_B^2/2g + P_B/\gamma$

$Z_B = 19$ ft
$V_B^2/2g = 0$ (Problem states to ignore the velocity head at point B)
$P_B = 3$ psi

$P_B/\gamma = 3 \times 144/62.4 = 6.92$ ft

Total head added by the pump =Head at point B – Head at point A + Friction head loss
Total head added by the pump = (19 + 0 + 6.92) – (0 + 0 + 0) + 0
Total head added by the pump = 25.92 ft (Ans B)

Discussion: Suction Head and Discharge Head:
Pumps typically pump water from a low point to a high point. In such situations, suction head and discharge head are shown in the figure below.

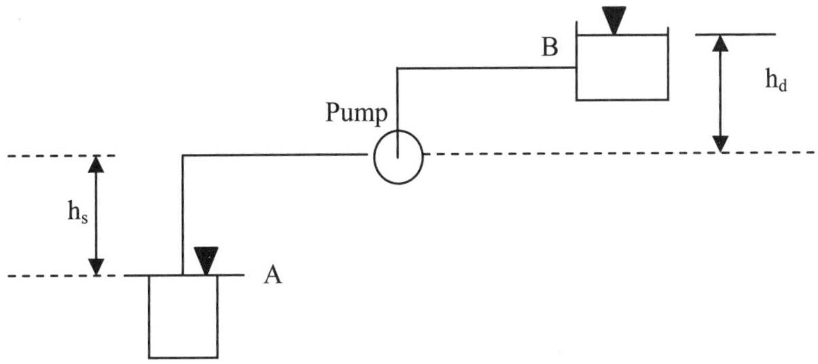

h_s = Suction head h_d = Discharge head

Pump has to suck water and then discharge to a higher elevation.

Problem 4.6): Parallel pipe system is shown in the figure.

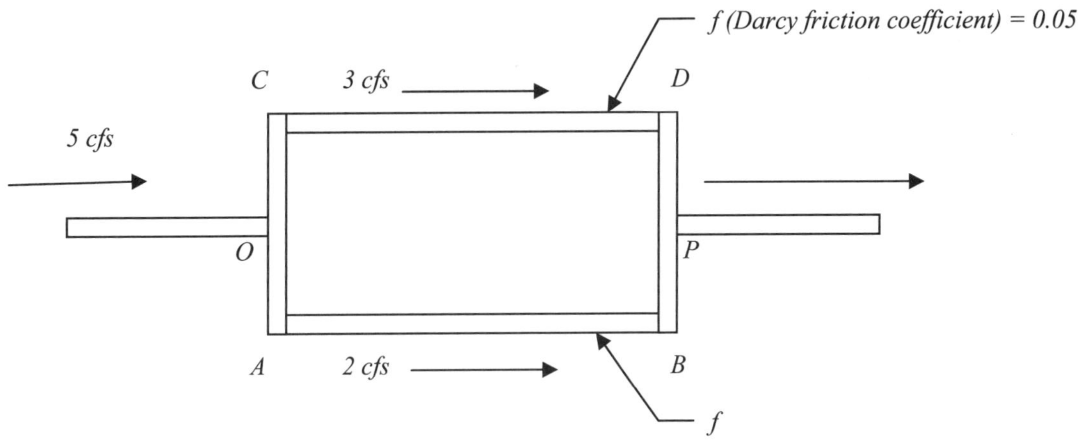

Length of pipe along OABP = 120 ft
Length of pipe along OCDP = 170 ft
Diameter of all pipes are 6 inches.
Find the Darcy friction coefficient of pipe AB.
A) 0.022 B) 0.114 C) 0.003 D) 0.159

Solution 4.6):

Following rule is needed to solve this problem.

In parallel pipes, friction head loss along all paths are the same

Friction head loss along CD = Friction head loss along AB

STEP 1: Find the friction head loss along pipe CD;

Head loss along pipe section CD = $f.L.v^2/2gd$

f = 0.05;
L = 170 ft

Velocity (v) = Flow/Area

Flow along CD = 3 cu.ft/sec
Area of pipe CD = $\pi \times d^2/4 = \pi \times 0.5^2/4 = 0.1963$ sq. ft

Velocity (v) = Flow/Area
v = 3/0.1963 = 15.28 ft

Head loss along pipe section CD = $\frac{f.L.v^2}{2.g.d} = \frac{0.05 \times 170 \times 15.28^2}{(2 \times 32.2 \times 0.5)} = 61.6$ ft

STEP 2: Find the head loss along pipe AB;

Head loss along pipe section AB = $\frac{f.L.v^2}{2.g.d} = \frac{f \times 120 \times v^2}{(2 \times 32.2 \times 0.5)}$

Velocity in pipe AB = Flow/Area

Area of pipe AB = $\pi \times d^2/4 = \pi \times 0.5^2/4 = 0.1963$ sq. ft

Velocity (v) = Flow/Area
Velocity in pipe AB = 2/0.1963 = 10.19 ft/sec

Head loss along pipe AB = $\frac{f.L.v^2}{2.g.d} = \frac{f \times 120 \times 10.19^2}{(2 \times 32.2 \times 0.5)} = 386.97$ f

STEP 3: Find "f";

Head loss along pipe AB = Head loss along pipe CD
386.97 f = 61.6 ft
f = 0.159 (Ans D)

Problem 4.7): Suction head of a pump is 8 ft and the discharge head is 9 ft. The flow is 300 gal/min. Find the horse power of the pump. Ignore the head loss due to friction and velocity head in pipes.

A) 5.3 HP B) 1.2 HP C) 0.3 HP D) 4.1 HP

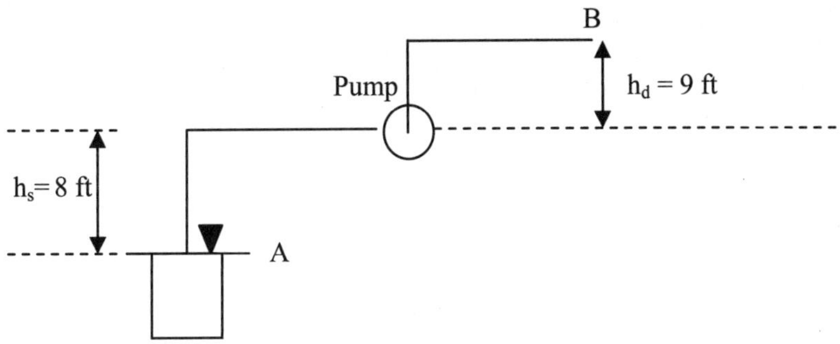

Solution 4.7):

Pump horse power is given by the following equation.

Pump horse power (HP) = h x m'/550

h = Head added by the pump.
m' = Flow in lbs/sec.
Note that in pump horsepower calculations, unit of flow is NOT cu.ft/sec but lbs/sec.

Typically flow is given in cu.ft per sec or gallons/minute. This has to be converted to lbs/sec.

Head added by the pump = Suction head + discharge head = 8 + 9 = 17 ft

Pump horse power (HP) = h x m'/550

Flow is given in gallons/minute. This has to be converted to lbs/sec.
Flow = 300 gal/min = 300/60 gal/sec = 5 gal/sec

1 gallon = 8.342 lbs; Hence 5 gal/sec = 5 x 8.342 lbs/sec.= 41.71 lbs/sec

h = 17 ft; m' = 41.71 lbs/sec

Pump horse power = h x m'/550 = 17 x 41.71/550 = 1.29 HP (Ans B)

Problem 4.8): A 6 in pipe with Tee runs, elbows and valves are shown below. There are two valves, one Tee run and five elbows. Energy loss at elbows, Tee runs and valves are given below. Flow is 200 gallons per minute and total length of pipes is 4,400 ft.

Energy losses (in ft of water):

Elbow = 2.1 ft;
Tee run = 2.8 ft;
Valve = 1.3 ft

Find the Darcy coefficient of the pipe line.

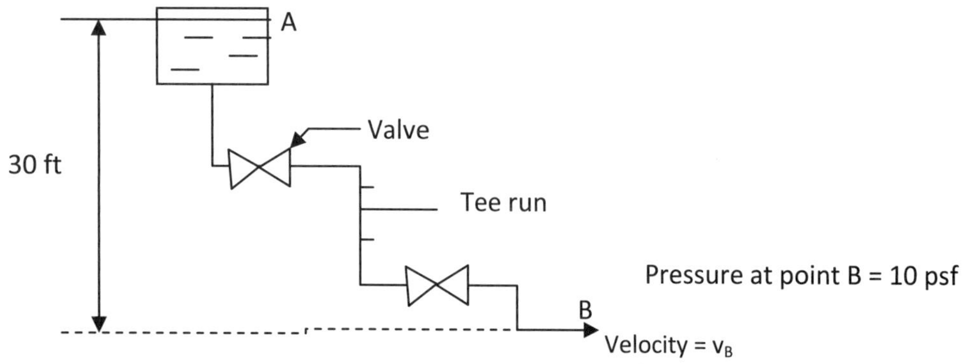

A) 0.020 B) 0.080 C) 0.002 D) 0.110

Solution 4.8):

Problem is to find the Darcy coefficient of the pipe line. When water flows from point A to point B, there is energy loss due to elbows, valves and tee runs as well as friction of the pipe.

Minor losses due to valves, Tee runs and elbows: Loss of energy in Tee runs, valves and elbows are given in ft of water. In some problems minor losses are given as equivalent length of pipe. In that case equivalents length of pipe should be added to the length of the pipe.

Energy loss due to elbows = 5 x 2.1 = 10.5 ft (There are 5 elbows)
Energy loss due to Tee runs = 2.8 ft (Only one Tee run)
Energy loss due to valves = 2 x 1.3 = 2.6 ft (There are 2 valves)

Total energy loss due to valves, elbows and tee runs = 10.5 + 2.8 + 2.6 = 15.9 ft

Energy at point A = $p_1/\gamma + v_1^2/2g + z_1$
$= 0 + 0 + 30 = 30$ ft
Pressure is zero at point A since it is open to atmosphere.
Velocity of a reservoir surface is negligible and considered to be zero.

Energy at point B = $p_2/\gamma + v_2^2/2g + z_2$
$= 10/62.4 + v_B^2/2g + 0$

When water flows from point A to point B, energy is lost due to friction in pipes and minor losses due to valves, Tee runs etc.
Hence following equation can be written down.

Energy at point A = Energy at point B + Energy losses due to valves, elbows and Tee runs + Energy loss due to friction in pipe

$$30 = 10/62.4 + v_B^2/2g + 15.9 + \text{Energy loss due to friction} \quad \text{------------------------------------} \quad (A)$$

Flow is given to be 200 gallons per minute.
(7.48 gallons = 1 cu.ft)
Q = 200 gallons per minute = 200/7.48 cu. ft/min
= 26.7 cu.ft/min
= 0.44 cu.ft/sec
Q = Area x velocity = $\pi. D^2/4 \times v = \pi. 0.5^2/4 \times v = 0.1963 . v$
Q is 0.44 cu. ft/sec
Hence 0.44 = 0.1963 v
v = 2.24 ft/sec

From above equation (A) $30 = 10/62.4 + 2.27^2/2g + 15.9$ + Energy loss due to friction in pipe (h_L)

Energy loss due to friction in pipe (h_L) = 13.86 ft

Energy loss due to friction in pipes is given by the Darcy formula.

Darcy Formula $h_L = f.L.v^2/2gD = \frac{f \times 4{,}400 \times 2.24^2}{(2 \times 32.2 \times 0.5)} = 685.6\ f$

$h_L = 13.86 = 685.6\ f$
$f = 0.02$ (Ans A)

Note: After any problem it is a good idea to see how a problem can be twisted in the exam in a different way.

In this case all the necessary parameters were given except friction coefficient (f). Since all the other parameters such as length, diameter, velocity and pressure are given head loss due to friction can be found. Then Darcy equation was used to find the friction coefficient (f). This same problem can be twisted. It is possible to give (f) and ask for the velocity at point B. Or it is also possible to give (f) and velocity and ask for the elevation difference between points A and B. Please go thru the example and see how to solve the problem in each case.

Discussion:

When water flows thru a pipe, energy is dissipated due to friction in the pipe. This is known as friction head loss. Head loss due to friction can be computed using Darcy Weishback formula or Hazen William formula. Also energy is lost due to elbows, Tee runs and valves. These losses are known as minor losses.

Friction and Minor Losses:

Energy in a fluid can be in three different forms. They are pressure, kinetic and gravity. Energy in a fluid can be lost due to friction. There are other causes for energy loss.

Elbows: There is considerable energy loss due to elbows. Energy losses due to elbows are given in ft.

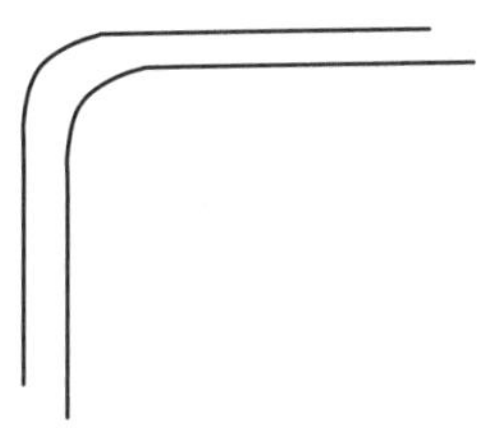

Valves: Valves are required to control the flow. Sometimes flow rate has to be reduced or completely stopped. There are many types of valves in use today. Ball valves, butterfly valves and gate valves are the most used.

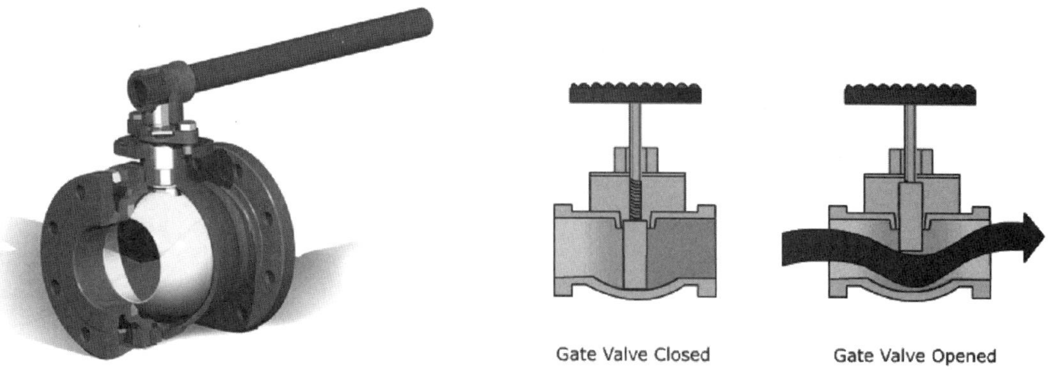

Ball valve on the left and gate valve on right.

Butterfly valve

In the case of a ball valves, there is a ball with a hole inside the pipe. In the case of butterfly valves, there is a plate that rotates. in the case of gate valves, there is a gate that can be lifted or lowered.

Tee Runs: Tee runs are needed to branch away flow from a pipe.

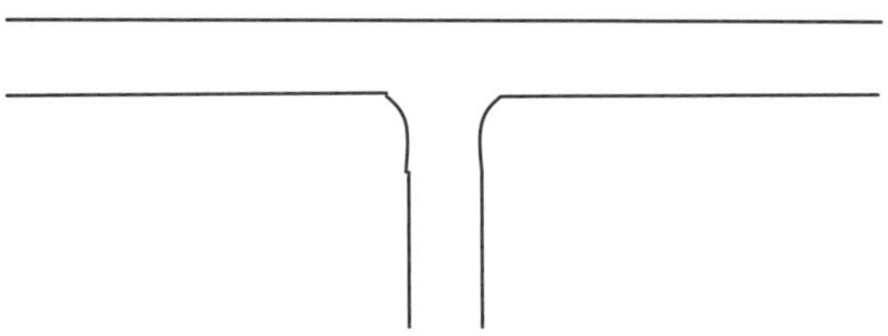

Symbols: In drawings typically following symbols are used.

Elbow *Tee run* *Valve*

Pipe Tee run

Problem 4.9): Approximate hydrograph for a stream is shown below.

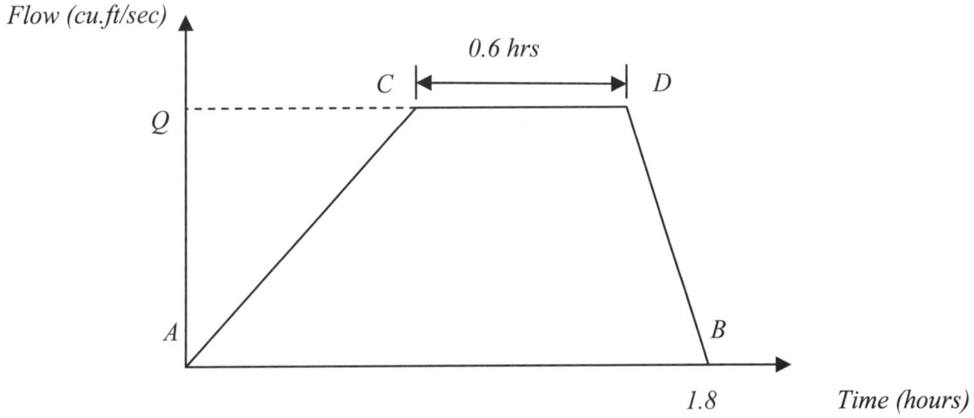

Area of the drainage basin was found to be 15 acres and the net precipitation was 1.3 inches.

Find the maximum flow (Q) in cu.ft/sec.

A) 16.4 cu.ft/sec B) 22.9 cu.ft/sec C) 76.7 cu.ft/sec D) 42.2 cu.ft/sec

Solution 4.9):

Area of the hydrograph provides the total flood volume. Total runoff is another name used for the total flood volume.

Area of the hydrograph = Runoff

Runoff is given by the following equation.

Runoff = Net precipitation x Area of the drainage basin

STEP 1: Find the area of the hydrograph;

Assume maximum flow to be Q cu.ft/sec.

Area of the hydrograph using geometry = (Top + Base)/2 x Height
= (AB + CD)/2 x Q

Time is given in hours. This needs to be converted to seconds.
AB = 1.8 hours = 1.8 x 3600 seconds;
CD = 0.6 x 3600 seconds

Area of the hydrograph = Runoff = (1.8 + 0.6) x 3,600 x Q/2 = 4,320 Q cu.ft

STEP 2: Find the runoff;

Runoff = Net precipitation x Area of the drainage basin
Net precipitation = 1.3 inches;
Area of the drainage basin = 15 acres
Runoff = 1.3 x 15 acre.in
Convert all units to ft.
Runoff = (1.3/12) x 15 x 43,560 cu.ft = 70,785 cu.ft

Area of the hydrograph provides the runoff.
Hence ;
4320 Q = 70,785 cu.ft
Q = 16.4 cu.ft/sec (Ans A)

Problem 4.10): Find the runoff for a unit hydrograph for the data given in the previous problem.

A) 54,450 cu.ft B) 64,950 cu.ft C) 12,512 cu.ft D) 23,887 cu.ft

Solution 4.10):

Net precipitation = 1.3 inches

Runoff = 70,785 cu.ft (Found in the previous problem).

Runoff for "1" net precipitation = 70,785/1.3
= 54,450 cu.ft (Ans A)

STRUCTURAL:

Problem 4.11) What is NOT an occupancy category with regard to seismic design
A) Buildings located in soft soils
B) Buildings that are not subjected to human occupancy.
C) Buildings with large number of occupants.
D) Buildings which are essential for post earthquake response work.

Solution 4.11):

Ans A

There are four occupancy categories. (ASCE 7.1)

Category I - Buildings normally not subjected to human occupancy. (ex. warehouses, storage buildings)

Category II - Buildings for ordinary occupancy (Residential buildings)

Category III - Buildings with large number of occupants (High rise buildings, sport arenas)

Category IV - Buildings that are essential for post earthquake response work. Also buildings that house hazardous material.

Problem 4.12) A building is located in a high groundwater area with loose sandy soil. The site class of the building in relation to seismic design is

A) Site class A or B
B) Site class C or D
C) Site class D
D) Site class F

Solution 4.12): Ans D (Site class F)

There are 6 site classes as per ASCE 7.1.

A - Hard rock
B - Rock
C - Very dense soil and soft rock
D - Stiff soil
E - Soft clay soil
F - Unstable soils that could liquefy or collapse due to an earthquake.

Loose sand in high groundwater areas can undergo liquefaction due to an earthquake. Hence it can be considered as unstable soil.

Problem 4.13) Seismic response coefficient (C_s) of a building was found to be 0.10. The effective seismic weight of a column is 300 tons. What is the seismic shear force acting on the column base.

A) 30 tons B) 20 tons C) 3,000 tons D) 200 tons.

Solution 4.13):

$$\text{Seismic base shear (S)} = C_s.\ W$$

C_s = Seismic response coefficient
W = The effective seismic weight

Seismic base shear (S) = 0.1 x 300 = 30 tons (Ans A)

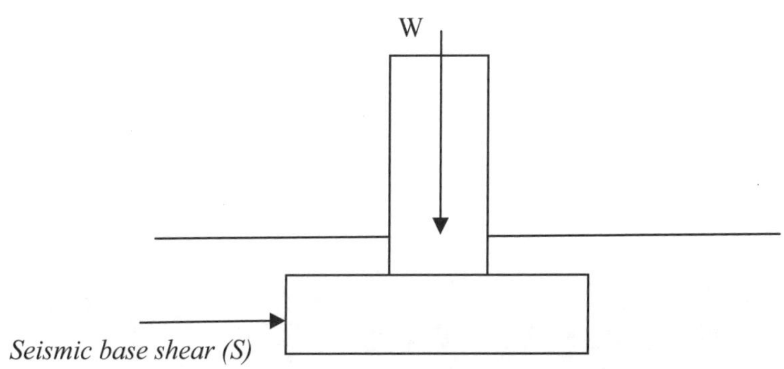

Problem 4.14): Find the area of steel required for a balanced section for the beam shown below.

Steel Young's modulus = 29 x 10^6 psi.
f_y= 60,000 psi
Concrete compression strength (f_c') = 4,000 psi
Concrete is assumed to crush when the strain in concrete is equal to 0.003. (ACI 318)

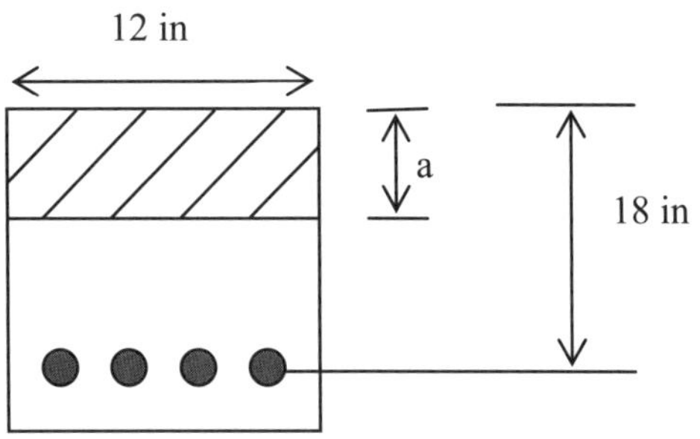

A) 7.08 sq. in B) 6.12 sq. in C) 1.24 sq. in D) 5.44 sq. in

Solution 4.14):

In a balanced section, yielding of steel and crushing of concrete occurs at the same time.

Steel yields when the steel strain = f_y/E

E = Young's modulus of steel
f_y = Steel yielding stress

This relationship is obtained by simple stress strain equation.

E = Stress/Strain

Strain = Stress/E = f_y/E

For a balanced section,
Concrete crushes at the same time steel start to yield.

STEP 1: Draw the strain diagram.

d = Depth (from top of the beam to center of gravity of rebars).
0.003 = Strain in concrete when concrete start to crush
f_y/E = Strain in steel when steel start to yield.
f_y/E = 60,000/29 x 10^6 = 0.0021

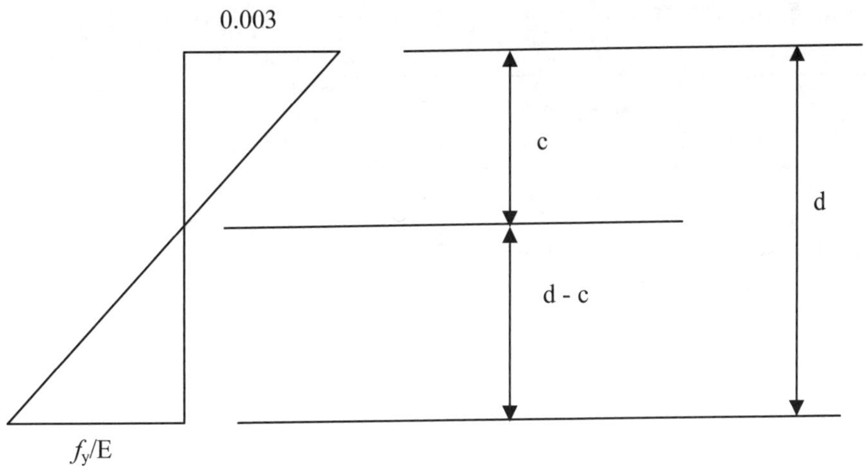

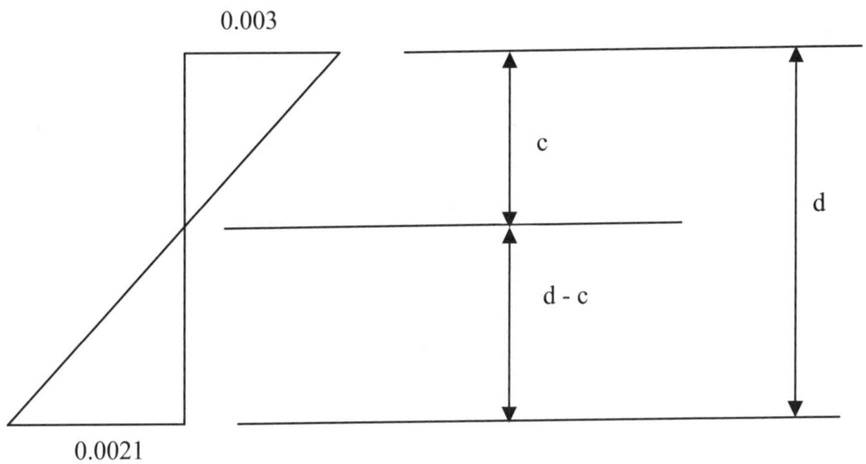

By proportioning;

$$\frac{0.003}{c} = \frac{0.0021}{(d - c)}$$

$0.003d - 0.003c = 0.0021c$
$0.003\ d = 0.0051\ c$
$c = 0.5882d$

$a = 0.85\ c = 0.85 \times 0.5882d = 0.5\ d$

$a = 0.5\ d = 9$ in.

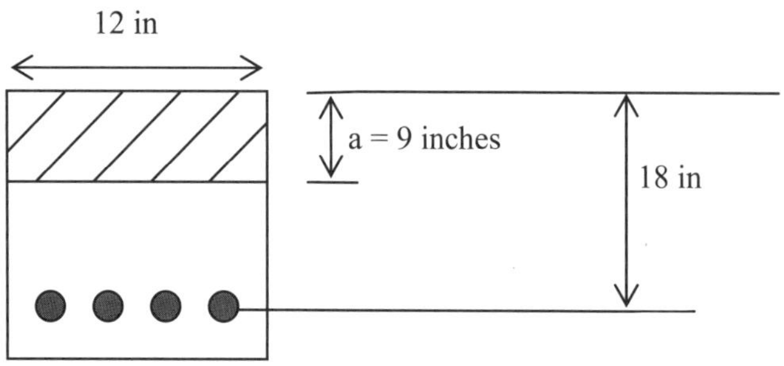

"d" is given to be 18 inches.
Hence;

a = 9 inches
A_c = a x b = 9 x 12 = 108 sq. in

STEP 2: Find the force in concrete for a balance section;

Force in concrete = 0.85 f_c' A_c
Force in concrete = 0.85 x 4,000 x 108
= 367,200 lbs. in

STEP 3: Equate force in concrete to force in steel;

Force in steel = f_y. A_s
Force in steel = Force in concrete
f_y. A_s = 367,200 lbs. in

60,000 x A_s = 367,200 lbs. in
A_s = 6.12 sq. in (Ans B)

Problem 4.15): Find the nominal moment of the T-beam given.

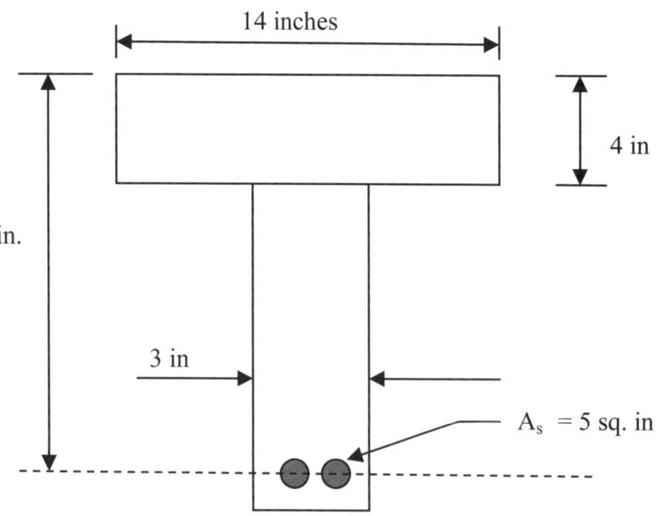

Steel yield stress = 60,000 psi
Concrete compressive strength = 4,000 psi
Steel area = 5 sq. in

A) 483 kip. ft B) 271 kip. ft C) 342 kip. ft D) 381 kip. ft

Solution 4.15):

STEP 1: Write the force balance equation.

Force in concrete = $0.85. f_c' A_c$

f_c' = Concrete compressive strength = 4,000 psi
A_c = Compressed area of concrete
(A_c is not known. A_c needs to be found).

Force in steel = $f_y. A_s$

f_y. = Steel yield stress = 60,000 psi
A_s = Area of steel = 5 sq. in

Force in concrete = Force in steel

$$0.85. f_c' A_c = f_y. A_s$$

$0.85. f_c' A_c = f_y. A_s$

$0.85 \times 4{,}000 \times A_c = 60{,}000 \times 5$

A_c = 88.2 sq. in

Area of the flange = 4 x 14 = 56 sq. in.

Compressed area of concrete goes below the flange.

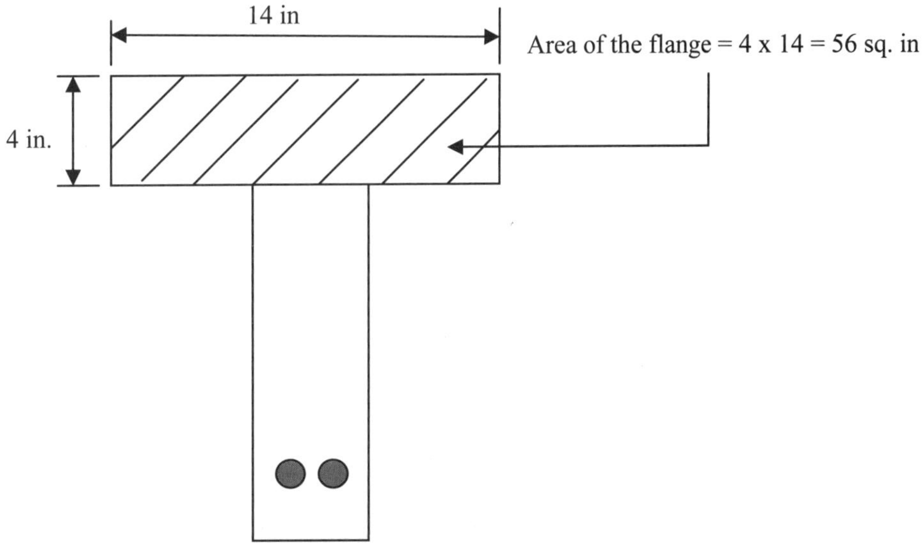

Area of the stressed section = 88.2 sq. in.

STEP 2: Draw the stressed section in concrete;

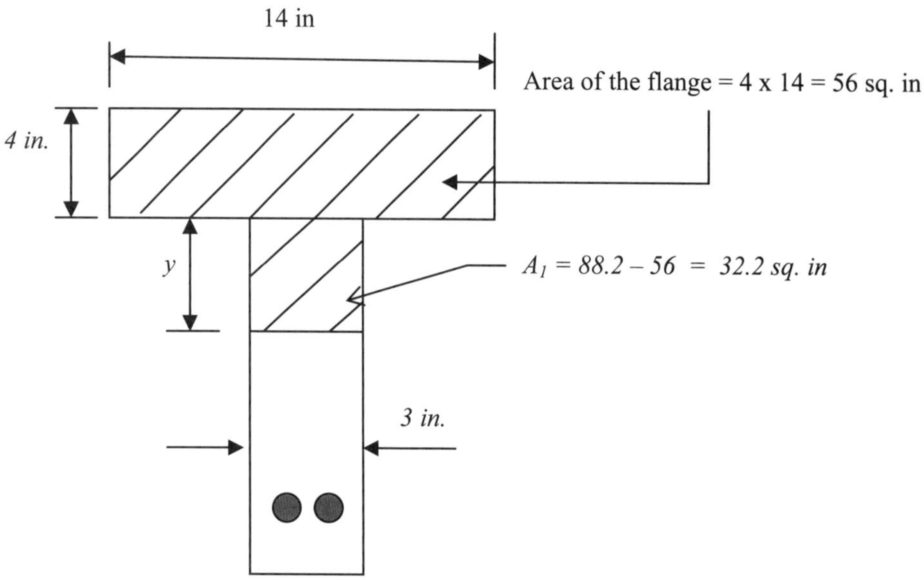

Area of the stressed section in concrete = 88.2 sq. in.
Area of the flange = 56 sq. in

Stressed area in the web = 88.2 – 56 = 32.2 sq. in
Distance "y" = 32.2/3 = 10.7

STEP 3: Find the center of gravity of the stressed section in concrete;

Assume the distance to center of gravity of the stressed section is "C" from the top.

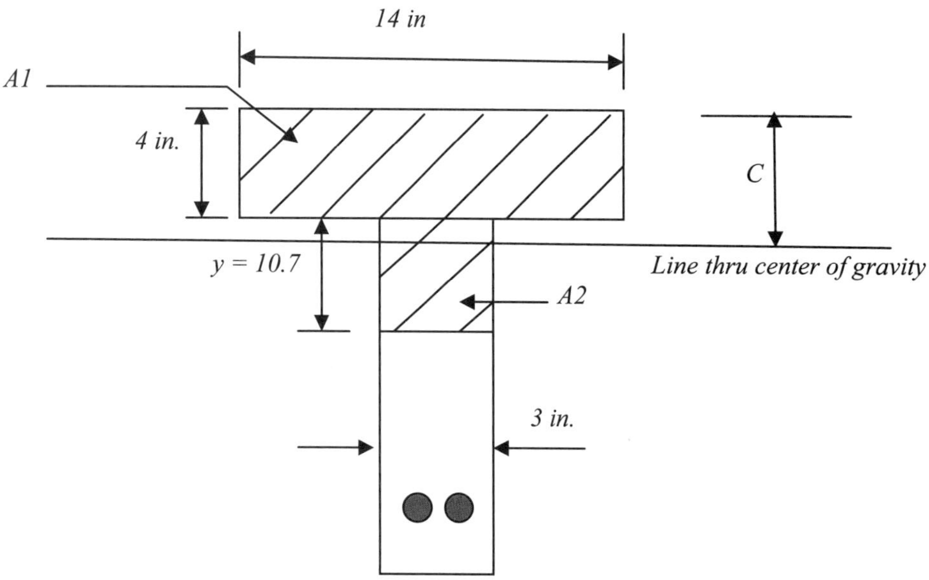

Equation to find the center of gravity;

Total stressed area x Distance to center of gravity = Area of flange x Distance to center of gravity of the flange + Stressed area of web x Distance to center of gravity of the stressed area in the web

Total stressed area = 88.2 sq. in
Distance to center of gravity measured from top = C
Area of flange (A1) = 4 x 14 = 56 sq. in
Distance to center of gravity of the flange from top = 4/2 = 2 inches
Stressed area of web = 10.7 x 3 = 32.2 sq. in
Distance to center of gravity of the stressed area in the web measured from top = 4 + 10.7/2

88.2 x C = (4 x 14) x 2 + 32.2 x (4 + 10.7/2)

C = 4.67
Center of gravity of the stressed area is below the flange.

STEP 4: Find the moment arm (Z);

Moment arm is the distance between center of gravity of concrete stressed area and steel area.

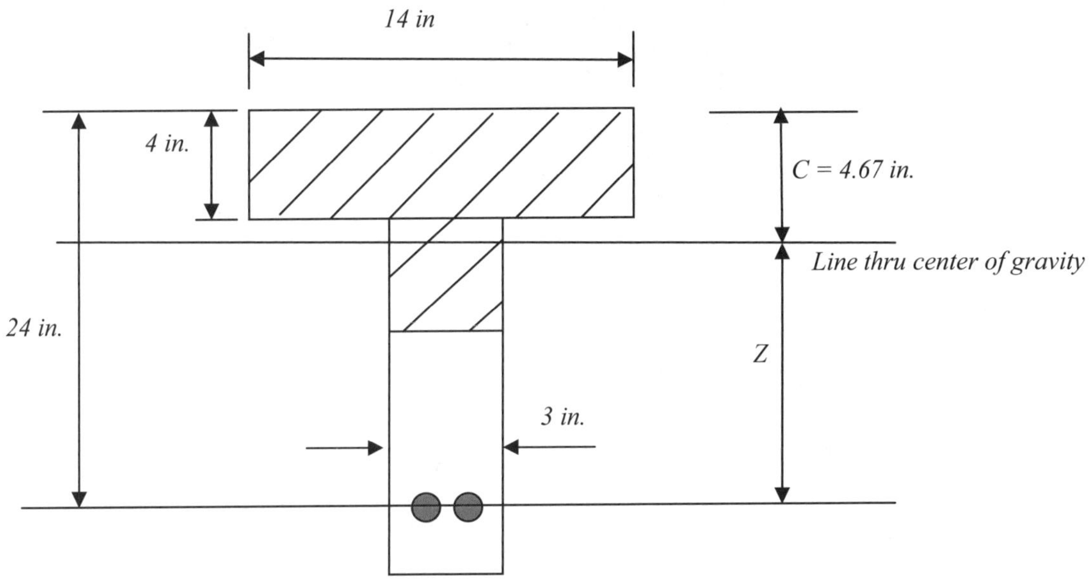

Z = 24 – 4.67 = 19.33 in.

STEP 5: Find the nominal moment of the section;

Force in steel = f_y. A_s

Since force in steel is equal to force in concrete one can also use force in concrete for the above equation.

Nominal moment = Force in steel x Moment arm

Nominal moment = f_y. A_s x Z

Nominal moment = 60,000 x 5 x 19.33
= 5,799,000 lbs. in
= 5,799,000/(12 x 1,000) kip. ft
= 483.3 kip. ft (Ans A)

Problem 4.16): Column to beam connection is shown in the figure. It can be considered as a
A) Moment connection B) Shear connection C) Can't tell D) Fixed end

Solution 4.16):

The beam can rotate due to horizontal forces laterally. Hence it is a shear connection. The connection cannot be considered as a moment connection or a fixed end. (Ans B)

This photograph shows a welded connection. Beam is welded to the column. Swaying is resisted in all directions and can be considered as a moment connection or a fixed end.

The beam is bolted and at the same time welded to the column. Hence swaying is resisted and can be considered as a moment connection or a fixed end.

Problem 4.17): Find the moment of inertia (I) thru the centroid of the T- beam shown.

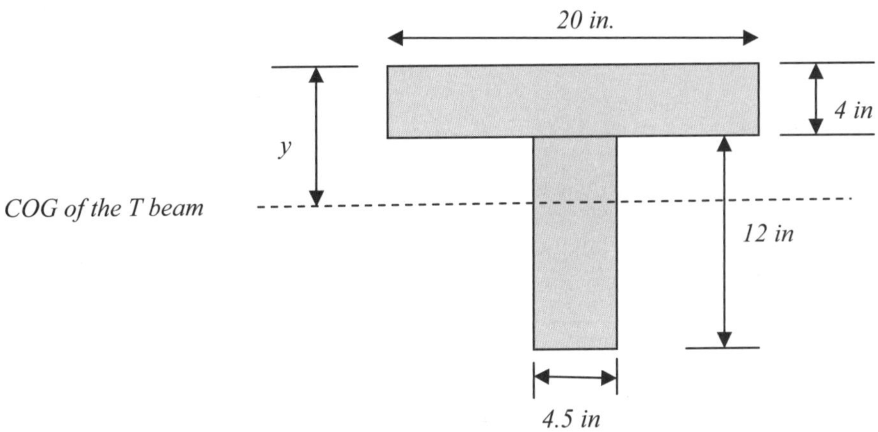

A) 2 ,143 in^4 B) 2,818 in^4 C) 986 in^4 D) 3,194 in^4

Solution 4.17):

STEP 1: Find the distance (y) to the center of gravity of the beam;

Divide the T beam to two sections. A1 and A2.
A1 is the flange and A2 is the web.

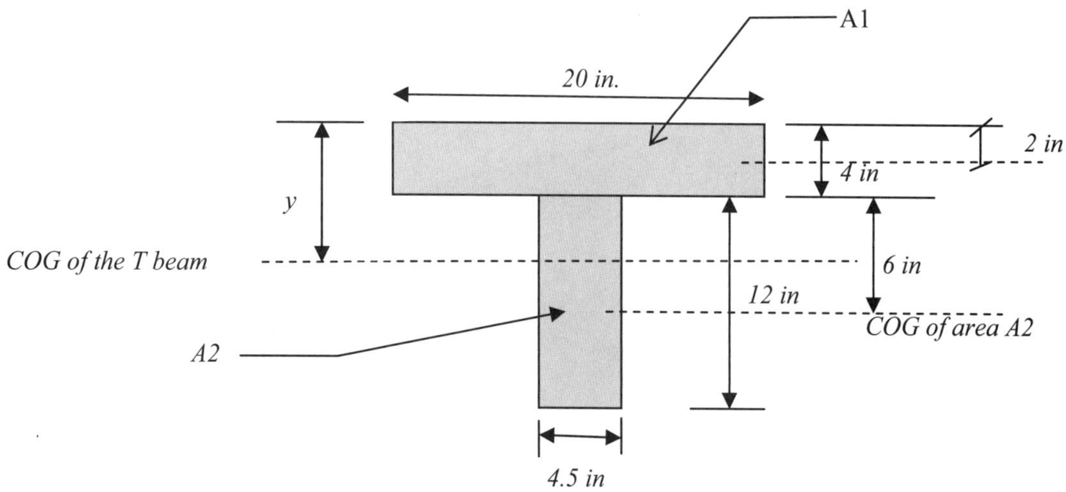

COG = Center of gravity

Center of gravity equation;

(A1 + A2) . y = (A1 x 2) + (A2 x 10)

Above "2" is the distance to center of gravity of area A1 from top.
Similarly "10" is the distance to center of gravity of area A2 from top. (6 + 4)

A1 = 20 x 4 = 80 sq. in
A2 = 4.5 x 12 = 54 sq. in

(80 + 54) . y = (80 x 2) + (54 x 10)

y = 5.22 in.

STEP 2: Find the moment of inertia (I) of A1 and A2 thru their own center of gravities.

Moment of inertia of a rectangular section is given by the following equation;

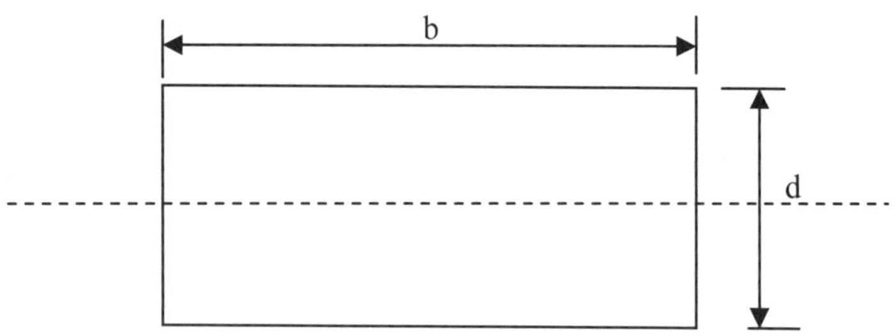

$$I = 1/12\ b.\ d^3$$

Moment of inertia of A1 thru its own center of gravity = I_{c1} = $1/12 \times 20 \times 4^3$
= $106.7\ in^4$

Moment of inertia of A2 thru its own center of gravity = I_{c2} = 1/12 x 4.5 x 12^3
= 648 in^4

STEP 3: Transfer the moment of inertia of A1 and A2 thru center of gravity of the T beam.

We have found the moment of inertia of A1 and A2 thru their own center of gravities.

Now we need to transfer them thru the center of gravity of the T-beam. Center of gravity of the T-beam is 5.22 in. below the top of the beam.

Equation to transfer moment of inertia;

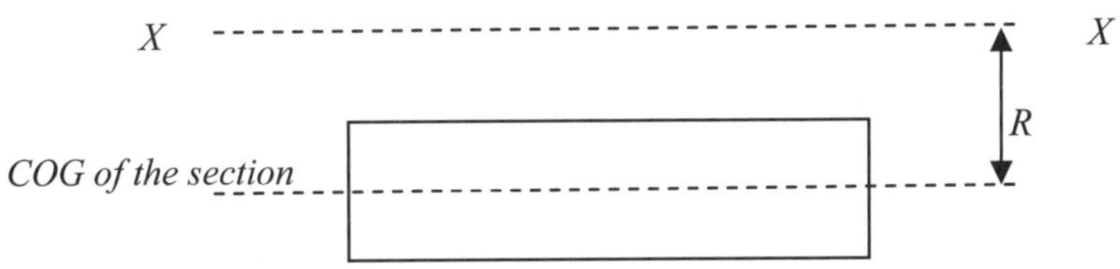

Moment of inertia thru COG of the section (center of gravity) = I_c

Moment of inertia thru XX = I_c + $A.R^2$

A = Area of the section
R = Distance between COG and line of reference

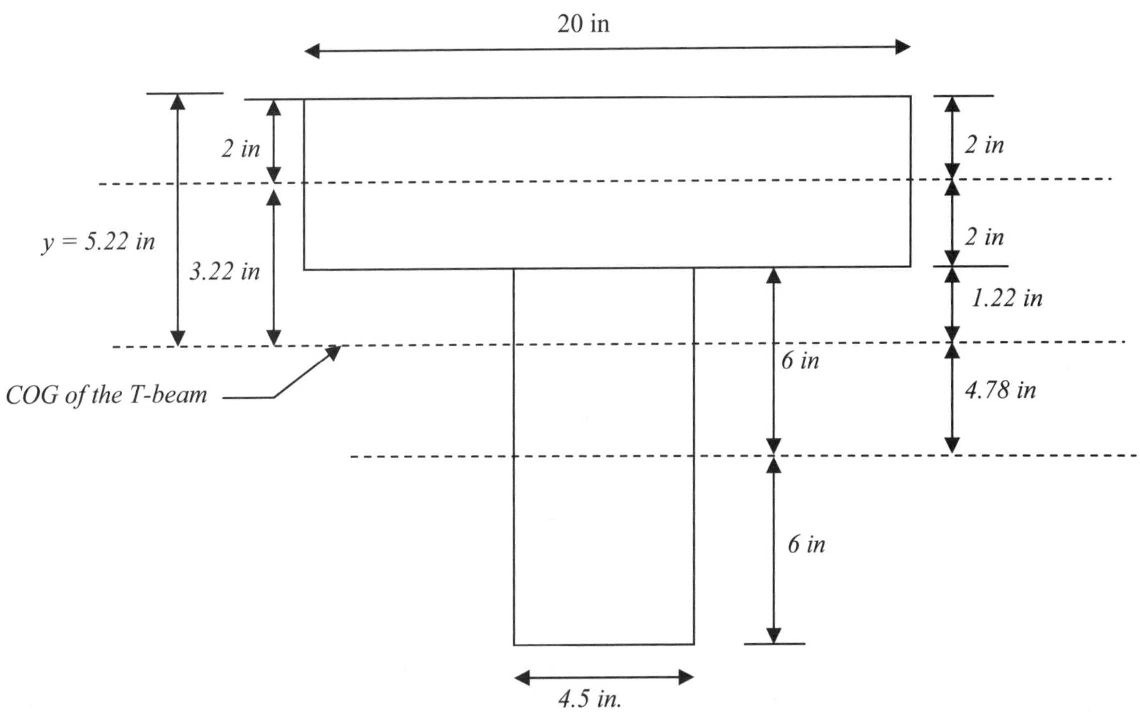

- Transfer moment of inertia of area A1 thru the COG of the T-beam.

  $I_1 = I_{c1} + A.R^2$

Moment of inertia of A1 thru its own center of gravity = I_{c1} = 1/12 x 20 x 4^3
= 106.7 in^4

R = 3.22 in.
"R" is the distance between center of gravity of A1 and center of gravity of the T-beam. (See the figure)

Hence;
$I_1 = I_{c1} + A1.\ R^2$
$I_1 = 106.7 + (20 \times 4).\ 3.22^2$
$I_1 = 936.2\ in^4$

- Transfer moment of inertia of area A2 thru the COG of the T-beam.

  $I_2 = I_{c2} + A2.R^2$

Moment of inertia of A2 thru its own center of gravity = I_{c2} = 1/12 x 4.5 x 12^3
= 648 in^4

R = 4.78 in. (See the figure)

"R" is the distance between center of gravity of A2 and center of gravity of the T-beam.

Hence;
$I_2 = I_{c2} + A2.R^2$
$I_2 = 648 + (12 \times 4.5) \times 4.78^2$
$I_2 = 1{,}881.8\ in^4$

Total moment of inertia (I) = $I_1 + I_2$
I = 936.2 + 1,881.8
I = 2,818 in^4. (Ans B)

Problem 4.18): W18 x 67 column has one end pinned and other end fixed. What is the nominal compression strength of the column in kips? Length of the column is 12 ft.

	Effective length factor (K)
Both ends pinned	1.0
One end pinned, other end fixed	0.8
Both ends fixed	0.65

Portion of AISC Table 4.1 (W 18 x 67)

K.L (with respect to least radius of gyration)	Axial compression strength (kips)
7	790
8	762
9	733
10	701
11	667
12	632
13	596

A) 596 kips B) 632 kips C) 667 kips D) 701 kips

Solution 4.18):

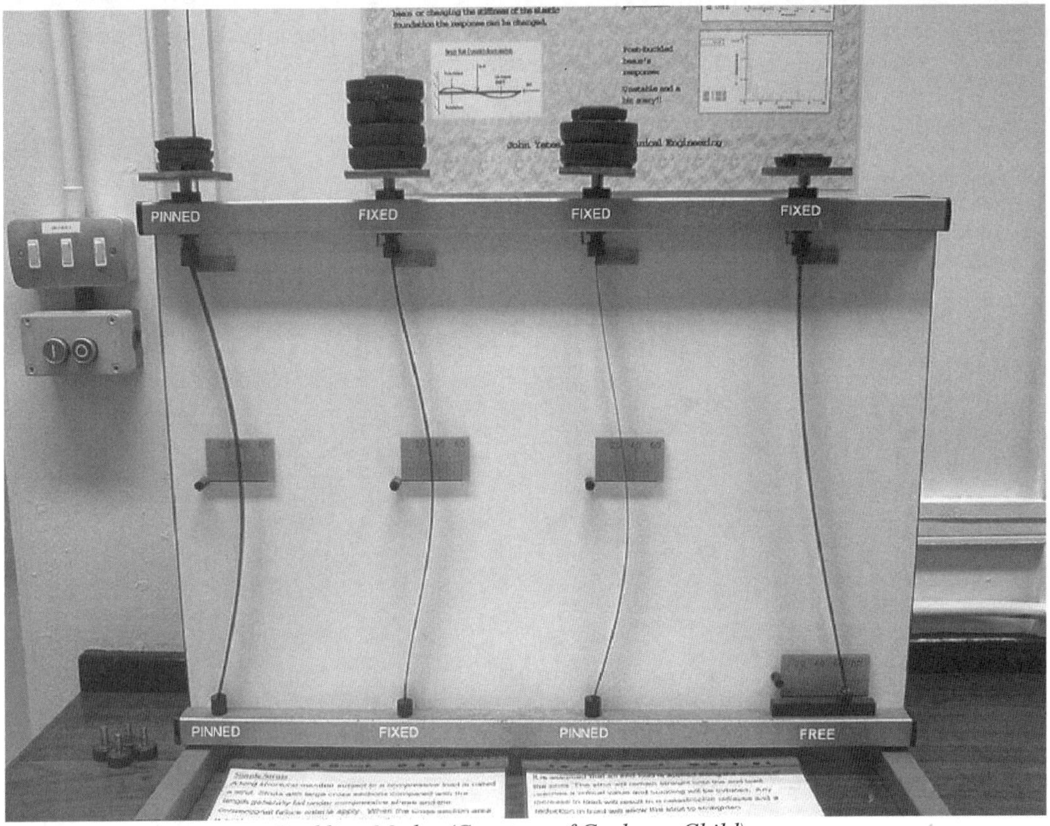

Buckling Modes (Courtesy of Grahams Child)

Left corner column is pinned from both ends.;
Both ends are fixed in the next one;
One end is fixed and the other end is pinned in the third one;
One end is free and the other end is fixed in the one on far right. (This is not a practical situation)
Second column from left can carry the most load. Though columns with fixed ends can carry more load, fixed end connections can be expensive. Typically fixed end connections are achieved by welding or by providing adequate amount of bolts.

When one end is pinned and other end is fixed, K = 0.8.

L = 12 ft
K.L = 0.8 x 12 = 9.6
To be conservative, use 10 for K.L.
Hence axial compressive strength = 701 kips (Portion of AISC table 4.1 is given in the problem).
(Ans D)

TRANSPORTATION:

Problem 4.19): Station at point of intersection (PI) is given to be 112 +70 and the degree of curve (arc method) is given to be 2.3^0. Find the station at PT. (Angle of intersection is 28^0 15' 08").

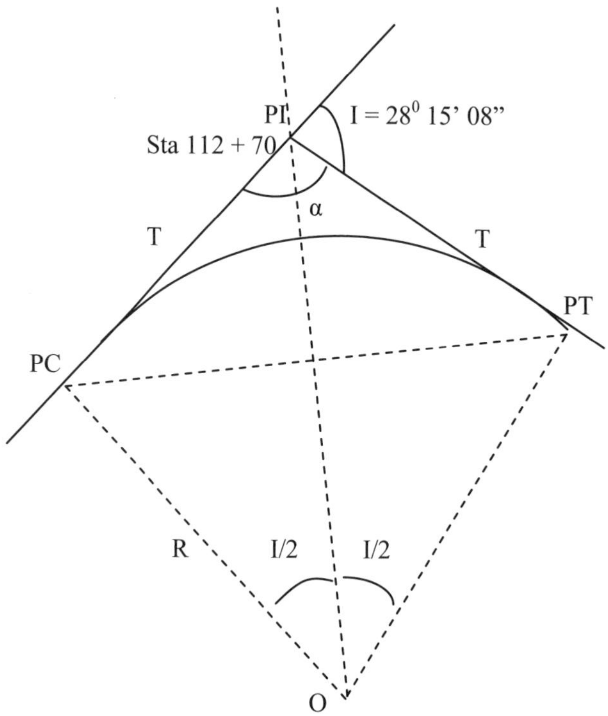

A) 118 + 71 B) 134+75 C) 220+67 D) 234+81

Solution 4.19):

In this problem, station at PT is required.

Station to PT is always measured **along the curve** from PC.

Never measure the station of PT from PI. That is wrong.
Always measure the distance to PT **along the curve** from PC.

As you could see radius of the circle is not given. Instead of giving the radius, degree of curve is provided.

There are two methods in highway surveying.

- Degree of curve (arc method)
- Degree of curve (chord method)

These methods were explained in sample exam 3.

Arc Method;

360^0 angle at the center of a circle will produce a full circle.

$360^0 \rightarrow 2\pi R$

1^0 degree angle at the center of a circle will produce an arc of

$1^0 \rightarrow 2\,\pi\,R/360$

D degree angle at the center of a circle will produce an arc of

$D^0 \rightarrow 2\,\pi\,R.D/360$

As per definition of degree of curve, D degrees will produce an arc of 100 ft.

Hence $2\,\pi\,R.D/360 = 100$ ft

$$D^0 = \frac{100 \times 360}{2\,\pi\,R} = \frac{5{,}729.5}{R} \quad \text{-----------------------------------} \quad (1)$$

In this problem degree of curve (D) is given to be 2.3^0.

Apply equation 1.

$$2.3 = \frac{5{,}729.5}{R}$$

$R = 2{,}491$ ft

As you could see, that the whole idea of giving the degree of curve is to obtain the radius of the circle.

STEP 1: Find the distance T

From trigonometry,
$T = R \tan I/2 = 2{,}491 \times \tan [(28^0\ 15'\ 08'')/2]$
$= 626.9$ ft

Station at PC = Station at PI – T
Station at PI = 112 + 70 = 11,270 ft
Station at PC = 11,270 – 626.9 = 8,929.2 = 10,643.09 = Sta. 106 + 43.09

STEP 2: Find the curve length from PC to PT:

Length of curve = $R \times \theta$
θ is the internal angle and has to be measured in radians.
Length along the curve from PC to PT = $R \times \theta$
Internal angle from PC to PT is $28^0\ 15'\ 08''$.
Hence $\theta = 28^0\ 15'\ 08''$
$\theta = 28^0\ 15'\ 08'' = 28.252^0$
This needs to be converted to radians.
Conversion from degrees to radians can be done using calculators. If not following equation can be used.

$\theta_{radians} = \theta_{degrees} \times \pi/180$
$\theta_{radians} = 28.252^0 \times \pi/180$
$\theta_{radians} = 0.4931$

Hence;

Length of curve from PC to PT = $R \times \theta$
Length of curve from PC to PT = $2491 \times 0.4931 = 1{,}228.3$ ft

STEP 3: Find the station at PT;

Station at PT = Station at PC + Length along the curve from PC to PT
Station at PT = Sta (106 + 43.09) + 1,228.3 = 106,43.09 + 1,228.3 = 11,871.38 = Sta. 118+ 71.38 (Ans A)

Problem 4.20) Radius of a horizontal curve is given to be 1,300 ft. Station at PC is given to be 187+34 and angle of intersection is 73^0. Find the deflection angle to station 190+00.

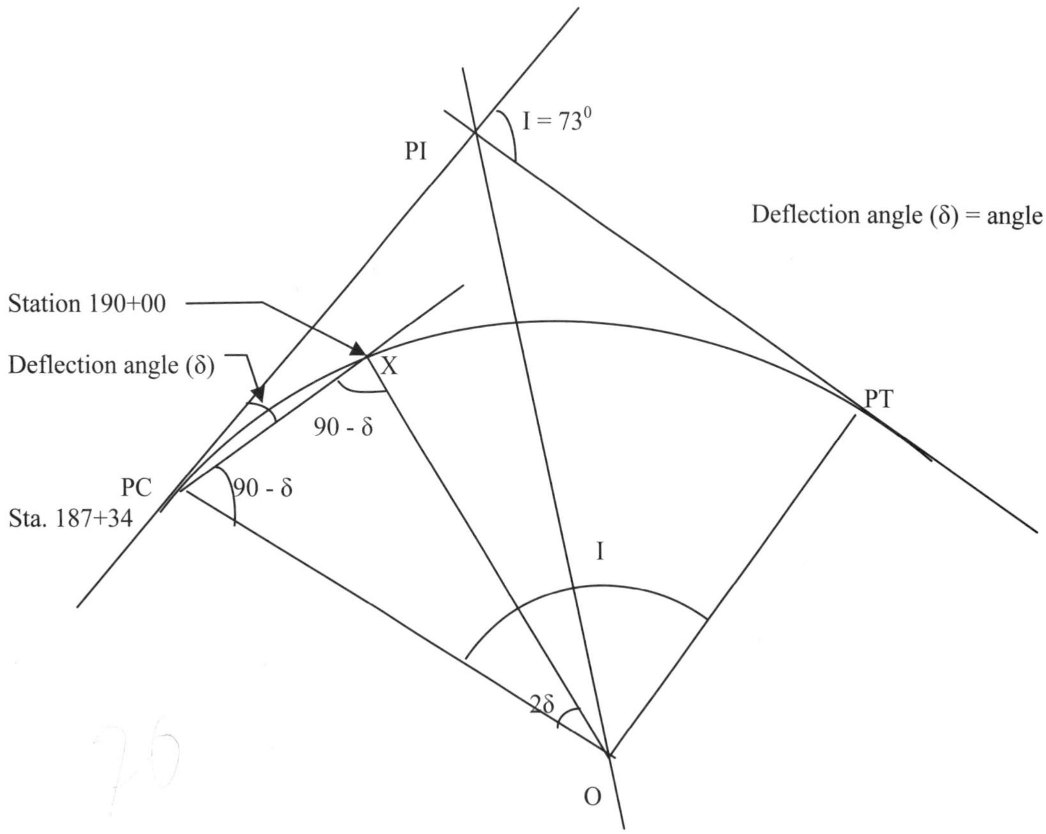

A) 3.95^0 B) 9.20^0 C) 5.86^0 D) 3.67^0

Solution 4.20):
Deflection angle (δ) is defined as the angle between line PC PI and line drawn to station of interest from PC.

By observation, angle X PC O = 90 - δ
Similarly due to symmetry PC X O = 90 - δ
Hence X O PC = 2 δ

Rule to remember

If deflection angle is δ, then the angle extended at the center is 2δ.

If deflection angle is δ, then the angle extended at the center is 2δ.

STEP 1: Find the length of the arc PC to X:

Station at PC = 187+34

Station at X = 190+00

Note that all stations are measured along the arc.
Length of the arc from PC to X = 190+00 - 187+34 = 266 ft

STEP 2: Find the length of the arc from PC to PT:

Previously we saw that if intersection angle is I, then the angle at the center of the circle encompassing PC to PT is also I.

Arc = Radius x θ
θ should be measured in radians.

Length of arc PC to PT = I (in radians) x 1,300

$I = 73^0$

This needs to be transformed to radians. You can do this using the calculator or use the following equation.

$\theta_{radians} = \theta_{degrees} \times \pi/180$

$\theta_{radians} = 73 \times \pi/180 = 1.274$

Length of arc PC to PT = 1,300 x 1.274 = 1,656.3 ft

STEP 3: Find the angle 2δ:

Length of arc between PC and X is 266 ft (Found earlier)

Length of an arc = Angle extended at the center measured in radians x Radius of the circle ---(1)
$266 = (2.\delta) \times 1{,}300$
$(2.\delta) = 266/1300 = 0.2046$ (in radians)

This needs to be converted to degrees. Again you can use the calculator directly or use the following relationship.
$(2.\delta)_{radians} = (2.\delta)_{degrees} \times \pi/180$
$0.2046 = \theta_{degrees} \times \pi/180$
$(2.\delta)_{degrees} = 11.72^0$
$2.\delta = 11.72^0$
$\delta = 5.86^0$ (Ans C)

Problem 4.21) Base free flow speed of a highway was assumed to be 70 mph. Following information is provided.
Lane width = 10 ft,
Right shoulder clearance = 3 ft,
Number of lanes in one direction = 3,
Two interchanges per mile.
Find the free flow speed.
Following parameters are given;
$f_{LW} = 6.6$, $f_{LC} = 1.2$, $f_N = 3.0$, $f_{ID} = 7.5$

A) 53.5 mph B) 51.7 mph C) 60.0 mph D) 54.3 mph

Solution 4.21):
For freeways;

$$FFS = BFFS - f_{LW} - f_{LC} - f_N - f_{ID}$$

Reference: Highway Capacity Manual

FFS = Free flow speed
BFFS = Base free flow speed
f_{LW} = Lane width adjustment (Exhibit 23-4, HCM)
f_{LC} = Adjustment for right shoulder lateral clearance (Exhibit 23-5, HCM)
f_N = Adjustment for number of lanes (Exhibit 23-6, HCM)
f_{ID} = Adjustment for interchange density (Exhibit 23-7, HCM)

$FFS = BFFS - f_{LW} - f_{LC} - f_N - f_{ID}$
$FFS = 70 - 6.6 - 1.2 - 3.0 - 7.5 = 51.7$ mph (Ans B)

Problem 4.22) Design traffic volume in a 6 lane freeway is 1,600 vehicles in both directions. It is reported that this highway has 3% truck traffic and 2% RVs (Recreational vehicles). The road is in rolling ground. Following vehicle flow rates were provided.

5:00 PM – 5:15 PM - 385
5:15 PM – 5:30 PM - 763
5:30 PM – 5:45 PM - 563
5:45 PM – 6:00 PM - 756

The traffic is mostly office commuters. Directional factor is 0.7.
Base free flow speed of the highway was assumed to be 70 mph.
Lane width = 10 ft, Right shoulder clearance = 3 ft, Number of lanes in one direction = 3,
Two interchanges per mile.
What is the level of service provided by the freeway?

$f_{LW} = 6.6$, $f_{LC} = 1.2$, $f_N = 3.0$, $f_{ID} = 7.5$
E_T = Truck equivalent factor = 2.5 for rolling ground
E_R = RV equivalent factor = 2.0 for rolling ground (obtained from table 21-8 (HCM)

Density vs LOS table from HCM;

	Level of Service (LOS)				
	A	B	C	D	E
Density (veh/mile)	11	18	26	35	45

A) LOS A B) LOS B C) LOS C D) LOS D

Solution 4.22):
To find the level of service (LOS), you need to find the density.

$$\text{Density} = \text{Flow/FFS}$$

FFS = Free flow speed (mph)
Density = Vehicles/mile
Flow = Vehicles/hr

Procedure to find LOS of a freeway:

STEP A: Find v_p (15 min passenger car equivalent service flow rate) from given flow data.
STEP B: Find free flow speed (FFS) from given data
STEP C: Find the density using the following equation

STEP 1: Passenger car equivalent service flow rate need to be found.

$$v_p = \frac{V}{PHF \times N \times f_{HV} \times f_P}$$

v_p = 15 min passenger car equivalent flow rate (pc/h/ln)
V = Hourly volume (veh/hr); This includes all vehicles including trucks, busses and RVs.
PHF = Peak hour factor;
N = Number of lanes
f_{HV} = Heavy vehicle factor (Heavy vehicle factor makes all vehicles to passenger car equivalent)
f_P = Driver population factor; This factor ranges from 0.85 to 1.0. For weekday office commuters this factor is 1.0. Office commuters are assumed to know where they are going and familiar with the road. During weekend, there would be more drivers who are driving for leisure and lower factor such as 0.85 should be used.

STEP 2: Find f_{HV};

$$f_{HV} = \frac{1}{1 + P_T (E_T - 1) + P_R (E_R - 1)}$$

$$f_{HV} = \frac{1}{1 + 0.03\,(2.5 - 1) + 0.02\,(2.0 - 1)} = 0.939$$

P_T = Fraction of trucks and busses = 0.03
P_R = Fraction of recreational vehicles = 0.02
E_T = Truck equivalent factor (obtained from table 21-8 (HCM) = 2.5 for rolling ground. (Given in the problem)
E_R = RV equivalent factor (obtained from table 21-8 (HCM) = 2.0 for rolling ground. (Given in the problem)

STEP 3: Find V
Volume is given for both directions.
V = 1,600 vehicles in both directions.

In commuter roadways, one direction would have more traffic in the morning and other direction would have more traffic in the afternoon. For an example, a certain freeway may have more traffic going east in the morning and more traffic going west in the afternoon.

Vehicles in one direction = Directional factor x Volume in both directions

Vehicles in one direction = 1,600 x 0.7 = 1,120

There are three lanes in one direction. (6 lanes total).

STEP 4: Find PHF

Find the maximum volume within 15 minute period.

Maximum volume within a 15 minute period occurs from 5:15 PM to 5:30 PM.

$V_{15minute\ max} = 763$

PHF (Peak hour factor) is given by the following equation.

$$PHF = \frac{\text{Hourly volume}}{4 \times V_{15minute\ max}}$$

Hourly volume = 385 + 763 + 563 + 756 = 2,467

Hence PHF = 2,467/(4 x 763) = 0.81

STEP 5: Find v_p;

PHF = 0.81; $f_{HV} = 0.939$; $f_P = 1.0$ (Weekday commuters); N = 3

$$v_P = \frac{V}{PHF \times N \times f_{HV} \times f_P} = \frac{1120}{0.81 \times 3 \times 0.939 \times 1.0} = 491 \text{ pc/h/ln}$$

N = 3 since there are three lanes in one direction and 1,120 is the flow in one direction.

Now you need to find FFS so that you can find the density. You need the density to find LOS.

STEP 6: Find FFS

For freeways;

$$FFS = BFFS - f_{LW} - f_{LC} - f_N - f_{ID}$$

FFS = Free flow speed

BFFS = Base free flow speed

f_{LW} = Lane width adjustment (Exhibit 23-4, HCM)

f_{LC} = Adjustment for right shoulder lateral clearance (Exhibit 23-5, HCM)

f_N = Adjustment for number of lanes (Exhibit 23-6, HCM)

f_{ID} = Adjustment for interchange density (Exhibit 23-7, HCM)

$FFS = BFFS - f_{LW} - f_{LC} - f_N - f_{ID}$

FFS = 70 – 6.6 – 1.2 – 3.0 – 7.5 = 51.7 mph

STEP 7: Find the Density;

Flow = Density x FFS
491 = Density x 51.7
Density = 9.5 veh/mile

Density vs LOS table from HCM;

	Level of Service (LOS)				
	A	B	C	D	E
Density (veh/mile)	11	18	26	35	45

Since 9.5 is less than 11, this freeway is LOS A. (Ans A)

Problem 4.23): Sag vertical curve of a roadway is shown in the figure. The roadway goes under an overpass.
Station of BVC = 115+45,
Elevation of BVC = 134.56 ft

Station of PVI = 120 + 34
Station of overpass = 123 + 13
Elevation of overpass = 143.25 ft

What is the maximum height of trucks that can be allowed in the roadway assuming 1 ft clearance between roof of trucks and the overpass.
Note that vertical curves are constructed in a manner so that PVI station is at the center of BVC and EVC.

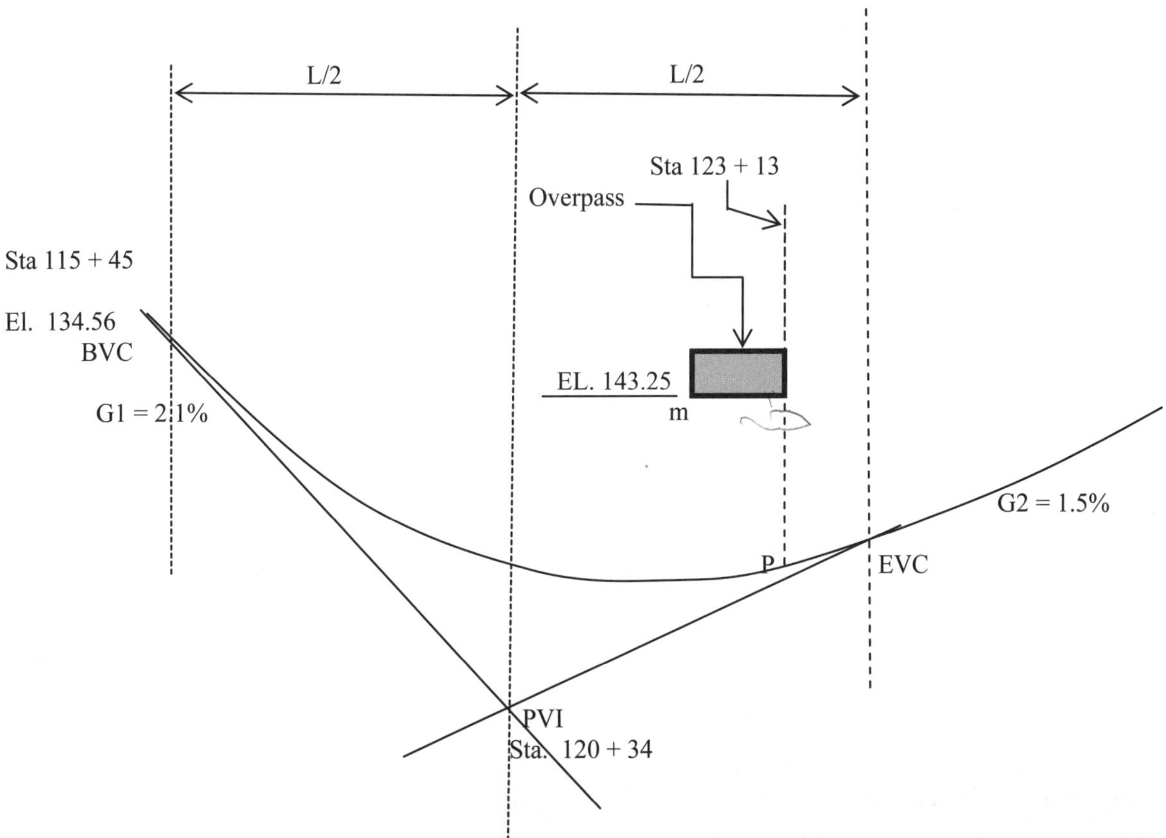

A) 13.45 ft B) 12.97 ft C) 14.56 ft D) 10.89 ft

Solution 4.23):

Vertical Curve Equation; (See sample exam 2 as well).

$$y = \frac{(G2 - G1)}{2L} \cdot x^2 + G1 \cdot x$$

y = Vertical height to point in the curve from BVC in ft
x = Horizontal distance between two points
G1 and G2 are gradients.
Upward gradient is taken to be positive (+) ve.
Downward gradient is taken to be (-) ve.
L = Horizontal distance between BVC and EVC in ft.

Let's write down the given parameters.
BVC station = 115 + 45
PVI station = 120 + 34
The horizontal distance between BVC and PVI = 12,034 – 11,545 = 489 ft

Note that PVI station is located midway between BVC and EVC.

Hence L = 2 x 489 = 978 ft

G1 = -0.021 (Downward gradient is considered to be negative)
G2 = 0.015 (Upward gradient is considered to be positive)
We need to find the horizontal distance between the overpass (point P) and BVC.
Horizontal distance to the overpass measured from BVC (x) = 12313 – 11545 = 768 ft

STEP 1: Find the elevation at point P:

$$y = \frac{(G2 - G1)}{2L} \cdot x^2 + G1\, x$$

$$y = \frac{(0.015 - -0.021)}{2 \times 978} \cdot 768^2 + (-0.021 \times 768)$$

y = 10.855 – 16.128 = -5.273

Note that "y" is measured from BVC.

Elevation at point P = 134.56 – 5.273 = 129.28

Elevation of the overpass = 143.25

The total clearance between the roadway and overpass = 143.25 – 129.28 = 13.97 ft

Clearance of 1 ft is needed between overpass and roof of trucks.
Hence maximum height of trucks = 12.97 ft (Ans B)

CONSTRUCTION:

Problem 4.24) A portion of surveyor's log book is shown below. What is the foresight reading could have been to point C.

Location	Elevation	Backsight Reading	Line of sight El.	Foresight Reading
Benchmark X	191.67	3.56		
Point A				2.81
Point B				4.34
Point C		10.89		
Point D	200.76			4.21

A) 2.46 B) 6.05 C) 13.56 D) 1.15

Solution 4.24): Fill the log book.

Location	Elevation	Backsight Reading	Line of sight El.	Foresight Reading
Benchmark X	191.67	3.56	191.67+ 3.56 = 195.23	
Point A	195.23-2.81=192.42		195.23	2.81
Point B	195.23-4.34=190.89		195.23	4.34
Point C		10.89		?
Point D	200.76		200.76+4.21= 204.97	4.21

Elevation of Benchmark X is given to be 191.67. Backsight reading to benchmark X is 3.56. Hence the elevation of line of sight = 191.67+ 3.56 = 195.23

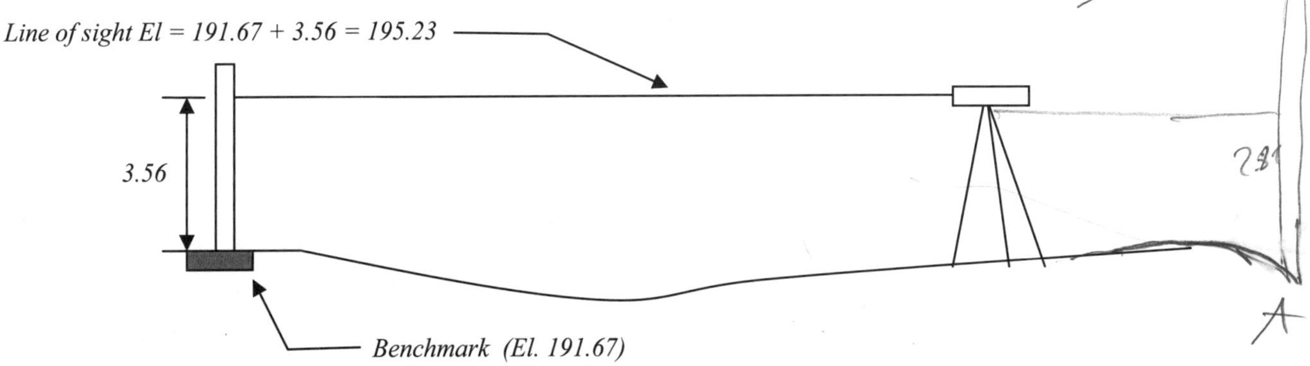

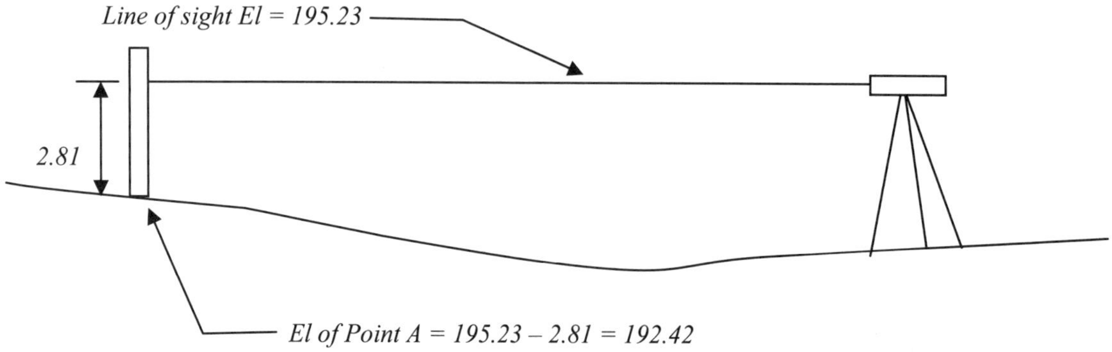

Elevation of point A = 195.23 - 2.81 = 192.42

Similarly,
Elevation of point B = 195.23 - 4.34 =190.89

Elevation of point C, cannot be found since foresight reading is not given.

After taking the reading at point C, the surveyor moved the instrument to a new location and obtained a reading to point D.
Elevation of point D is given to be 200.76 and the foresight reading to point D is given to be 4.21.

Hence we can establish the elevation of new line of sight.
Elevation of new line of sight = 200.76 + 4.21= 204.97

Now we can find the elevation at point C.

Elevation of point C = Elevation of new line of sight – backsight reading = 204.97 – 10.89 = 194.08

Remember that the foresight reading to point C was taken from the old location of the instrument prior to moving it to the new location.

Hence,
Elevation of point C = Elevation of line of sight (old) – foresight reading to point C

Elevation of line of sight (old) = 195.23
Elevation of point C was found to be 194.08.
Elevation of point C = Elevation of line of sight (old) – foresight reading to point C
Hence

194.08 = 195.23 – foresight reading to point C

Foresight reading to point C = 195.23 – 194.08 = 1.15 (Ans D)

Location	Elevation	Backsight Reading	Line of sight Elevation.	Foresight Reading
Benchmark X	191.67	3.56	191.67+ 3.56 = 195.23	
Point A	195.23 - 2.81 =192.42		195.23	2.81
Point B	195.23 - 4.34 =190.89		195.23	4.34
Point C	204.97 - 10.89 =194.08	10.89		195.23-194.08 **=1.15**
Point D	200.76		200.76+4.21= 204.97	4.21

Bench mark by US Geological Survey. Elevation is shown to be 4,070 feet above sea level

Discussion: Why surveyors move the instrument?

Surveyors would be more than happy to take all elevation readings from one location. Unfortunately it is not possible to find one location where all the points are visible, unless you are surveying in a desert or in a grass field. In most cases, there are buildings, trees, roads, sign boards etc are in the way. Hence surveyors are forced to move their leveling instruments.

Typical Procedure in Leveling:

STEP 1: Find a benchmark with a known elevation.

STEP 2: Setup the instrument at a comfortable location and take a reading to the benchmark. Since the elevation of the benchmark is known, elevation of the line of sight can be established.
STEP 3: Then take readings to points of interest. Since the surveyor knows the elevation of line of sight, he can establish the elevations of these points. In this case, elevations of points A, B and C were found from the first setup.
STEP 4: Now he finds some points are not visible from the location that he is right now. He has to move the instrument. Move the instrument to a location where the points of interest are visible.
STEP 5: Obtain a reading to a point that was surveyed before. In this case, a reading was taken to point C. Elevation of point C is known. Hence he can establish the elevation of the line of sight.
STEP 6: Now the elevation of line of sight is established, obtain readings to points of interest.

Problem 4.25): Bearing of line AB is given to be S 45^0 14' 13"W. Interior angle of ABC is 80^0 23' 11". Find the azimuth of line CB.

A) 265^0 14' 13"   B) 144^0 51' 2"   C) 175^0 14' 33"   D) 134^0 24' 33"

Solution 4.25) Draw line AB. Line AB is south to west as shown. Then draw line BC. Interior angle of ABC is given. Hence line BC can be drawn. Azimuth of line CB needs to be found. Azimuth is always measured from North.

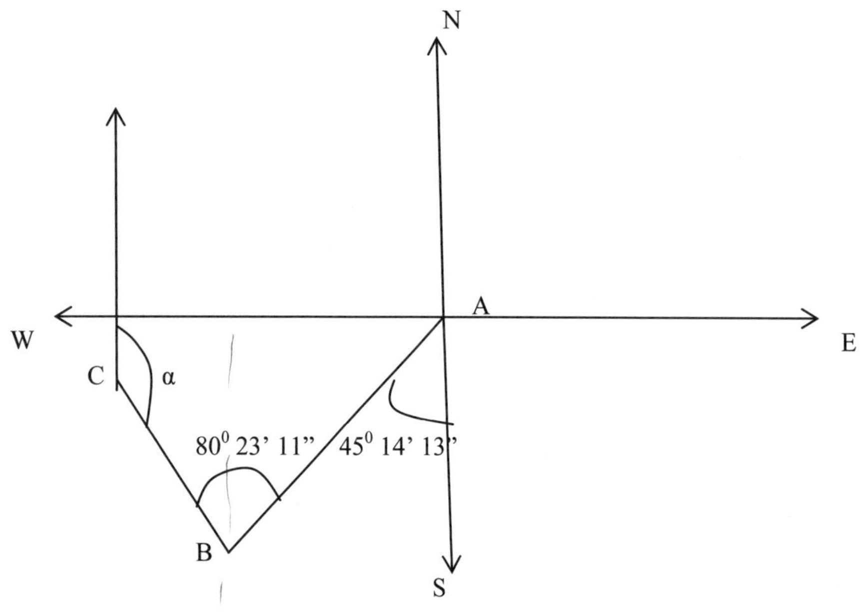

STEP 1: Draw line AB. AB is 45^0 14' 13" measured from south to west.

STEP 2: Draw line BC with interior angle ABC to be 80^0 23' 11".

Azimuth of line CB is the angle between North and line CB measured in clockwise direction. (α).

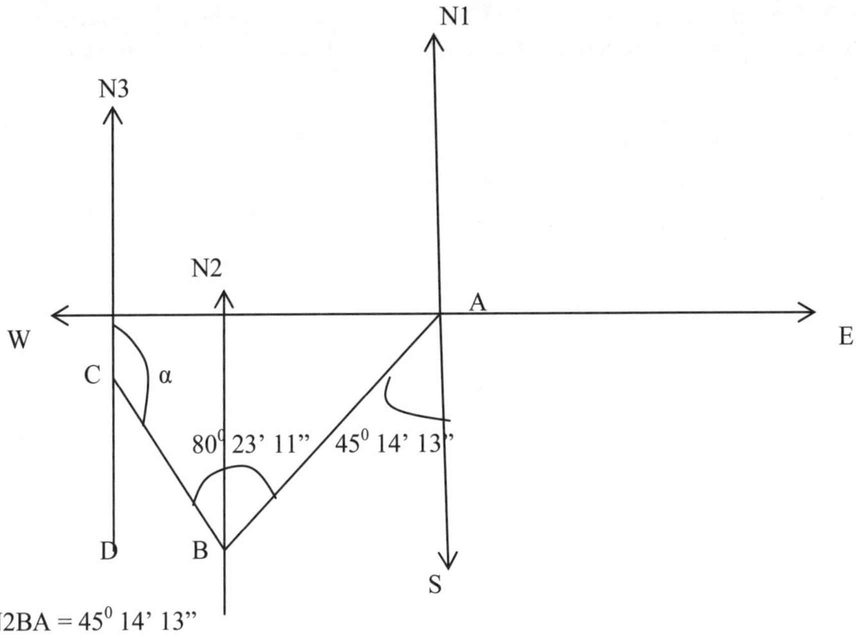

Angle N2BA = 45^0 14' 13"
Angle N2BC = 80^0 23' 11" - 45^0 14' 13" = 35^0 8' 58"
Angle N2BC = Angle DCB = 35^0 8' 58"
Azimuth of line CB = α = 180^0 - 35^0 8' 58" = 144^0 51' 2" (Ans B)

Note that N1, N2 and N3 are north direction . Different symbols (N1, N2 and N3) are given for clarity.

Use the network given in problem 4.26 to solve problems 4.27 and 4.28.

Problem 4.26) Activity on arrow diagram is shown below. Find the duration of the project.

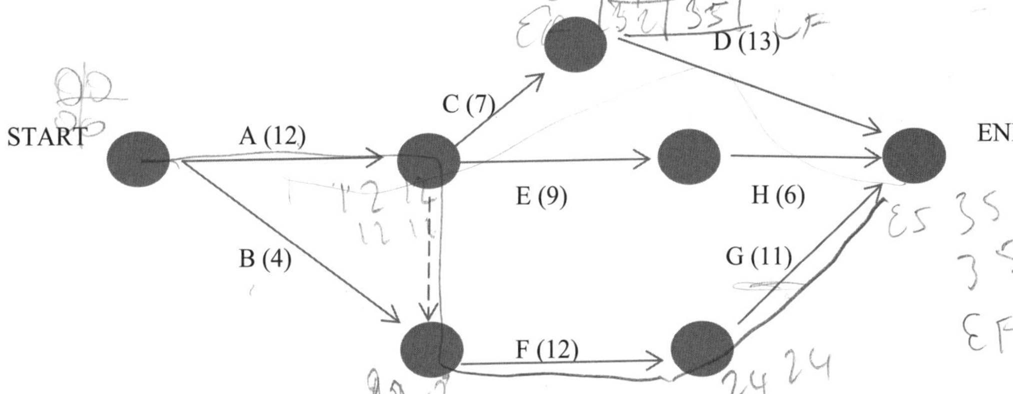

A) 32 days B) 25 days C) 27 days D) 35 days

Solution 4.26): Activities are shown on arrows in activity on arrow diagrams

In activity on arrow diagrams, arrows are used to show activities.
In activity on arrow diagrams, nodes are called events.
First node is always the "START" event and last node is the "END" event.
On the other hand, in activity on node diagrams, activities are shown on the node.

Dummy Activities: Broken arrows are used to indicate dummy activities. Activity F cannot be started until activities A and B are completed. If the dummy activity is not shown, activity F can be started as soon as activity B is completed.

There are number of paths exist from start to end.

Duration of path A, C, D = 12 + 7 + 13 = 32
Duration of path A, E, H = 12 + 9 + 6 = 27
Duration of path A, F, G = 12 + 12 + 11 = 35
Duration of path B, F, G = 4 + 12 + 11 = 27

To complete the project, longest path needs to be completed.

Duration of the project = 35 (Ans D)

Problem 4.27) Find the early start and late start of activity D for the project given in the previous problem.

A) ES = 12, LS = 15 B) ES = 13, LS = 22 C) ES = 19, LS = 22
D) ES = 22, LS = 24

Solution 4.27):

Activity D, cannot be started until activity C is completed.

Early start of activity D = 12 + 7 = 19 (ES = 19)

Find the late start (LS) of activity D

Finding late start of an activity can be tricky.

We found that project duration to be 35 days.
Now we need to find what is the latest day that activity D can be started without delaying the project?

Latest day that activity D can be started without delaying the project = 35 – 13 = 22
35 is the duration of the project and 13 is the time need to complete activity D.
If activity D is started on day 22, there would not be any delay to the project. LS = 22
(Ans C)

Problem 4.28) Find the total float of activity D.

A) 1 B) 3 C) 9 D) 2

Solution 4.28):

Total float of an activity is given by the following equation;

Total Float of an activity = LS – ES

LS = Late start; ES = Early start

LS of activity D = 22
ES of activity D = 19
Float = LS – ES = 3

Problem 4.29) What type of cement is known as general purpose cement?

A) Type I B) Type II C) Type III D) Type IV

Solution 4.29) Ans A

There are five types of cements. They are;

Type I (General purpose cement): Type I cement is known as general purpose cement. When there are no special conditions to be considered type I cement is used.

Type II Cement: Type II cement can withstand moderate heat and moderate amount of sulfates.

Type III Cement: Special property of Type III cement is that it develops high early strength. If forms have to be removed earlier to progress concrete work, type III cement may be used.

Type IV Cement: Special property of type IV cement is that it generates less heat during hydration. Hydration is the process of reaction between water and cement. Heat is generated during hydration.
High level of heat generation affects the strength of concrete. In large dams, heat does not discharge quickly from center and large amount of heat can be generated inside. This would affect the strength of the concrete. Type IV cement can be used in such situations.

Concrete in the center could generate large amount of heat

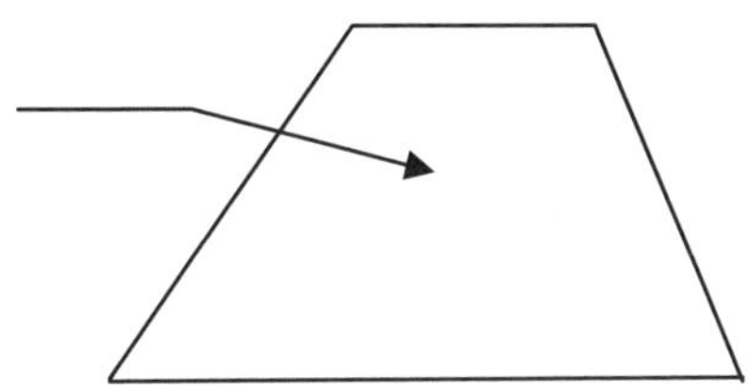

Type V Cement: Special property of type V cement is that it's resistant to sulfate attack. Sulfates are present in certain marine environments. In such situations type V cement can be used.

Above mentioned 5 types of cement vary in their properties and cost.

Problem 4.30): Concreting has to be done in cold weather. What type of cement is not desirable?

A) Type I/II B) Type III C) Type IV D) Type V

Solution 4.30): Ans C

Type IV cement generates less heat during hydration and desirable in hot weather conditions. Type IV cement is not desirable for low temperature environments.

Problem 4.31): Topography of a proposed road is shown below.

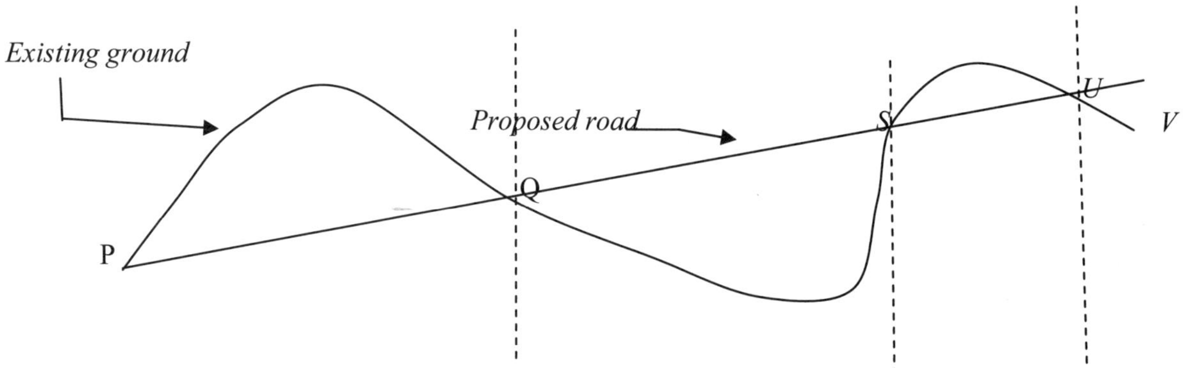

What is the correct mass-haul diagram for the above project?

A)

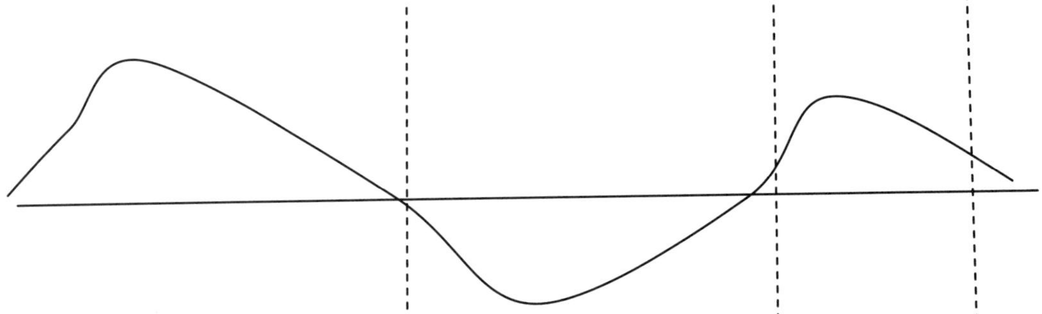

B)

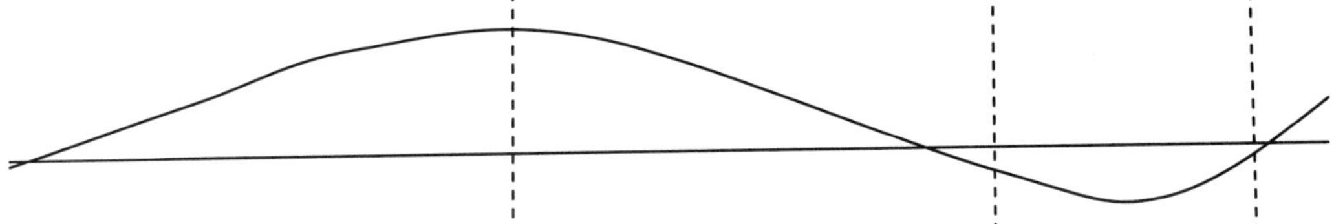

C)

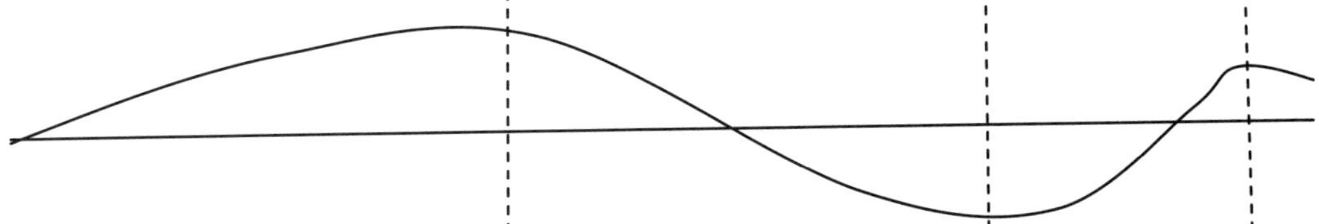

D)

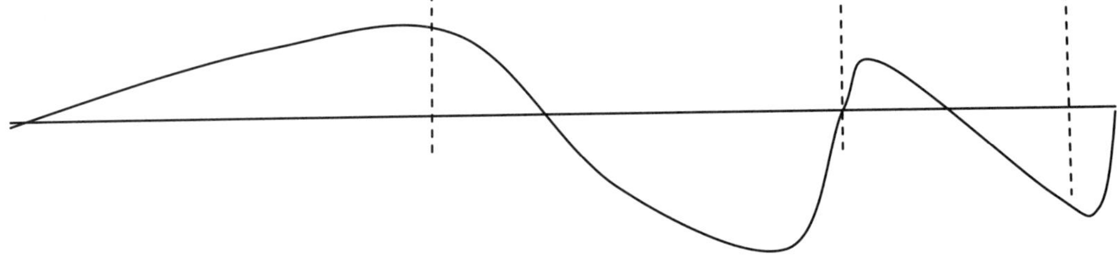

Solution 4.31): (Ans C)

In a mass haul diagram X axis represents the distance and the Y axis represents the cumulative cut.
Mass Haul diagram is drawn to find the net cut. Cut volume keeps increasing from P to Q.

P to Q = P to Q is all cut. Net cut should keep increasing from P to Q. Mass haul curve should head upwards from P to Q.

Q to S = Q to S is all fill. Net cut should decrease from Q to S. From Q to S, mass haul curve should head downwards.

S to U = S to U is all cut. S to U mass haul curve should head upwards.

U to V = U to V is all fill again. U to V mass haul curve should be directed downwards.

(Ans C)

Cut section of a road is shown in the figure. See two figures shown below for explanation

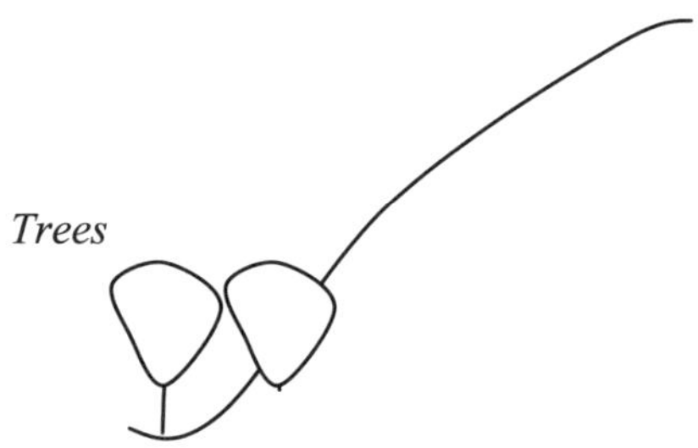

Original ground

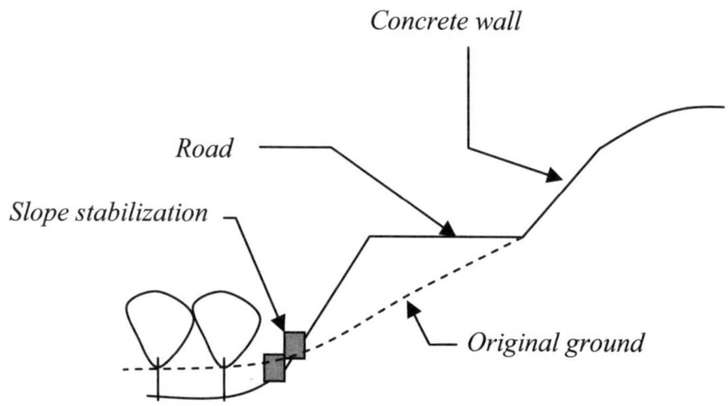

Ground was cut to build the road.

Problem 4.32): Painting crew consists of 3 painters and 2 helpers. Wages of a painter is \$60/hr and of a helper is \$40 per hour. What is the labor hour rate (LH).

A) 55 B) 52 C) 77.2 D) 34.7

Solution 4.32):

$$\text{Labor Hour (LH)} = \frac{\text{Crew Hour}}{\text{No. of workers in the crew}}$$

STEP 1: Find the crew hour rate;

Crew hour rate = 3 x \$60 + 2 x \$40 = \$260

Every hour, \$260 has to be paid for the painting crew.

STEP 2: Find the labor hour (LH):

Labor Hour = Crew Hour/No. of workers in the crew
Labor Hour = 260/5 = \$52 (Ans B)

GEOTECHNICAL:

Problem 4.33): Site with gray sandy silt is shown below. Original groundwater was at 5 ft below the ground surface. Due to a severe drought, groundwater level drops down to 9 ft below the ground surface.
What would happen to the effective stress at point A when groundwater level drops down to 9 ft below ground surface. Total density of soil is 110 pcf.

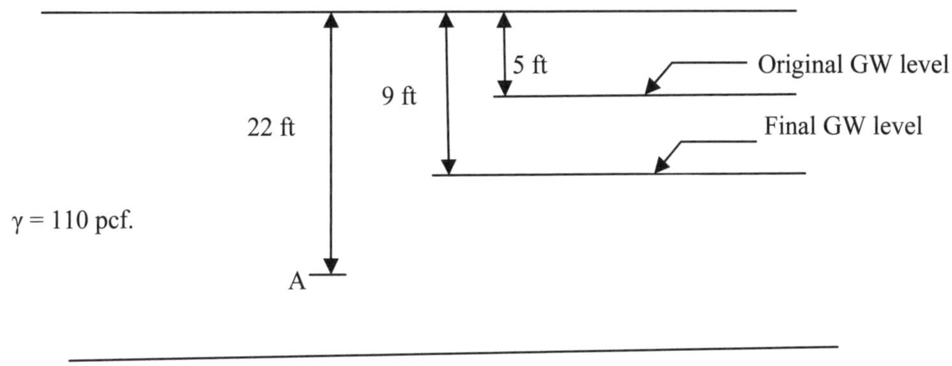

A) Vertical effective stress increases by 249.6 psf.
B) Vertical effective stress decreases by 249.6 psf.
C) Vertical effective stress increases by 169.6 psf.
D) No change

Solution 4.33): When calculating vertical effective stress following rules should be used.

- Density above groundwater is taken to be total density.
- Density below groundwater is taken to be buoyant density.

$$\text{Buoyant density} = \gamma - \gamma_w$$

$$\gamma_w = \text{Density of water}$$

STEP 1: Find the vertical effective stress at point A when groundwater is 5 ft below surface;

$$= 110 \times 5 + 17 \times (110 - 62.4) = 1{,}359.2 \text{ psf}$$

STEP 2: Find the vertical effective stress at point A when groundwater is 9 ft below surface;

$$= 110 \times 9 + 13 \times (110 - 62.4) = 1{,}608.8 \text{ psf}$$

Vertical effective stress increases by 249.6 psf.
(Ans A)

Problem 4.34): Mechanism of secondary compression of a clay soil is best described;

A) Secondary compression is caused by expulsion of air particles due to applied load.
B) Secondary compression is caused by expulsion of water particles due to applied load.
C) Secondary compression is caused by rearranging of soil particles due to applied load.
D) Due to combination of above A and B.

Solution 4.34): Clay soils have water inside the pores. When the soil is loaded, this water would get squeezed out. Hence the structure would undergo settlement. This is known as primary consolidation.

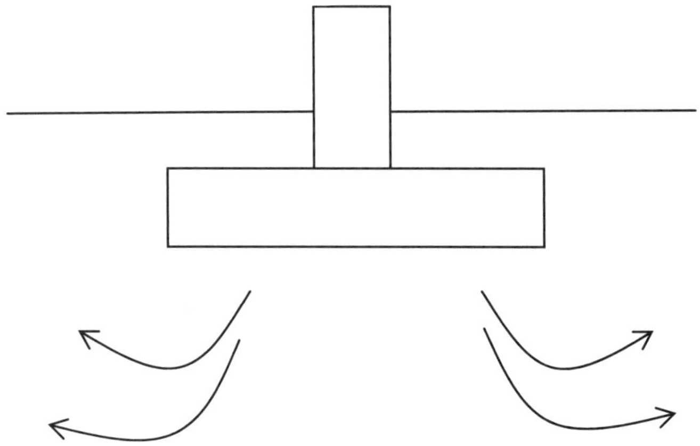

When loaded, water squeeze out. This process is known as primary consolidation.

Researchers have found even after all the pore water is squeezed out, settlement to continue. This settlement is known as secondary compression. Secondary compression is due to rearranging of soil particles. (Ans C)

Problem 4.35): Clay soil is loaded in a consolidation test apparatus. The void ratio was measured during loading and plotted and shown below.

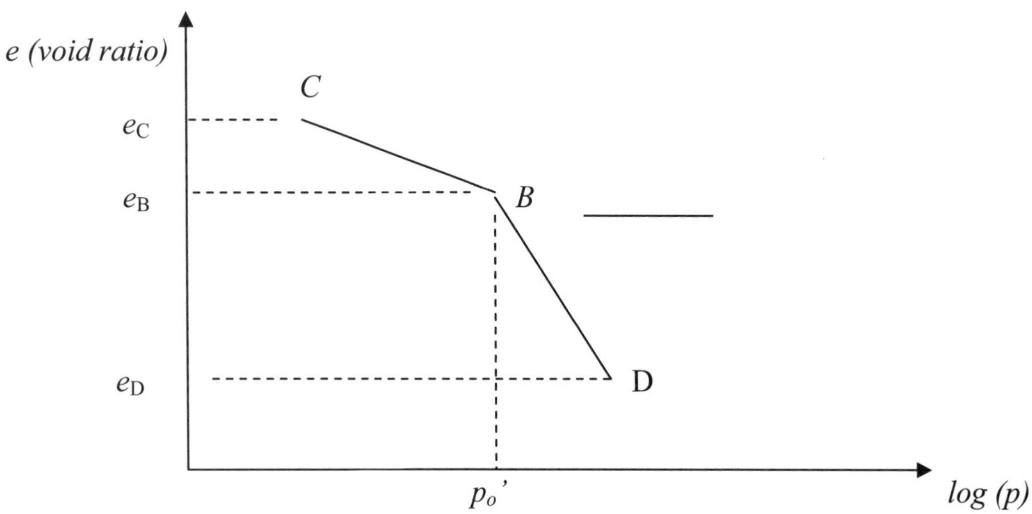

X axis represents log pressure and Y axis represents void ratio.

This clay is a

A) Normally consolidated clay
B) Overconsolidated clay
C) Silty clay
D) Organic clay

Solution 4.35):

When a normally consolidated clay is loaded following graph is obtained.
Void ratio would decrease proportional to the log of the pressure applied.

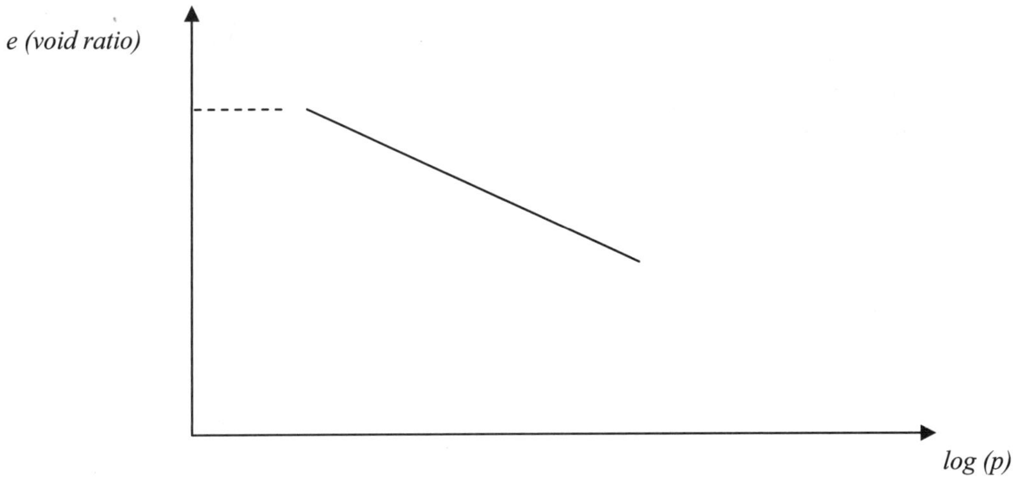

When an overconsolidated clay is loaded, following graph is obtained.

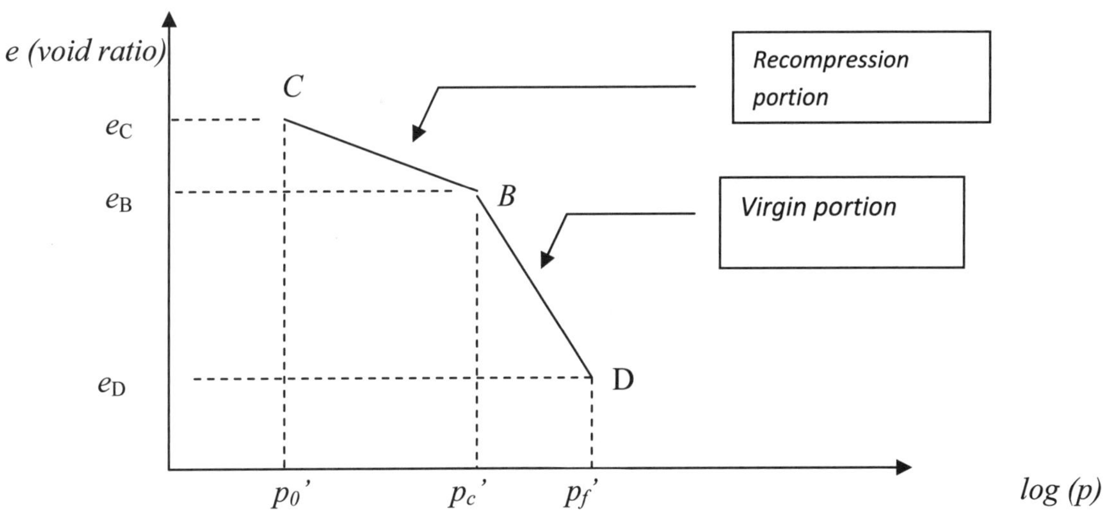

Overconsolidated clays have been subjected to high pressures in the past. Highest stress the soil had subjected to in the past is p_c'. C to B is known as recompression portion.

Point C = Present stress. Normally represented by p_0'
Point B = Highest stress soil had subjected to in the past. This is known as preconsolidation pressure. (p_c')
Point D = Stress in soil after the footing load is applied. (p_f')

C to B = Recompression curve
B to D = Virgin curve

(Ans B)

Problem 4.36): In the previous problem, point C to B is known as

A) Recompression curve
B) Virgin curve
C) Overconsolidated curve
D) None of the above

Solution 4.36): Ans A

Point C to B is known as the recompression curve
Point B to D is known as the virgin curve.

Problem 4.37): Engineers are planning to place a shallow foundation as shown.

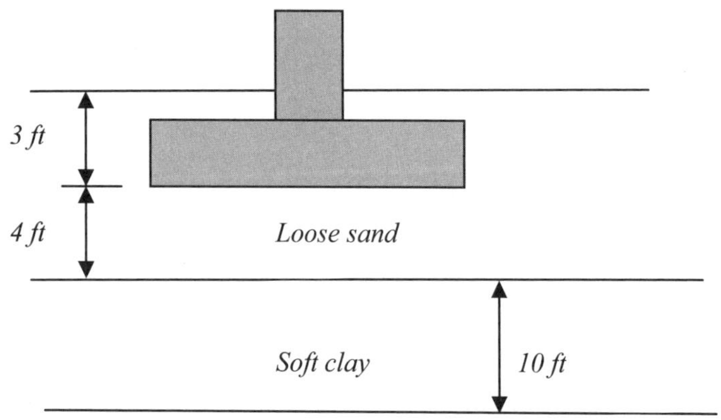

Engineers have decided that loose sand and soft clay are not capable of supporting the foundation without been treated. What is the best ground modification method for this site?

A) Conduct dynamic compaction to treat soft clay and conduct pressure grouting to treat loose sand.
B) Conduct dynamic compaction to treat loose sand and conduct surcharging to treat soft clay.
C) Conduct vibrocompaction to treat soft clay and conduct pressure grouting to treat loose sand.
D) Conduct pressure grouting to treat soft clay and conduct surcharging to treat loose sand.

Solution 4.37):

First let's look at ground modification methods.

Dynamic Compaction:

Dynamic compaction is used to densify loose sand. Large weight, typically in the range of 10 tons been dropped on loose sand.

Dropping a large weight on soil

Dynamic compaction is best suited for sandy soils. This method may not work for soft clay.

Pressure Grouting:

A hole is drilled and grout is injected to the ground with high pressure. Grout would fill the voids and strengthen the soil. This method is best suited for sandy soils. It is also can be used for clay soils. Pressure grouting is an expensive method to stabilize ground. Pressure grouting is also known as jet grouting.

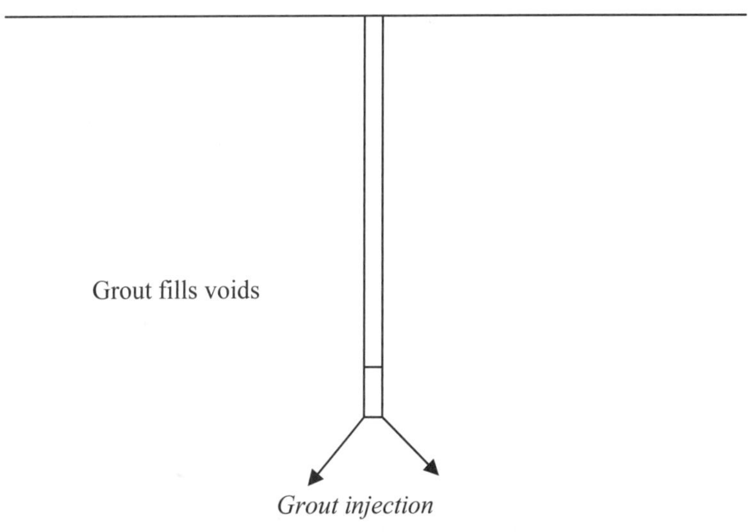

Vibroflotation (Vibrocompaction): Soil is vibrated using a vibrating device. When the soil is vibrated, loose soil would get densify. Stones are introduced to fill the voids. This method is best suited for sandy soils.

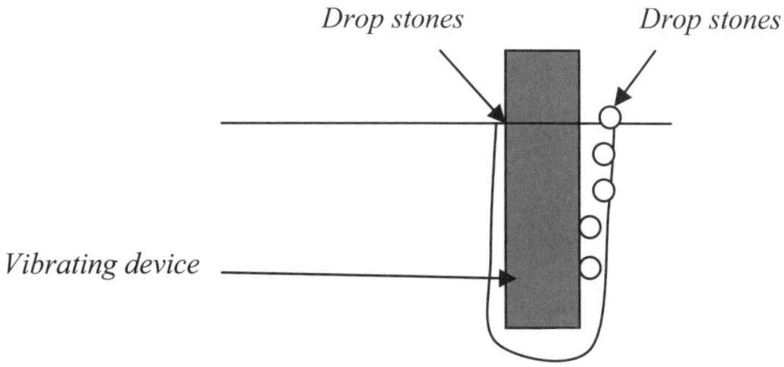

Surcharging: Large soil mountain is placed on soft clay. Over time due to the pressure, soft clay would consolidate. This method is typically used for clay soils.

Soil surcharging

Problem 4.38): Find the length of the footing required to carry a load of 120 tons with a factor of safety of 2.0.

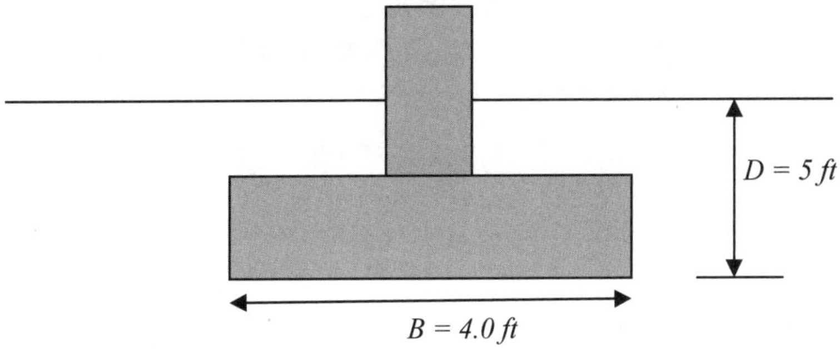

Following parameters are given;

Width of the footing = 4.0 ft
$N_c = 12.9$
$N_q = 4.4$
$N_\gamma = 2.5$
Shape factors S_c, S_q and S_γ;
$S_c = 1.2$
$S_q = 1.0$
$S_\gamma = 0.8$
Total density of soil (γ) = 120 pcf
Cohesion of the soil (c) = 400 psf
Factor of safety = 2.0

Use the Terzaghi Bearing Capacity Equation:

Bearing Capacity in psf = $c.N_c\, S_c + q.\, N_q\, S_q + \frac{1}{2}\, B.\, \gamma.\, N_\gamma\, S_\gamma$

Ignore depth factors.

A) 11.34 ft B) 8.27 ft C) 4.45 ft D) 13.77ft

Solution 4.38):

STEP 1: Write down the Terzaghi Bearing Capacity Equation;

$$\text{Bearing Capacity in psf } (q_{ult}) = c.N_c\, S_c + q.\, N_q\, S_q + \tfrac{1}{2}\, B.\gamma.\, N_\gamma\, S_\gamma$$

q_{ult} = Ultimate bearing capacity
c = Cohesion of soil (psf) (c = 400 psf)
N_c = Bearing capacity factor for cohesion. ($N_c = 12.9$)
Tables are provided by various researchers to find this factor. In this problem N_c is given to be 12.9
$q = \gamma.\, D$
"q" is known as overburden.
γ = Density of soil. In this problem γ is given to be 120 pcf.

D = Distance from top of ground to bottom of footing. (D =5 ft)
N_q = Terzaghi bearing capacity factor for overburden. (N_q = 4.4)
B = Width of the footing. (B = 4.0 ft).
γ = Density of soil (γ = 120 pcf)
N_γ = Terzaghi bearing capacity factor for soil density. (N_γ = 2.5)
Shape factors S_c, S_q and S_γ;
$S_c = 1.2$
$S_q = 1.0$
$S_\gamma = 0.8$

STEP 2: Apply the Terzaghi bearing capacity equation;

Bearing Capacity in psf (q_{ult}) = $c.N_c\, S_c + q.\, N_q\, S_q + ½\, B.\, \gamma.\, N_\gamma\, S_\gamma$

$q = \gamma.D = 120 \times 5 = 600$ psf

Bearing Capacity in psf (q_{ult}) = (400 x 12.9 x 1.2) + (600 x 4.4 x 1.0) + (½ x 4 x 120 x 2.5 x 0.8)
Bearing Capacity in psf (q_{ult}) = 6,192 + 2,640 + 480
Bearing Capacity in psf (q_{ult}) = 9,312 psf

STEP 3: Find q_{net};
The third factor in the Terzaghi bearing capacity equation assumes that the footing is placed on top of the ground surface. Hence correction needs to be done to account for the fact that the footing is placed 5 ft below the ground surface.

$q_{net} = q_{ult} - \gamma.D$
$q_{net} = 9{,}312 - 120 \times 5 = 8{,}712$ psf

STEP 4: Total bearing capacity of the footing;

Bearing capacity of the footing (tons) = Bearing capacity (psf) x Area/2,000

Length of the footing = L (*unknown*)
Width of the footing = 4 ft
Area of the footing = 4 x L

Bearing capacity of the footing (tons) = 8,712 x (4 x L)/2,000 ------------------------------------(1)

Load on the footing = 120 tons.
The problem says to use a factor of safety of 2.0.
Hence, bearing capacity required = 2 x 120 tons
= 240 tons

From (1)
240 tons = 8,712 x (4 x L)/2,000

L = 13.77 ft (Ans D)

Problem 4.39) Soil sample has a porosity of 0.55 and specific gravity of 2.65. Moisture content of the soil sample was found to be 33%. What is the degree of saturation of the soil sample?

A) 43% B) 71% C) 12% D) 34%

Solution 4.39):

STEP 1: Survey the equations and see which equations have given parameters.

Porosity (n), specific gravity (G_s), moisture content (w) are given.
Degree of saturation (S) needs to be found.

n = 0.55
G_s = 2.65
w = 0.33

STEP 2: Use the equation between void ratio and porosity.

Void ratio = e

e = n/(1 - n) --------------------------------(1)
e = 0.55/(1 – 0.55)
e = 1.222

STEP 2:

Use the following equation;

S.e = G_s.w --------------------------------(2)

S = Degree of saturation;
e = void ratio (e = 1.222)
G_s = Specific gravity of soil (G_s = 2.65)
w = Moisture content (w = 0.33)

S x 1.222 = 2.65 x 0.33

S = 0.71 (Ans B)

Problem 4.40): There is a weak soil layer 12 ft below an existing column footing (5 ft x 5 ft) as shown in the figure. The maximum additional stress that can be applied at the midpoint of the weak soil layer is 700 psf. Any stress more than 700 psf would cause excessive settlement.
What is the maximum load that can be applied on the footing?
Assume stress distribution to be 1:1.

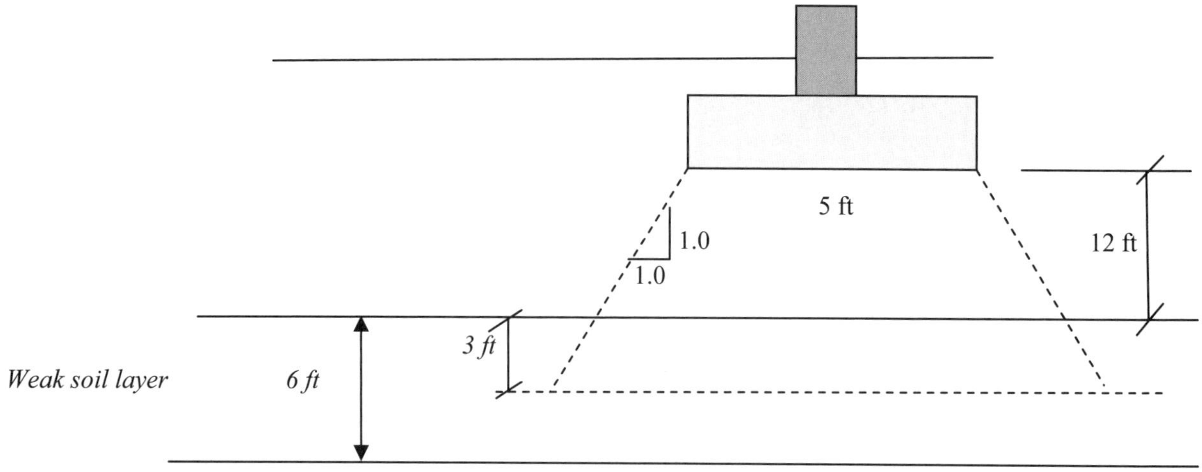

A) 341 tons B) 171 tons C) 428 tons D) 487 tons

Solution 4.40):

STEP 1: Draw the stressed zone. We are concerned only of the midpoint of the weak soil layer. Midpoint of the weak soil layer is 15 ft below the footing.

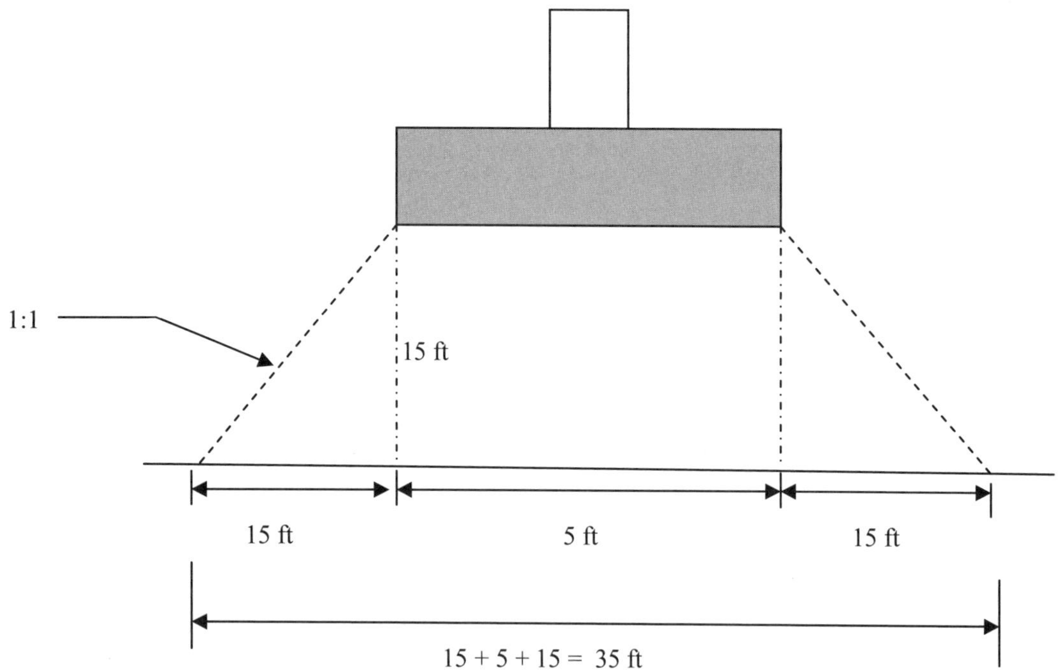

STEP 2: Draw the plan view of the footing and stressed zone.

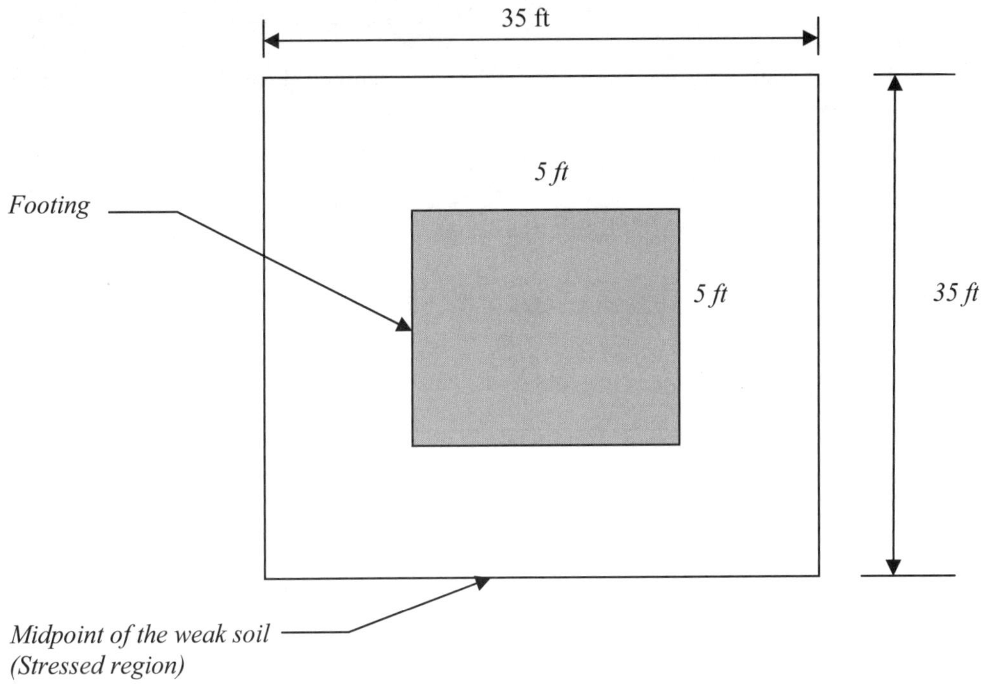

STEP 3:

Total area of the stressed region = 35 x 35 = 1,225 sq. ft

Allowable stress in the midpoint of the weak soil = 700 psf

Total load allowed in the midpoint of the weak soil layer = 700 x 1,225 lbs
= 857,500 lbs
= 428.75 tons

This is same as the total load on the footing.
Total load on the footing will be ultimately transferred down.
(Ans C)

APPENDIX GEOTECHNICAL:

Phase Relationships:

Soil consists of solids, air and water. Solids are soil particles. Soil matrix can be schematically represented as shown below.

Volume ***Mass***

V_a $M_a = 0$ (Mass of air is taken to be zero).

V_w M_w (Mass of water)

V_s M_s (Mass of solids)

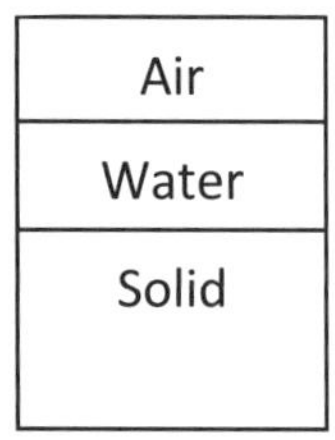

Soil phase diagram

M_a = Mass of air = 0 (Usually taken to be zero)
V_a = Volume of air (Volume of air is **not** zero)

M_w = Mass of water
V_w = Volume of water

M_s = Mass of solids
V_s = Volume of solids

M = Total mass of soil $= M_s + M_w$
V = Total volume of soil $= V_s + V_w + V_a$
V_v = Volume of voids $= V_a + V_w$

Density of Water (γ_w) = Density of water can be expressed in many units.

$\gamma_w = M_w/V_w$

SI Units: $\gamma_w = 1$ g/cm^3 = 1,000 g per liter = 1,000 kg/m^3 = 9.81 kN/m^3.
fps Units: γ_w = 62.42 pounds per cu. feet (pcf)

Total Density of Soil γ_t, γ_{wet} or γ:

Some books use γ_{wet} and some other books uses γ_t or simply γ to denote total density of soil. Total density (also known as wet density) is simply mass of soil (including water) divided by the volume of soil.

$\gamma_{wet} = M/V$
$M = M_w + M_s$ and $V = V_w + V_a + V_s$
{M = Total mass of soil including water}
{V = Total volume of soil including soil, water and air}

Dry Density of Soil (γ_d): $\gamma_d = M_s/V$

M_s = Mass of solid only.
V = Total volume of soil including soil, water and air
$V = V_w + V_a + V_s$

Density of Solids = M_s/V_s

Specific Gravity (G_s): Specific gravity is defined as density of solids divided by the density of water. Density of solids represented by G_s or by simply G.

Specific Gravity (G_s) = $M_s/(V_s \cdot \gamma_w)$

Void Ratio (e): Void ratio (e) is defined as the ratio of volume of voids to volume of solids.

$$e = V_v/V_s$$

V_v = Volume of voids (Volume of water + Volume of air) = $V_a + V_w$

V_s = Volume of solids

Moisture Content (w): Moisture content (w) = M_w/M_s

M_w = Mass of water; M_s = Mass of solids

Porosity (n):

$$n = V_v/V$$

V_v = Volume of voids V = Total volume = $V_s + V_w + V_a$

What does porosity means? If you look at the top term V_v, basically tells us how much voids are there in the soil. The ratio between voids and total volume is given by porosity. In other words porosity gives us an indication of pores in a soil. Soil with high porosity would have more pores than soil with low porosity. It is reasonable to assume that soils with high porosity would have a higher permeability.

Degree of saturation (S): $S = V_w/V_v$

V_w = Volume of water

V_v = Total of volume of voids

When total volume of voids is filled with water S = 100%.

Degree of saturation tells us how much water is in the voids.

Some Relationships to Remember:

Relationship 1:

$$\gamma_d = \gamma_{wet}/(1 + w)$$

This is an important relationship that needs to be remembered. This relationship appears in soil compaction section as well.

It can be proven as follows.

$\gamma_{wet} = M/V$, hence $V = M/\gamma_{wet}$

$\gamma_d = M_s/V$;

Replace V with M/γ_{wet}

$\gamma_d = M_s/(M/\gamma_{wet}) = M_s \text{ x } \gamma_{wet}/M$

$M = M_s + M_w$ (mass of air is ignored)
$\gamma_d = M_s \times \gamma_{wet}/(M_s + M_w)$

Divide both top and bottom by M_s.

$$\gamma_d = \gamma_{wet}/(1 + w)$$

Relationship 2:

$$S.e = G_s.w$$

This relationship can be shown to be true as below.
$S = V_w/V_v$; $e = V_v/V_s$

$S.e = V_w/V_s$------------------------------------ (1)
$G_s = M_s/(V_s. \gamma_w)$; $\gamma_w = M_w/V_w$; Replace γ_w in the equation.
$G_s = M_s/(V_s) \times (V_w/M_w)$

Hence $G_s = (M_s \times V_w)/(V_s. M_w)$;
$w = M_w/M_s$

$G_s.w = [(M_s \times V_w)/(V_s. M_w)] \times (M_w/M_s)$
By simplification;
$G_s.w = V_w/V_s$------------------------------------ (2)
Equations (1) and (2) are equal. Hence $S.e = G_s.w$.

Relationship 3:

$$n = e/(1 + e)$$

Proof:

Replace "e" with V_v/V_s in the above equation.
$n = e/(1 + e) = (V_v/V_s)/[1 + V_v/V_s]$

Multiply top and bottom by V_s.

$e/(1 + e) = (V_v)/[V_s + V_v]$
$V_s + V_v = V$
$e/(1 + e) = V_v/V = n$ (V_v/V is porosity)

Relationship 4:

$$e = n/(1 - n)$$

Proof: From relationship 3;

$n = e/(1 + e)$
$n + ne = e$
$n = e - ne$
$n = e (1 - n)$

$e = n/(1 - n)$

Relationship 5:

$$\gamma_d = \gamma_w . G_s/[1 + (w/S)G_s]$$

Proof:

$$\gamma_d = \frac{\gamma_w . G_s}{[1 + (w/S)G_s]}$$

$$\gamma_d = \frac{\gamma_w . M_s/(V_s . \gamma_w)}{[1 + (M_w/M_s/(V_w/V_v) \times M_s/(V_s . \gamma_w)]}$$

$$\gamma_d = \frac{\cancel{\gamma_w} . M_s/(V_s . \cancel{\gamma_w})}{[1 + (M_w/\cancel{M_s}/(V_w/V_v) \times \cancel{M_s}/(V_s . \gamma_w)]}$$

$$\gamma_d = \frac{M_s/V_s}{[1 + (M_w/(V_w/V_v) \times 1/(V_s . \gamma_w)]}$$

$$\gamma_d = \frac{M_s/V_s}{[1 + (\cancel{\gamma_w} . V_v) \times 1/(V_s . \cancel{\gamma_w})]}$$

$$\gamma_d = \frac{M_s/V_s}{[1 + (V_v) \times 1/(V_s)]}$$

Multiply top and bottom by V_s

$$\gamma_d = \frac{M_s}{[V_s + V_v]}$$

$$V_s + V_v = V$$

$$\gamma_d = \frac{M_s}{V}$$

Relationship 6:

$$\gamma_{wet} = \frac{\gamma_w . G_s \times (1 + w)}{[1 + e]}$$

Proof:

Write down relationship 5.

$$\gamma_d = \frac{\gamma_w . G_s}{[1 + (w/S)G_s]}$$

Substitute for γ_d and S.

$\gamma_d = \gamma_{wet}/(1 + w)$ and $S.e = G_s.w$

Hence $S = G_s.w/e$

$$\gamma_{wet}/(1 + w) = \frac{\gamma_w . G_s}{[1 + (w.e/G_s.w)G_s]}$$

$$\gamma_{wet} = \frac{\gamma_w . G_s \times (1 + w)}{[1 + e]}$$

S.e = G.w

0.7 x e = 2.65 x 0.6

e = 2.27

A

B

C

D

L

M

N

O

P

R

S

T

U

V

W

Y